THE
UNSEEN
AQUITANIA

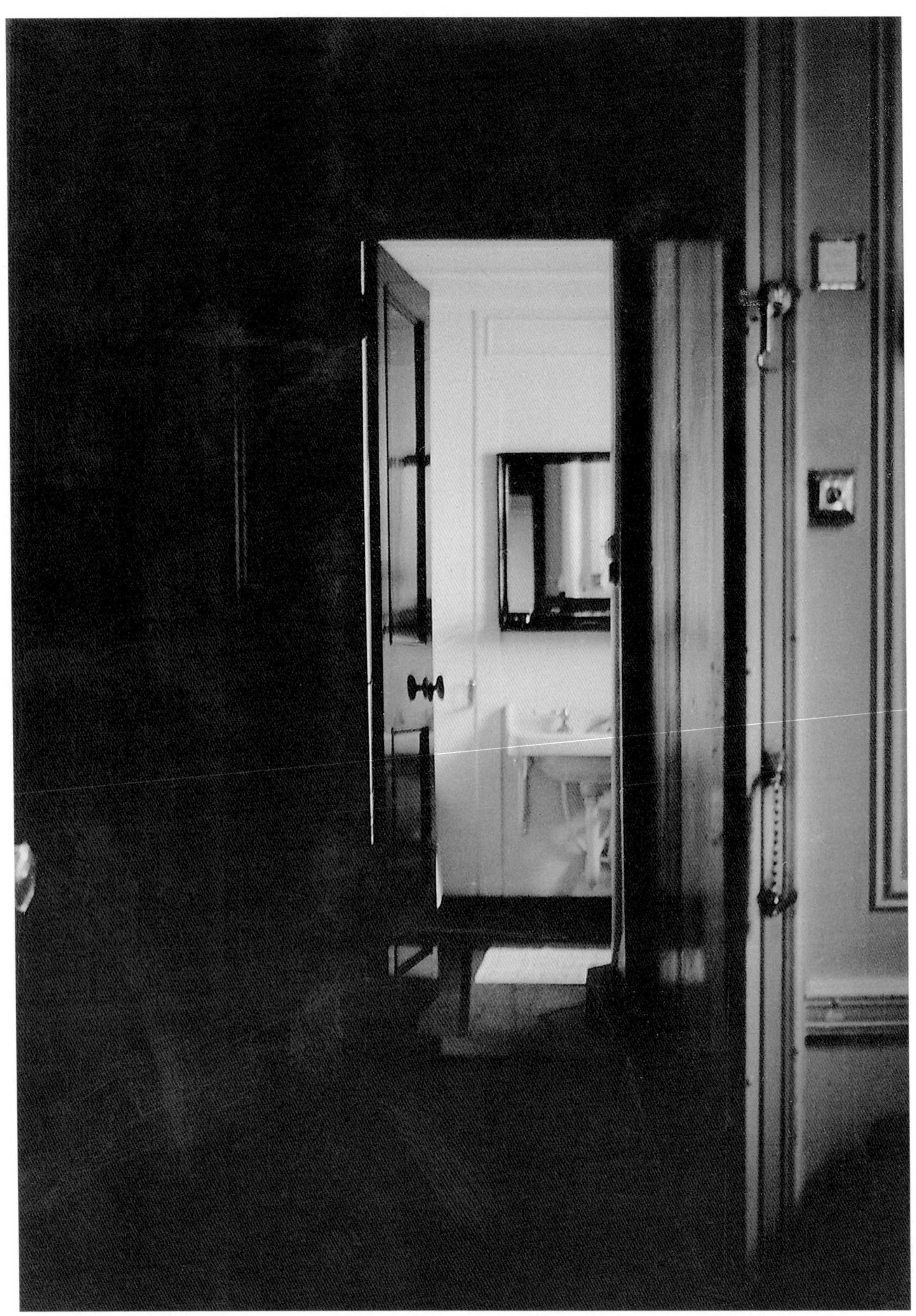

This artistically composed portrait shows the interior of one of *Aquitania*'s suites. In the distance is the sink and mirror for the suite's private washroom. (Steven B. Anderson Collection)

THE UNSEEN AQUITANIA

THE SHIP IN RARE ILLUSTRATIONS

J. KENT LAYTON & TAD FITCH

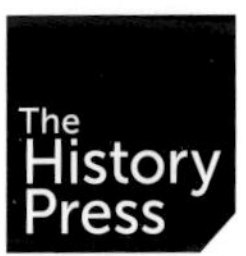

The History Press

Front cover: The *Aquitania* in 1937. (Clyde George Collection)
Back cover: The *Aquitania* enters the Mersey in Liverpool for the first time on 14 May 1914. (Clyde George Collection)

First published 2016
This paperback edition first published 2024

The History Press
97 St George's Place, Cheltenham,
Gloucestershire, GL50 3QB
www.thehistorypress.co.uk

© J. Kent Layton & Tad Fitch, 2016, 2024

The right of J. Kent Layton & Tad Fitch to be identified as the
Authors of this work has been asserted in accordance with the
Copyright, Designs and Patents Act 1988.

British Library Cataloguing in Publication Data.
A catalogue record for this book is available from the British Library.

ISBN 978 1 80399 586 1

Typesetting and origination by The History Press
Printed and bound in India by Thomson Press India Ltd

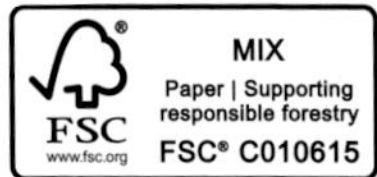

Trees for Life

CONTENTS

INTRODUCTION

In 1958, eight years after she had been withdrawn from service, *Aquitania* made an appearance in the classic film *A Night to Remember*. The grainy footage showed her at the start of one of hundreds of crossings she made over decades of service. Another scene showed *Lusitania*. Only fourteen 'four stackers' were ever built: *Titanic* is undoubtedly the most famous and yet others met unfortunate ends, too. *Lusitania* was struck by a German torpedo and sank in May 1915; *Britannic* fell victim to a mine in November 1916. People often assume that any steamship with four funnels must be the doomed *Titanic*.

What distinguishes *Aquitania* from *Titanic* and her other contemporaries is the success she enjoyed over such a long period. Cunard were discussing the new ship five years before the First World War broke out and yet she was not withdrawn from service until five years after the end of the Second World War. She existed in one form or another – whether as the first keel plates or the remnants of her hull – during five different decades.

There was a key connection between them, however. The early years of the twentieth century were a period of fierce rivalry between Britain's Cunard and White Star Lines. Cunard benefited enormously from government assistance to build *Lusitania* and *Mauretania*, which entered service in 1907. And yet, a year later, construction had begun on White Star's *Olympic* which would surpass them in size and luxury; *Titanic* was not far behind.

Cunard's own response was a form of flattery, because in *Aquitania* they had a far larger ship whose operating model was far more similar to the White Star ships than *Lusitania* or *Mauretania*. She was larger, slower, and far more luxurious – financed by commercial means rather than government support. They aimed to learn all the lessons from their competitors and their naval architect, Leonard Peskett, even travelled on *Olympic* to understand the competition and see which concepts might work on *Aquitania*.

Hundreds of thousands of passengers crossed on her, and she became a favourite of many repeat travellers. Their experiences form a fascinating social history. Arguably, a single volume is insufficient to do justice to every aspect of *Aquitania*'s extraordinary career. What we see here is a remarkable compilation of images, each of which has a story behind it or a perspective to contribute through the photographer's lens. If we study them carefully and give them the attention they deserve, we learn a lot.

The exceptional images within this fine book give us an insight into the life and times of 'The Ship Beautiful'.

Mark Chirnside

AUTHORS' NOTE

They still fire our imagination. One would think that their time had come and gone, and that – particularly in this fast-paced world that we live in, where the 'next big thing' in technology or industry is right around the corner, and only keeps our attention for a few moments – we would long since have forgotten them. Yet the Atlantic liners live on in our collective consciousness. Why?

It's difficult to say. Perhaps it has something to do with the alluring nostalgia of the past, for the most famous of the liners lived in much simpler and, many would argue, better times … times when there was actual collective optimism regarding the future, and when it seemed there was good reason for such. Perhaps it was the simplicity of the age. What better thing to reminisce over   in these hectic times when electronic devices are all clamouring for attention, and when 'unplugging' comes with fears of missing something important – than a time when people could step off shore, away from life, and set out across the open sea on a journey wherein the biggest news of the day was who won in the ship's mileage pool? On these voyages, little was vying for attention and, when the weather and ocean was not threatening the continued existence of one's ship and one's very life, or when one was not suffering the malaise of seasickness, those days at sea were eminently relaxing.

Perhaps it has something to do with the adventure and romance of those voyages. There were those who loved simply being at sea, who 'connected' with their favourite liner, entrusting their lives and the lives of their family and friends to 'her', and who watched with fascination – and more than a tinge of adventure – as that liner battled the elements. There was also more than a touch of romance in the air while on a sea voyage aboard an Atlantic liner; it seemed that love was just waiting to cast its spell on the unsuspecting.

With some ships, such as the *Titanic* or the *Lusitania*, there is the grim allure of disaster. Yet the majority of the liners lived long lives, and still captivate us today. This is certainly true of the *Titanic*'s sister *Olympic*, as well as the Cunard Line's legendary *Mauretania* of 1907. It is also true of one of Cunard's longest-lived and most successful liners: the *Aquitania*. With a service career lasting thirty-four-and-a-half years, she was nearly able to match the longevity of the *Queen Elizabeth 2*, which was in Cunard's employ for an astonishing thirty-nine-and-a-half years.

Aquitania also had the distinction of being borne of those golden years of the Edwardian Era – a time which technically ended in 1910 with the death of King Edward VII of England, but which really seemed to continue until the outbreak of the First World War. During that time, technological progress led to the creation of unprecedented ocean liners. Though her life would be in an era far different than those foreseen when she was being constructed, she adapted well to changing times and served happily in peace and both World Wars.

While some superb books about this great and beautiful ship have been written before, this is, to our knowledge, the first time that an original book has been given this type of large, high-quality photographic format that comes with the prestigious *Unseen* series of books. We are proud to take you aboard one of the great legends of the Atlantic through words and pictures.

Bon Voyage,
J. Kent Layton & Tad Fitch

ACKNOWLEDGEMENTS

Although our names are on the cover of this book, it was impossible for us to pull together everything that sits between the front and back cover of this volume without the assistance of many other individuals. This is our opportunity to thank everyone who has helped us so much with the project.

First, we would like to offer a special thank you to Amy Rigg and the staff at The History Press, who made this book possible. Significant photographic and illustrative contributions were made by the following individuals: Steven B. Anderson, Clyde George, Ioannis Georgiou, Brian Hawley, Michael Poirier, Eric Sauder, Kalman Tanito, Alexander Warton, Richard Weiss, Randy White, and Russ Willoughby. Your help and generosity is what makes this book stand out.

A special thank you is owed to Mark Chirnside, not only for the wonderful introduction, but also for his ongoing support and willingness to share his research with us. An additional thank you to Siân Wilks at the University of Liverpool for his assistance. Finally, both authors would like to thank their family members and friends for their encouragement and support as we worked on this project.

If, in the course of this list, we have managed to forget anyone, please let us know so that we can correct this omission. Lastly, while great effort was taken to ensure that the information contained in these pages is accurate and tells *Aquitania*'s complete story, if any errors have been made, they are the authors' alone.

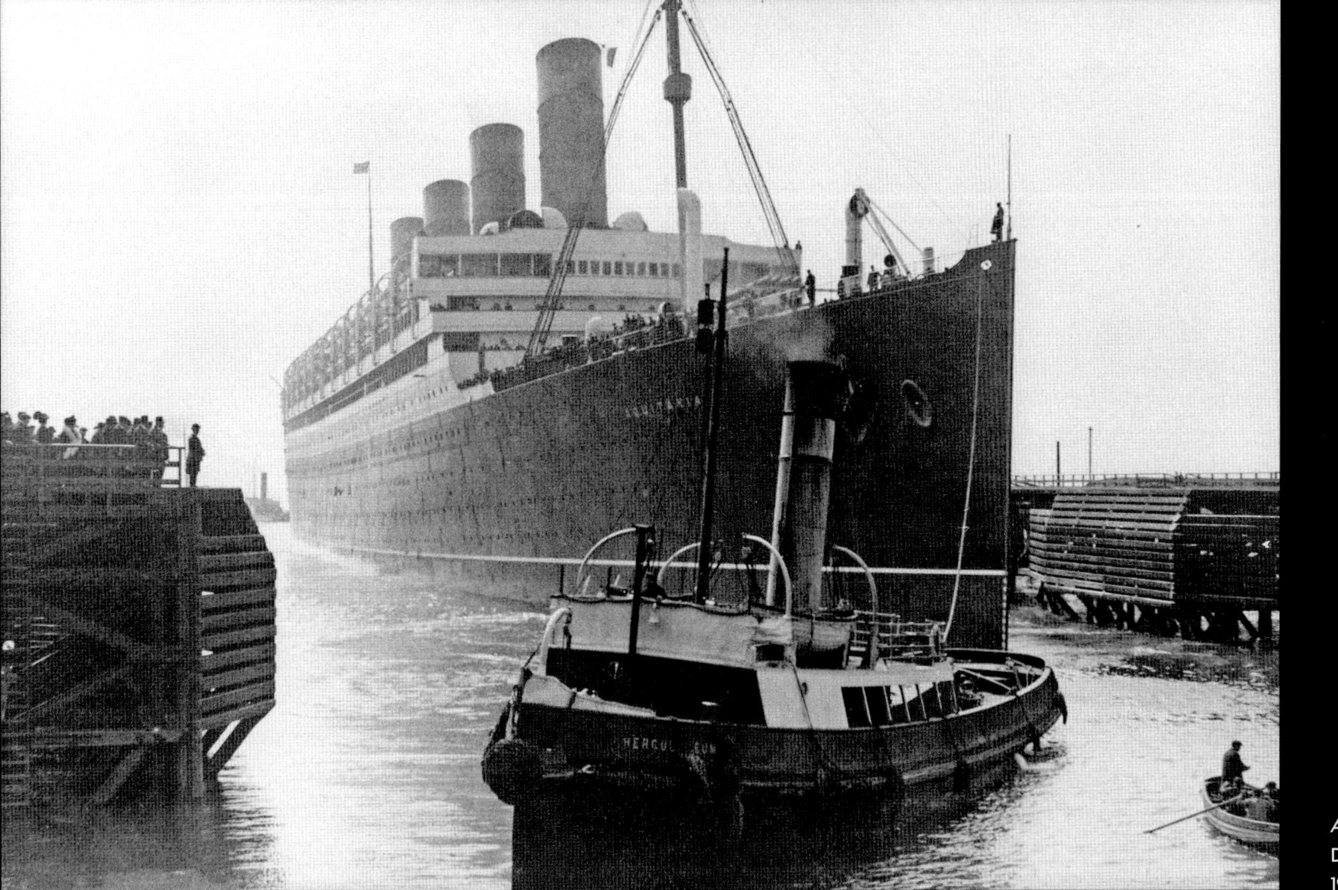

Aquitania enters the Gladstone Dry Dock in Liverpool on 15 May 1914. (Clyde George Collection)

It was December 1909, and the Cunard Steamship Company had just managed to pull off the impossible. Barely seven years before, the great line had been on the very precipice of financial oblivion, despite their prestigious six-decade history. Although they had been the first company to establish a regularly scheduled transatlantic service, competition had sprung up in the intervening years. Some of it had foundered – sometimes quite literally – on an abysmal safety record; against these, the ever-cautious Cunard company were able to boast over their record of never having lost a life. However, other competitors, like the White Star Line, or the German Norddeutscher Lloyd and Hamburg-Amerika (or HAPAG) Lines, had more sticking power.

Yet, as the nineteenth century drew to a close, Cunard had found themselves in an increasingly precarious position. The *Campania* and *Lucania*, their crack liners of 1893, had quickly been eclipsed by larger, faster and more luxurious ships from White Star, Norddeutscher Lloyd and the HAPAG Line.

These new ships really pushed the boundaries of what was previously thought possible on the Atlantic: while the German ships took and then retained the speed title for a decade beginning in 1897, the title of what ship was 'most luxurious' or 'largest' changed hands regularly. During this period, Cunard placed into service three intermediate steamers – *Ivernia*, *Saxonia* (both 1900) and *Carpathia* (1903). They followed these up with two more substantial vessels, the *Caronia* and *Carmania* (1905), but even then their fleet was hardly in a position to tackle White Star and the Germans head-on.

When their less-than-ideal situation was compounded by a rate war that broke out on the North Atlantic, Cunard found themselves on the brink of calamity. It looked like it would be difficult for them to remain viable; some thought that another shipping company might buy them out, but considering the ships in their fleet, it was unlikely that the offers would be good. Indeed, Cunard's situation was so bad that American financier J.P. Morgan, who was then building a sea-going monopoly under the banner of the International Mercantile Marine, had shown very little interest in Cunard, even as he bought out the White Star Line and made alliances with the German steamship lines.[1]

Fortunately, the right man was at the helm of the company at what appeared to be their darkest moment. Lord Inverclyde approached the British Government and secured financial backing to build liners that would prove, by a good measure, to be the largest, most luxurious and also the fastest that the world had ever seen. He and his company selected two shipyards for the project, and construction commenced.

It was a tenuous time, for the new liners would be the pushing the boundaries of technology in ways that no previous ship had. If anything went wrong technically, the fallout would be catastrophic – for the shipbuilders *and* for the Cunard company. Yet if the situation had looked tenuous up to that point, everything turned around once the ships had actually entered service. In short order, the *Lusitania* and *Mauretania* had claimed the Atlantic Blue Riband speed records in both directions. They were instantly greeted with great acclaim in the British and American maritime technical journals and the press. The travelling public thought very highly of them and, more importantly, booked passage on them in droves. They fast became the darlings of the Atlantic; in the busy seasons, they were frequently booked to capacity.

The *Lusitania* and *Mauretania* had instantly eclipsed all other ocean liners on the North Atlantic. Cunard was now in the lead, and suddenly it was their competitors who were trying to catch up. Yet that lead was a fragile one.

On 30 April 1907, the White Star Line formally requested that the Belfast shipbuilding company Harland & Wolff design a pair of much larger ships to pit against the new Cunarders.[2] Even before this formal request, it was clear that the order was in the works: in 1906, the Belfast shipyard had already demolished three of their slipways to accommodate two behemoth new ones, on which the shipbuilder could simultaneously build a pair of extraordinarily large ships. Two years before that, construction had begun on a new drydock in Belfast that could accommodate enormous liners even longer than the *Lusitania* and *Mauretania*.

In the end, the two new White Star liners were named *Olympic* and *Titanic*, and they would enter service in 1911 and 1912, respectively. Each was roughly 50 per cent larger, in terms of gross tonnage, than the Cunard speedsters.[3] While they would be larger and more luxurious, they would not be as fast: White Star opted for a hybrid powerplant – partly consisting of traditional reciprocating engines; but with the addition of a single low-pressure turbine driven by their exhaust – that would propel them at a slated service speed of 21 knots. The *Mauretania*'s best average speed for a crossing then stood at 26.06 knots. White Star clearly felt that many of their prestigious passengers would opt for a more comfortable, if slightly more leisurely, passage rather than an all-out dash in a slightly smaller ship. At the same time, maintaining a slightly lower service speed also meant significant savings in fuel expenses on each trip. White Star retained an option to build a third ship of this class if they were satisfied with the first, or the first two. They exercised this option and ordered a third liner at the conclusion of the *Olympic*'s maiden voyage in June 1911.

Meanwhile, across the English Channel, the German Hamburg-Amerika Line began to formulate plans to quickly build three new superliners. These monsters would successively become the largest ships in the world, and would hold that title for about two decades. They were to sport turbine engines, similar to those on the Cunard speedsters, which would drive them at speeds faster than the *Olympic* and *Titanic*. They were to be named *Imperator*, *Vaterland* and *Bismarck*.

Each of these companies was attempting to implement a balanced and reliable weekly service on the Atlantic. At 21–26 knot service speeds, it was impossible for two ships alone to maintain a weekly schedule. Pairing an older, smaller, slower ship with two primary liners meant that, at times, only a less desirable ship would be available for finicky prospective passengers. Having three well-matched ships running together – sister ships, or at the very least all prestigious ships offering similar amenities – was ideal. Even the swift *Lusitania* and *Mauretania*, or the enormous *Olympic* and *Titanic*, would be incomplete without another entrant in the service.

To be fair, Cunard were doing remarkably well even if their two crack liners were still teamed up with a smaller, slower vessel. Indeed, the company's 1910 profits, reported in early 1911 – including a sum of £6,988 brought forward from the previous year – amounted to £994,149 3s. 8d. Their total receipts for passage, freight and mail for that year had reached £3,068,425.[4]

However, Cunard's Board of Directors knew that more work remained: they had to add a third superliner to their fleet as soon as they could. Because the new White Star and Hamburg-Amerika liners were going to be so much larger, building a third of the *Lusitania/Mauretania* type would not be enough. Astonishingly, the *Lusitania* and *Mauretania*, the 1907 leviathans of roughly 30,000 tons each, were now beginning to seem rather small when compared with their successors then under construction.

The 1907 sisters, designed for high speed, suffered from some drawbacks: they were prone to vibration, which was extremely uncomfortable for passengers; there were also occasions when their arrow-like bows – designed to slice through a large wave rather than pitch up over it – made for a less comfortable ride than that which the latest ships from competitors would eventually offer. Additionally, to drive them at peak speeds, the engines of the *Lusitania* and *Mauretania* consumed large quantities of fuel. The terms of the original government loan which funded their construction had provided them with an operating subsidy of £150,000 per annum, or £75,000 for each ship, which offset this potential for financial deficiency, but that would not apply to the new liner.

By building a ship roughly the size of the new White Star vessels, and aiming for a slightly more moderate speed, Cunard could add extra luxuries and comforts. They could also carry more paying passengers on each trip, while burning less fuel. All of this would ensure that the new liner would help to maintain Cunard's edge against their competitors. One period journal described the various factors that they were balancing in this way:

The chief problem was a commercial one ... It was desired that the new vessel should earn as much per voyage without a subsidy as the older boats did with their subsidies, and that she should be able to sail regularly from each terminal port once in every three weeks.

To make up for the absence of the subsidy it was decided that the new vessel should have a lower speed and a larger passenger carrying capacity than her forerunners. In order to avoid the necessity for an expensive and hurried turning round at New York and Liverpool it was not desirable that the ship should take longer than 5½ days on the voyage from port to port. This necessarily restricted the amount by which the speed could be lowered and practically fixed it at 3 knots less than that of the older vessels, namely, at 23 knots. The passenger accommodation had therefore to be increased until the saving of coal, upkeep costs etc., represented by the lower speed, together with the extra passenger earnings, equalled the subsidy earned by the sister ships. To provide this additional accommodation the new vessel was given an extra deck and her length was increased by over 100ft. For the sake of stability corresponding increases had to be made in the beam and depth.[5]

Working from these basic parameters, a rough concept was formed, and then several variant plans were prepared and submitted for consideration. By July 1910, Cunard had preliminary discussions with three shipbuilding firms about finalising the ship's configuration, and obtaining tenders for construction. The companies were the Tyne-based Swan, Hunter & Wigham Richardson, which had built the *Mauretania*; the Barrow-in-Furness yard of Vickers Sons & Maxim; and the Clydeside firm of John Brown & Company, which had built the *Lusitania*.

Vickers had initially been in the running to build either the *Lusitania* or *Mauretania*, but had dropped out early on. Swan, Hunter had catapulted to prestige because of the success of the *Mauretania*. In fact, in

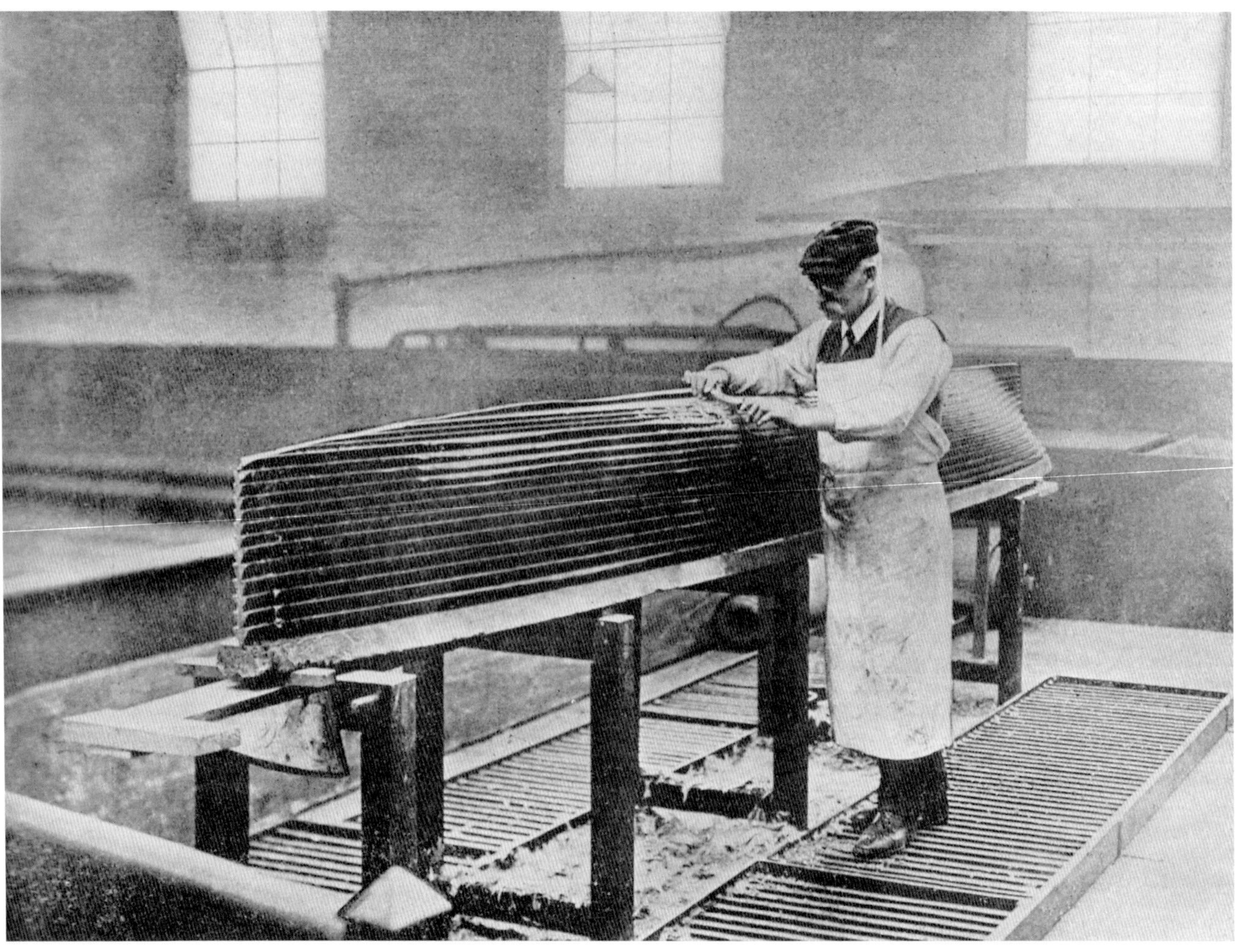

A craftsman shapes a wax model of the proposed *Aquitania*'s hull. Many different shapes and forms were tried and refined before the liner's design was finalised. (Authors' collection)

early 1910 they had beaten John Brown for the contracts to build the *Laconia* and *Franconia*.[6] By September 1910, Cunard formally invited these three firms to submit a tender for construction of the new liner – at a fixed price, and with a fixed delivery date. The fixed-price proviso was a change from the cost-plus agreement which had been used for construction of the *Lusitania* and *Mauretania*, and which had allowed for nearly catastrophic cost overruns.[7]

On 6 December 1910, John Brown's tender was submitted, and their price was £1,414,222. Six days later, they were awarded the contract, which was signed on 31 January 1911.[8] The name of the new ship had been selected at a meeting of the Cunard Board of Directors on 19 January: she was to be called *Aquitania*.[9] This name was in keeping with Cunard tradition. First, it ended in the same traditional *-ia* suffix that so many of their ships bore. It also harmonised well with the names of her two swift predecessors, *Lusitania* and *Mauretania*, named after ancient Roman provinces located in Spain and Africa, respectively. The Roman province Gallia Aquitania, or Aquitaine Gaul, had been located in the south-west of modern-day France. Hence all three of Cunard's premiere liners would bear names from a common lineage.

Meanwhile, a great deal of work was being done by both Cunard and John Brown staff, to finalise *Aquitania*'s design. A number of lengths, breadths and hull forms were proposed and model-tested, then refined to alleviate deficiencies found during the experiments. The British Admiralty was also involved in design decisions, under the terms of their previous agreement with Cunard on the funding of the *Lusitania* and *Mauretania*. As *The Shipbuilder* pointed out:

> The *Aquitania* is liable to be called upon at any time for naval service, and her design allows for the carrying of a number of quick-firing guns on the bridge and shelter decks ... the positions arranged for them are specially strengthened. Two of these ... are situated on the port and starboard sides respectively of the boat deck amidships, two on the port and starboard sides of the same deck aft, two on each side of the shelter deck forecastle head, and two on each side of the shelter deck aft. The

sub-division of the ship and the arrangement of the bunkers on each side of the boiler rooms afford protection as far as possible against damage caused by shells, torpedoes or ramming.[10]

Cunard also wanted to ensure that the new liner had a sufficient metacentric height – which is a technical measurement of a vessel's inherent stability. *Aquitania*'s final design ensured that she would provide her passengers with a forgiving ride in all but the most violent weather.

When it came to the question of stability, most ships of the period utilised what were called bilge keels – or long fins attached to the underwater surfaces of their hulls – in order to help prevent rolling. However, in recent years Dr Frahm of the German Blohm & Voss shipyard had been advocating a new type of anti-rolling countermeasure. This system consisted of tanks, partially filled with water, located on the sides of a ship and connected by piping, through which the water of a tank on one side could 'slosh' into its counterpart on the opposite side. The Frahm anti-rolling tanks were being adopted on the new *Imperator* and *Vaterland*, and Cunard had tested them with success in their *Laconia*. Since its inception in 1840, the Cunard company had often proved very conservative in their thinking. However, they had showed great forward-thinking with the adoption of turbine engines in the *Lusitania* and *Mauretania*. In this instance, Cunard opted to adopt Frahm anti-rolling tanks rather than bilge keels. Yet in an interesting twist, their conservatism still seemed to rear its head on the point, for it was reported:

> There seems to be still some doubt regarding the methods which will be adopted ultimately in order to reduce the rolling of the vessel. At first bilge keels were included in the design, and these were actually made and fitted but not riveted on the vessel. Since the proposal was made that Frahm antirolling tanks should be used it has been decided to dispense with bilge keels, as both devices should not be necessary ... It is understood, however, that they will be kept, so that they may be fitted at any time if they are found necessary.[11]

In external appearance, the *Aquitania* would closely resemble her two running mates, but on a larger scale. Her final length overall would be 901ft 6in, while her hull's maximum breadth would be 97ft. While the German liners were being built with only three funnels apiece – the four-funnel trend was coming to an end – the *Aquitania* would still sport four. Each of these funnels would be given an attractive dousing of the house livery: black tops capping off orange-red segments that were separated by narrow black bands. The fourth funnel was connected to the No. 4 Boiler Room, but because that compartment would have only three boilers instead of the six in each of the other three boiler rooms, it would vent less exhaust gas. Instead of making it smaller than the other three, it was decided to make it the same size, and then to use it to help ventilate the central Turbine Engine Room.[12]

Aquitania's long bows and forecastle would project some 200ft from the forward face of her superstructure; in comparison, Cunard's first ship, the *Britannia* of 1840, was only some 207ft in length. This meant that if *Aquitania*'s forecastle had been completely open and devoid of equipment and other encumbrances, *Britannia* could comfortably have ridden on the deck forward of her bridge.

The *Aquitania* would be comprised of nine decks in total, with the lowest of these being only a partial deck that ran forward and abaft of the machinery spaces amidships. The topmost deck was the Boat Deck, beneath which were located: Promenade Deck (A), Bridge Deck (B), Shelter Deck (C), Upper Deck (D), Main Deck (E), Lower Deck (F), Orlop Deck (G) and at last the partial deck, designated the Lower Orlop Deck (H). The floor of B Deck would be the uppermost strength deck, while the decks above that would be superstructure decks of much lighter construction; these would, more or less, ride atop the primary hull of the ship. The Boat and A Decks were built to overhang the side of the ship by about 2ft on either side, thus creating extra space for lifeboats and strolling passengers. A series of deckhouses, built atop the Boat Deck, would allow for an even higher spot where passengers could enjoy sea views or participate in games. All of this made for a structure that projected above the water by about one deck's height more than the *Lusitania* and *Mauretania*; her quartet of funnels were slated to rise a little over 10ft higher out of the water than theirs. Like her predecessors, a separate segment of superstructure was built astern of the primary one. This detached 'block' of the Boat and Promenade Decks would house a number of Second-Class public rooms, in an extremely similar arrangement to the older ships.

At the forward end of the *Aquitania*, where the superstructure rose from the forecastle, the original idea seems to have been to design a curved face that resembled the gracefully curved forward superstructures of the first two ships. Some early artist's renderings, fairly detailed and accurate in other respects to the ship's finished form, showed this; at least three of these renderings were released as early publicity material. A curved face made sense, for during the design phase of the *Lusitania* and *Mauretania*, and especially in the case of the latter ship, much testing had been done to determine the effect of wind resistance on various shapes of the forward superstructures, and curved faces offered less resistance.

Yet when built, the forward superstructure of the *Aquitania* looked less aesthetically pleasing, and less graceful. Its face was curved, but nearly imperceptibly; from many angles, each deck's forward bulwark presented a nearly 'flat', and somewhat ungainly, appearance. The overall effect worked from some angles, but did not work well at all from others; comparing this area of her structure to that of the older speedsters, the *Lusitania* and *Mauretania* had a clear advantage.

Atop the ship's bridge was located a secondary, open-air bridge, with a rather graceless side-to-side bulwark covering the forward face of the upper bridge wings, perched awkwardly above the primary bridge. The need for this odd open-air secondary bridge sprang from the ship's extraordinarily long forecastle. Officers on the bridge, particularly during docking manoeuvres or in the narrow confines of harbours, would not be able to see clearly over the prow in order to observe anything close to the liner's bow – be that a tug or a dock wall. The shortcomings of her original bridge location were an oddity that would be the site of continually evolving updates and additions throughout the liner's career.

Although John Brown & Co. had built the *Lusitania*, the largest ship in the world when she entered service less

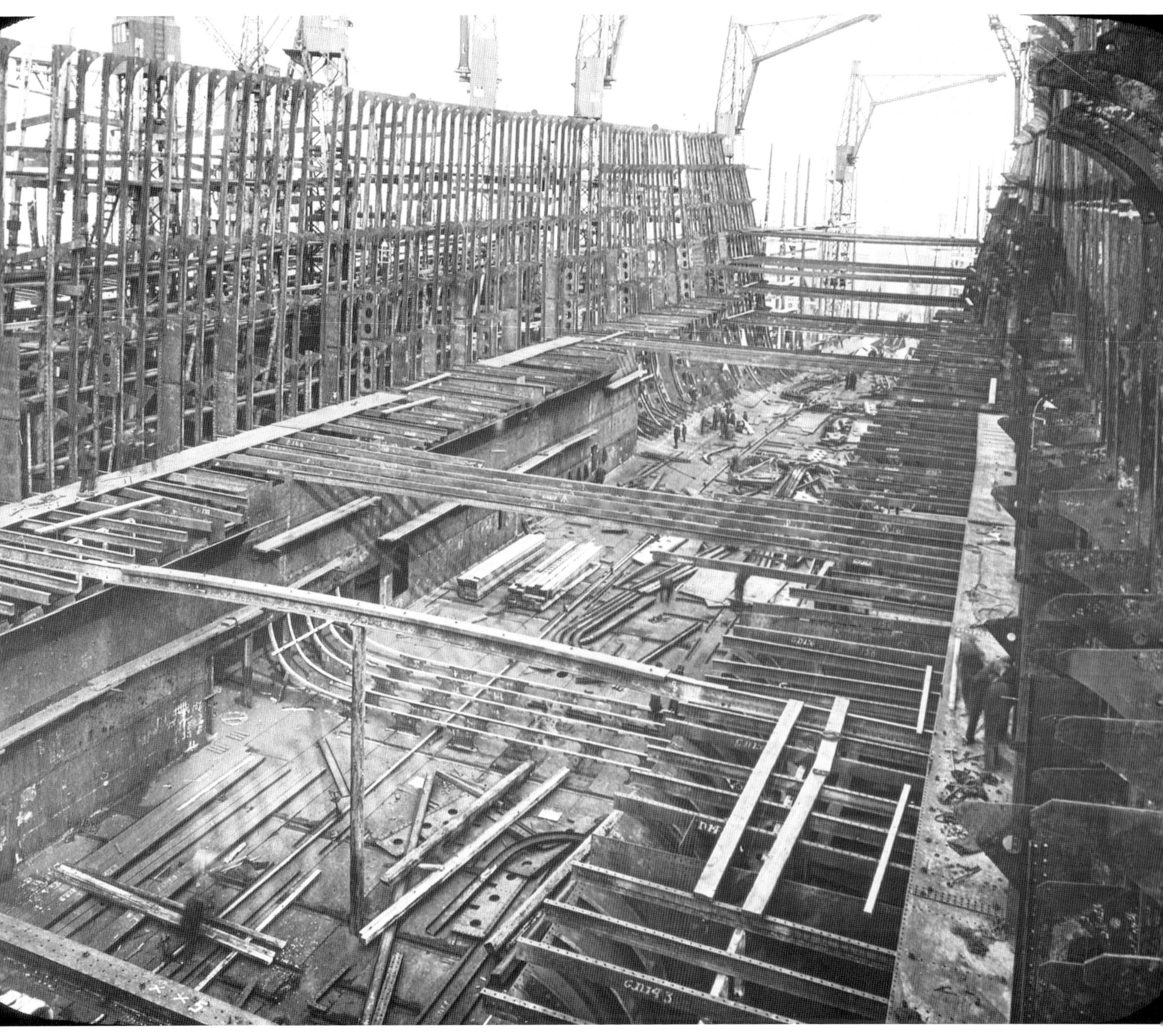

A fantastic view of the *Aquitania* in mid-construction. The ship's frames, or ribs, are visible at the sides, as are the longitudinal bulkheads which formed the bunkers that held the ship's fuel. (Brian Hawley Collection)

than four years before, the *Aquitania* was an enormous leap forward in terms of size. A great deal of work had to be done to prepare the shipyard for construction. It was said:

> The berth on which the *Aquitania* is being constructed, while virtually that on which the *Lusitania* was built, was subject to some extension and consolidation for the much larger and heavier ship. Before laying the keel blocks in March, 1911, it was found necessary to lengthen considerably the berth and alter its angle to the river, principally to secure a freer launching run. To deal effectively with the work of construction the builders also installed a very complete arrangement of electric jib cranes having lattice girder standards.[13]

This was no small job. Another account gave further details:

> Before the construction of the ship was begun the whole face of the yard had to be changed to make room for laying down the skeleton. Several large buildings had to be demolished.
>
> The area of the sliding ways for the *Aquitania* will be about 10,000 square feet and the pressure about 2.6 tons per square foot. Though the ship has been built on the berth used by the *Lusitania*, the ground has had to be remade and lengthened. To bear the great load of steel heaped upon it, the ground has been piled and cross-piled. Over the cross piles have been placed layers of steel plates, then quantities of cement, especially toward the wayends, where the pressure when the vessel is partly waterborne will be terrific. Of course it is essential that the ground should not yield an inch.[14]

Even with the more acute angle to the river, the demolition of existing buildings in the yard was necessary because the new liner's prow was going to penetrate an extra 100ft ashore than the *Lusitania*'s had. The new cranes on the slip were electrically powered, and built by Sir William Arrol & Co. John Brown spent no less than £27,750 to prepare the new slip.[15] Meanwhile a great deal

of work had to be done to prepare the River Clyde to accept her. In April 1913 it was said:

> The Clyde Lighthouse Trustees have accepted the tender of Mr. C.H. Campbell, dredging contractor, of London, for the widening and deepening of the Clyde off Greenock and Port Glasgow, and the work is now proceeding day and night.
>
> Completion is necessary in twelve months to permit of the safe passage out of the Clyde of the great Cunarder *Aquitania*.[16]

John Brown ended up contributing £10,000 to this task.[17] A newspaper editorial from shortly before the liner was launched gave some interesting details on this work:

> Under their act of 1904 the Clyde Navigation Trustees widened a part of the river opposite the yard for the purpose of accommodating the *Lusitania*. Since then, in anticipation of the launching of the *Aquitania*, the widening has been continued along the north side of the Newshot Isle. Quite a large piece of land has been taken away, and the dredging operations to increase the depth of water in front of the slip are nearing completion. Special provision has had to be made in order that when the ship 'dips' on finally leaving the ways there will be depth enough for her. The depth at this spot will be abnormal, as when the *Aquitania* makes her bow ... which will indicate that she is completely waterborne, she will go down by the head a rather long way. But the abnormal depth is necessary over only a comparatively small area.
>
> The builders' fitting-out basin at Clydebank has also had to be dredged to accommodate the liner when she is towed alongside for completion.[18]

Only once the liner's final specifications and designs had been produced, and all of these improvements to the shipyard were completed, could actual construction begin. Considering the scale of the project, the time from the signing of the contracts to the commencement of work was rather short. On 5 June 1911, the first keel plate of John Brown Hull No. 409, the *Aquitania*, was laid down.

The Cunard company kept a tight wrap on the particulars of their new steamer, most especially concerning her final overall length and her intended gross tonnage. *The Engineer* was moved to say, a year after the construction project had begun, in May 1912:

Although little has been said about it, it is common knowledge that the Cunard Line is having a mammoth steamship built for its Transatlantic service by John Brown and Co., Limited, of Clydebank.[19]

The unusual level of secrecy over the ship's particulars is more understandable when one takes into account the context of the period: a number of very large ships were then being built by Cunard's competitors, and each firm wanted to have the world's longest or largest ship. Placing into context the construction and entry into service of the *Aquitania* can help us to understand their secrecy.

On 31 May 1911, or five days before the *Aquitania*'s keel was laid, the new White Star liner *Olympic* departed from the Harland & Wolff shipyards in Belfast, where she had been built. This ship was 882ft 9in in her overall length, and was then registered with a gross tonnage of 45,324 – nearly 50 per cent larger in internal volume than the *Lusitania* and *Mauretania*. Her maiden voyage would commence on 14 June, eleven days after construction on the *Aquitania* had commenced.

Also on 31 May 1911, the second *Olympic*-class liner, *Titanic*, was launched at the Harland & Wolff shipyard. Contrary to popular opinion, she was identical in overall length and beam to her older sister, *Olympic*; thus she did not become the largest moving object in the world at the time of her launch. Because of some modifications to her internal configuration, however, when she was finished, the *Titanic*'s gross tonnage was registered at 46,329, roughly 1,004 tons more than her sister. These small internal changes and the resulting alterations to her tonnage calculations would, upon her registry in 1912, give *Titanic* the right to claim the title of 'world's largest ship'.

Meanwhile, in Germany, another behemoth was being built at the A.G. Vulkan shipyard at Stettin; her keel had been laid on 18 June 1910. Designated only by her yard number, No. 314, she would eventually be named *Imperator*. Launched on 23 May 1912, her gross tonnage was, when completed, 52,117. Her maiden voyage would begin on 10 June 1913. *Imperator*'s finished dimensions were also shrouded in secrecy. As it turned out, her hull was 909ft in overall length, or 7½ft longer than the *Aquitania*'s. However, because Cunard was playing their hand 'so close to the vest', Hamburg-Amerika was clearly nervous that their new ship would not be the longer of the two. As a result, they added a ghastly eagle figurehead to the prow of their ship; this preposterous, scowling bird added to the ship's overall length, making it 918ft 8in.

By the time that the *Imperator* had been launched, a second of the class, eventually named *Vaterland*, had been laid down at the Blohm & Voss shipyard in Hamburg. She was to be even longer and larger than the *Imperator*, at 950ft and 54,282 tons, and would enter service on 14 May 1914.

Back in Belfast, construction on the third *Olympic*-class entrant commenced on 30 November 1911. Her dimensions were also the subject of some secrecy and exaggeration in the press, but in the end she was identical in length to her two older sisters, namely 882ft 9in. Even before her construction commenced, it was decided to increase her hull's width from 92ft 6in to 94ft. This meant that when she was finally registered, she would be the largest of the three White Star ships, with a gross tonnage of 48,158.

Early Cunard publicity estimated the *Aquitania*'s tonnage to be '47,000', a figure somewhat inflated over her final 45,647. Thus, her gross tonnage was greater than that of *Olympic* in 1911, but not of the *Titanic*'s 46,329. By 1913, *Olympic*'s gross tonnage had been raised to 46,359. *Aquitania*'s tonnage certainly didn't come close to any of the German ships, which were all larger than 50,000 tons, and neither did she eclipse the overall length of any of them. However, she was longer than the three White Star liners. Although she would never hold the title of 'world's largest' or 'world's longest' ship, she was the 'longest British-built ship' on the Atlantic until 1936.

Meanwhile, on 10 April 1912, just over ten months after the *Aquitania*'s keel had been laid, the second of the White Star giants, *Titanic*, began her maiden voyage from Southampton to New York. Her voyage ended ignominiously off the Grand Banks of Newfoundland on the night of 14/15 April, and she took 1,496 of her passengers and crew to a watery grave.[20]

The *Titanic* disaster shook the maritime world badly. Up until that point, complacency had been growing. *Titanic* was not the only ship dubbed either 'unsinkable' or 'practically unsinkable'. It was not uncommon for engineering journals, a steamship company's publicity material or the general press to simply pronounce a ship to be 'unsinkable'. Simultaneously, the British Board of Trade had complacently failed to follow the rapid growth of ships, failing to increase their requirements on the number of lifeboats a large ship carried. All of this changed the morning after the *Titanic* disaster. The advantages and disadvantages of different styles of watertight subdivision were suddenly hotly debated, and lifeboats for all were quickly required.

Fortunately, the design of the *Aquitania*'s watertight bulkheads was so thorough that it cost Cunard less than £7,000 in order to make the necessary alterations to the *Aquitania*. Significant improvements in her system of longitudinal watertight subdivision were also made over that found in the *Lusitania* and *Mauretania*. Detailed analyses of extreme flooding scenarios for the *Aquitania* were published in technical journals, helping to demonstrate that she could withstand anything that she might encounter in either peace or war. In fact, her design was so stringent that the day before her maiden voyage commenced, one newspaper – either audaciously or foolishly in that post-*Titanic* world – said:

> The *Aquitania* is two ships in one, for the inner shell is separated from the outer shell by fifteen feet, which with a new system of watertight compartments, is reckoned to make the vessel unsinkable.[21]

In addition, *Aquitania* would be provided with enough lifeboats for everyone aboard. Twenty-two standard wooden lifeboats and fifty-eight collapsible lifeboats would be rounded out by a pair of 30ft motorboats. Each of these latter craft was fitted with a battery-powered wireless set to request assistance or guide rescuers to the location of lifeboats. Together, these boats gave her a lifesaving capacity of 4,584, a comfortable margin above her certified capacity of 4,202.

While her safety systems were being sorted out, and construction continued, there were other details to tend to. One of the most important of these, from the perspective of making the liner attractive to potential passengers, was ensuring that her scheme of interior decoration was of the highest calibre and appeal. Cunard, by that point, had received a good deal of general feedback on the interior decoration of their previous liners. Cunard could also glean much about what did, and did not, work on some of their competitors' new ships.

Before the new White Star liner *Olympic* made her maiden voyage, she was opened for public inspection in Liverpool, on 1 June 1911. Cunard's chairman, Alfred Booth, their general manager, A.D. Mearns, and their resident naval architect, Leonard Peskett, took the opportunity to tour the liner and record their impressions. John Currie and Alexander Galbraith, from Cunard's engineering department, also visited the ship, and were shown around her engineering spaces by none other than the liner's Chief Engineer, Joseph Bell. On 13 June, the day before *Olympic*'s maiden voyage, Arthur Davis of the firm Mewès & Davis, along with some other key personnel who would be involved in providing the internal decorations for the *Aquitania*, toured her in Southampton. In August 1911, Leonard Peskett even booked passage on the *Olympic* to study her.

Peskett's report seemed rather biased about the *Olympic*. He felt that the combination reciprocating engine/turbine powerplant, even if economical, took up too much space in the hull, reducing the number of berths available; the lifts were in a bad location; the backlit stained-glass windows in the First-Class Smoking Room were very poor; there was too much furniture in the staterooms, and not enough space. His list of negatives ran on and on, and it is clear that he thought he could do better. One ray of sunshine among his otherwise glum take on the *Olympic* was the numerous portholes behind the casement windows in her First-Class Dining Saloon, since they allowed more light to enter the room; however, he felt that the scheme of decoration in that room was a failure.

Despite his apparent bias against the White Star liner, on several points he had more or less hit the nail on the head in his criticisms. The fact that the First-Class Dining Saloon was only a single deck in height was a decision that

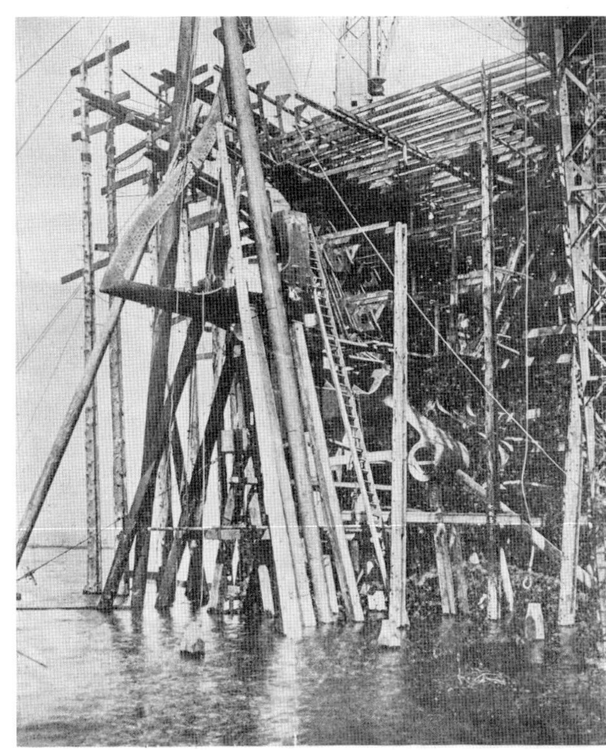

Left: This photograph shows the liner's after framing, as well as the stern posts and the brackets which would hold the propeller shafts. Although very little of the ship's final form is discernible at this stage of construction, the curve of her stern frame is immediately identifiable. (Authors' Collection)

Below: A workman uses a pneumatic cutting machine to create a porthole opening in the side plating of the *Aquitania*. (Authors' Collection)

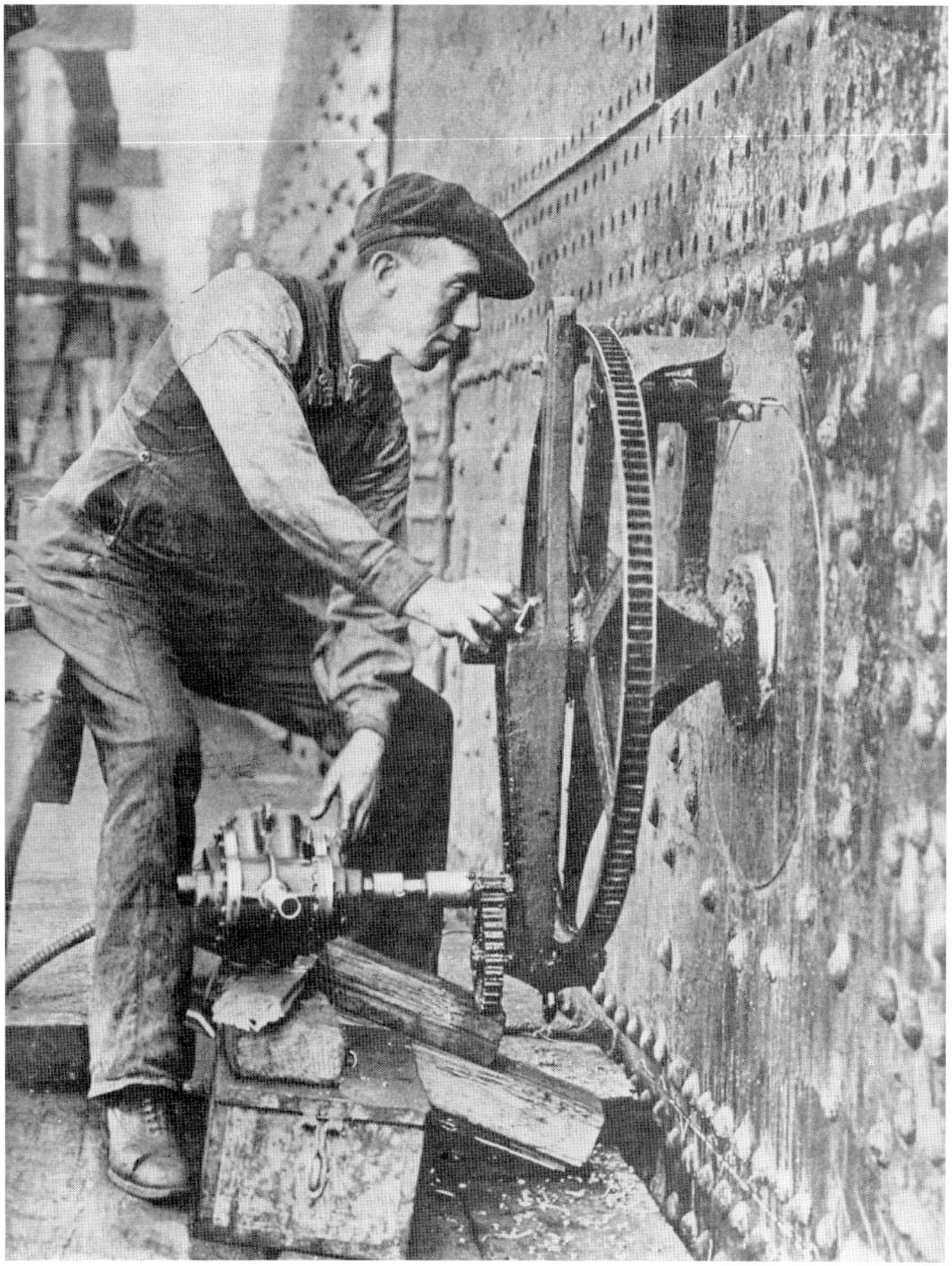

was questioned by many; not only did it make the room feel smaller, despite its great footprint, but it also failed to give hot air anywhere to escape, helping to make the room uncomfortably stuffy, particularly in the warm summer season. Also, Peskett noted that the layout of the Turkish Baths was not ideal. Both of these details, and some of the others he frowned on, were things that White Star and Harland & Wolff would eventually work to address.

Yet Peskett's criticisms seem unusually harsh. Many of his fellow Cunard employees, and those who would be involved in creating the *Aquitania*'s décor, clearly felt differently. Indeed, when the *Aquitania* entered service, there were many ways in which it was clear that her design had been influenced by the *Olympic*. One of these features was actually an expansion of a concept on the *Olympic*: on the White Star liner, there was a well-thought-out corridor which ran aft from the Lounge to the aft Entrance and Grand Staircase, from which access was given to the Smoking Room. High praise for this functional corridor was given by Cunard's people. A similar, but far larger and grander corridor, which was termed the Long Gallery, was eventually installed on *Aquitania*.[22]

Back at the shipyard, work on the new Cunarder continued. One account of her construction read:

Day by day this great spine [the keel] was added to by the ribs of framing. This delicate and difficult job consisted of something more than joining parts made to fit one another as one puts together the pieces of a jig-saw puzzle. With the aid of almost human cranes … the job certainly looked easy. But each process was the work of a group of specialised men who had grown up with the work, and it was due to their high efficiency and technical skill that the difficult task was performed so smoothly and rapidly.

Armies of workmen were now busying themselves at the bottom of the ship and working up into position the side framing, first amidships and then gradually extending in either direction towards the stem and to the stern, until eventually the ship presented the appearance of the skeleton of an enormous animal lying on its back. The work of putting on the flesh and the skin crept on, and a few weeks later the ship took more shape. … Before the plating was proceeded with, of course, the enormous beams which span the inside had to be put into place to take the decks.

To finish off, the framing of the ship, a most intricate operation, had to be faced. That involved the erection of the massive stem and stern frames, made of solid cast steel.[23]

The liner's stern frame, main and smaller brackets, and rudder, were constructed by the Darlington Forge Company, Ltd, which had built counterpart components for the *Lusitania*, *Mauretania*, *Olympic* and *Titanic*. The *Aquitania*'s stern frame was made of special Siemens-Martin mild steel, and had a length of 44ft, a height of 42ft and an extreme width of 12ft 6in. The total weight of the rudder alone was about 70 tons. Once these items were completed, it was necessary to ship them to the Clydebank shipyard for installation, which posed something of a problem:

Although the main piece of the stern frame was loaded … upon a railway trolley wagon having a rated capacity of 54 tons, its dimensions were of such size as seriously to exceed the railway load gauge, the 'wing' of the frame overlapping the wagon side by 10ft 10in., while the overhang on the abutment side was 5ft 6in. The total width was 15ft 2in., and the distance from rail level to the top of the load was 13ft 6in. It was therefore quite impracticable to transport the castings from Darlington to Glasgow by rail, and the only alternative was to convey them to Middlesbrough to be shipped from there by special steamer to Clydebank.

… The railway portion of the journey was successfully performed on Sunday last, April 28th. The main piece of the stern frame formed a very unsymmetrical load, and its position on the wagon required very careful handling … A maximum speed of about six miles per hour was observed, and all traffic on the adjoining lines was meanwhile suspended, whilst a temporary programme involving alterations in the existing passenger and goods service was introduced to admit of the safe transit of the load.[24]

Meanwhile, plans were moving ahead to engage the appropriate firms for creating the *Aquitania*'s grand interior spaces. Harold Peto had been engaged for the *Mauretania*, while James Miller had been hired to create the *Lusitania*'s interior spaces. Peto's price had been extravagant, and while the *Mauretania*'s interiors were a triumph of exotic, intricately carved woods, the *Lusitania*'s were less expensive and had a fresh, light, airy appeal. More than one firm was engaged to design and build the interior spaces of the new ship, and the contracts for this work were signed in October 1912.

One of the most sought-after architectural firms of the day was that of Mewès & Davis. Charles Mewès was a French architect, who had formed a partnership with British architect Arthur Davis to create the firm. The pair had executed the fashionable interiors of the London Ritz Hotel, which opened in May 1906, among their many popular projects. Albert Ballin, of the Hamburg-Amerika Line, had been so taken with their work that he had engaged Mewès on some of his own ships, including the upcoming *Imperator*-class behemoths.

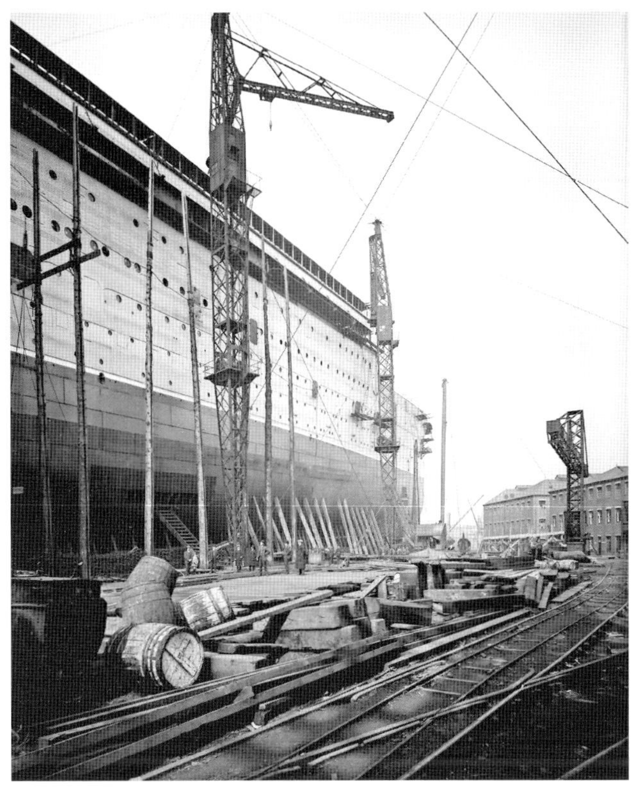

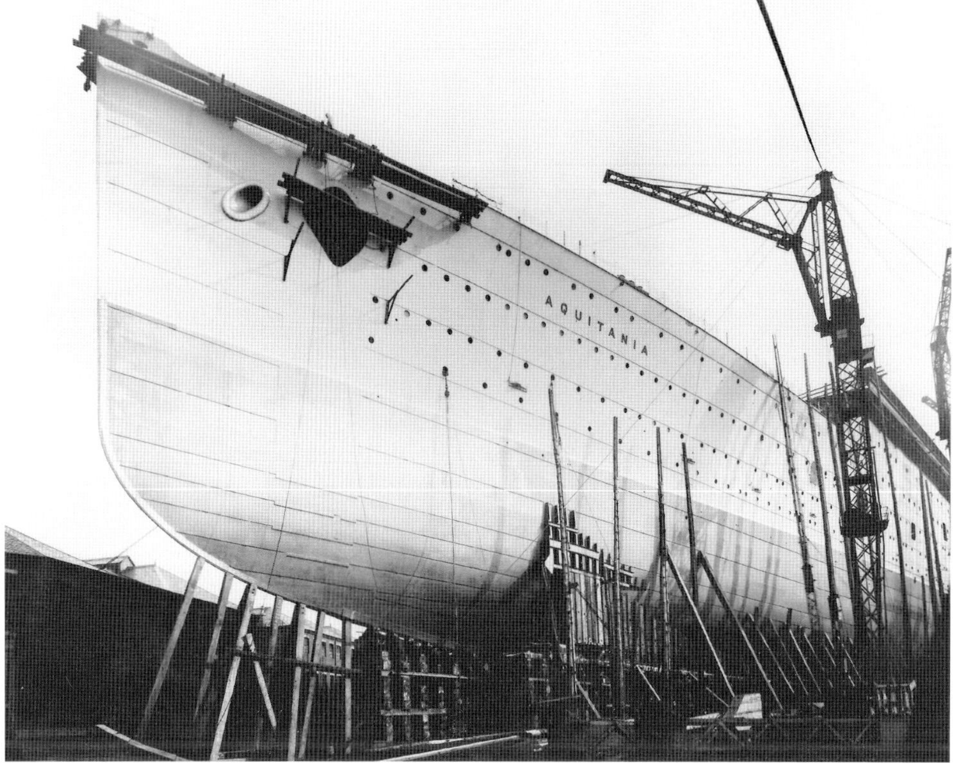

Clearly, Mewès was not only preoccupied with work on the new German ships, but to ask him to work on a new Cunard liner would also create a conflict of interests. However, Mewès' partner, Arthur Davis, was not then engaged on the German ships, and Cunard and HAPAG agreed that Davis could be engaged on the *Aquitania* project. The proviso, which was arguably a rather silly one, was that Mewès and Davis could not share with each other the details of their work. In the many years since, Davis has retained the lion's share of the credit for the *Aquitania*'s interior décor. However, there were others:

As is only to be expected in a ship of the great dimensions of the *Aquitania*, the accommodation for first, second and third-class passengers has been arranged on the most generous and elaborate scale. ... [The] excellent results achieved are unsurpassed in any vessel afloat, either in general utility or beauty of decoration. Messrs. Mewès & Davis, of London, have been the architects for the decoration, while Mr. James Miller, A.R.S.A., has designed the public rooms ...

The following firms, among others, have carried out work by contract in the passenger accommodation: M. Marcel Boulanger, Paris (main stairways and entrance); Messrs. P. H. Rémon & Sons, Paris (first-class restaurant decoration); Messrs. W. & E. Thornton-Smith, London (first-class smoking room); Messrs. Lenygon & Co., Ltd., London (first-class lounge); Messrs. Waring & Gillow, Ltd., London and Liverpool (first-class smoking room gallery, garden lounges, first-class secondary staircase, and the second-class verandah café); Messrs. Hoskins & Son, Ltd., Birmingham (furniture for the first-class staterooms on D deck); Messrs. Wylie & Lochead, Ltd., Glasgow; and Messrs. Trollope & Colls, Ltd., London.

Behind the scenes, all of these architects and firms were working to finalise the internal layout of the ship. In the end, the decorating work was so successful that the White Star Line, in the midst of construction of the *Britannic*, decided that upgrades to her accommodations needed to be made in order to help her remain competitive.

Top right: The extraordinarily long prow of the ship is still surrounded by workers' scaffolds. Interestingly, the upper hull of the ship was painted in light colours for the launch, just as the *Mauretania* built on the Tyne River was for that same event. The *Lusitania*, although built on the same slip as the *Aquitania* just a few years before, bore a darker hull colour for her launch. (Clyde George Collection)

Top left: The starboard side of the *Aquitania* towers over the John Brown shipyard. (Brian Hawley Collection)

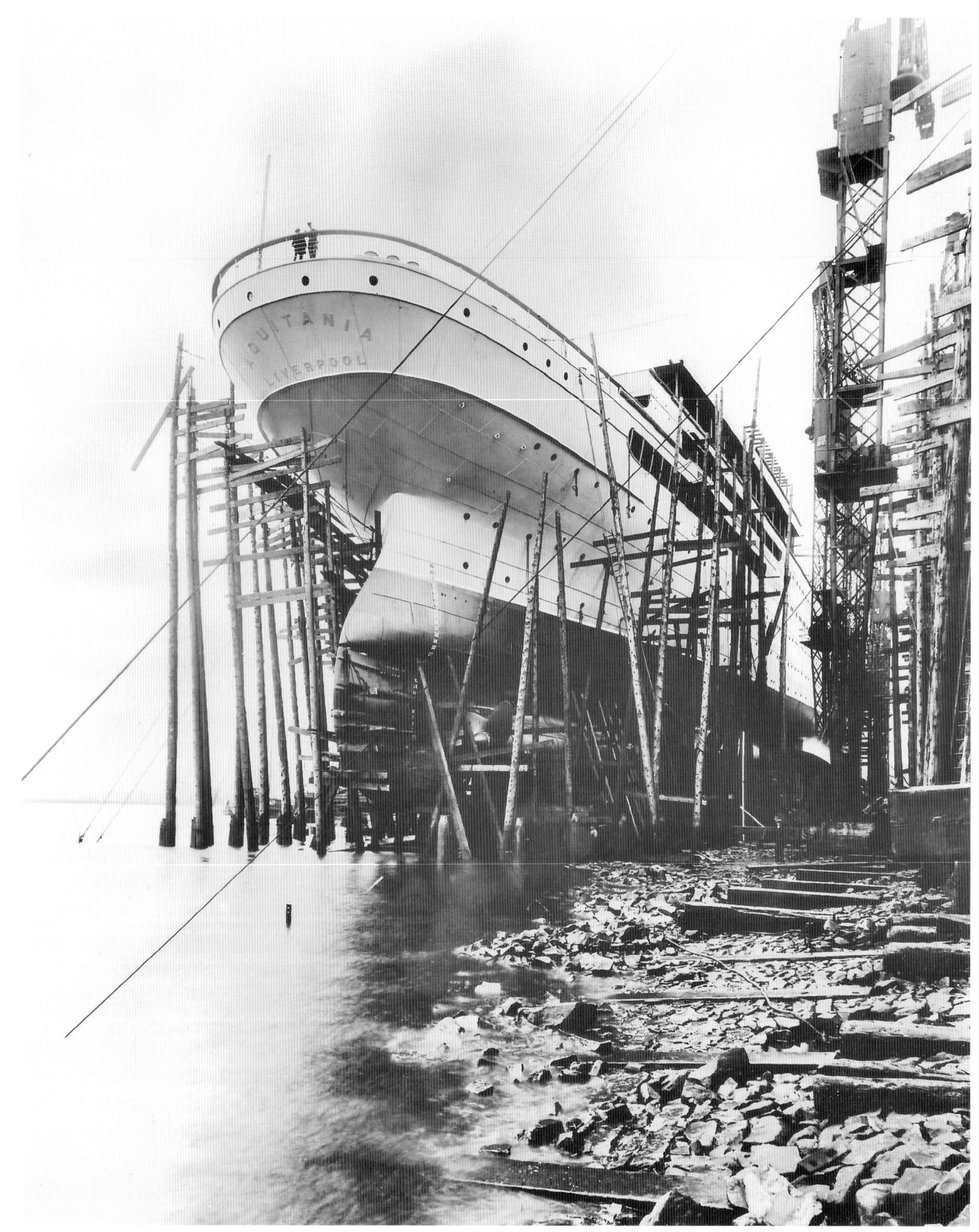

At the stern, it is clear that the liner's hull is complete, and only a little more work remains to be done on the upper works. Although scaffolds still shroud the hull, it is obvious that she will soon be ready to take to the water for the first time. (Clyde George Collection)

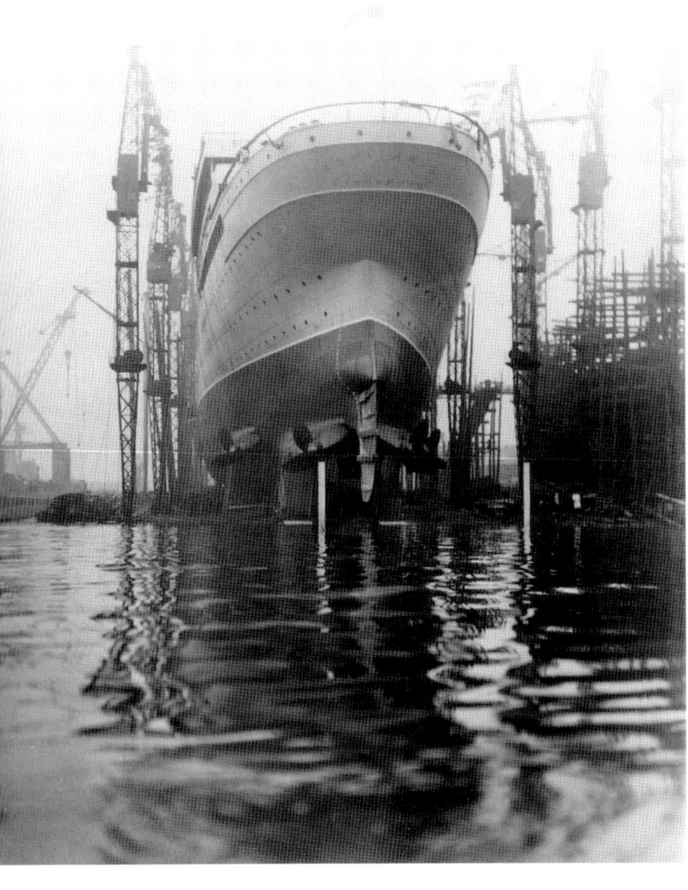

This photo was taken very shortly before the *Aquitania* was launched. Her hull is free of all encumbrances, and a flag flies from her taff rail. (Clyde George Collection)

While architects were designing the interiors, the actual ship was being prepared for launch. By January 1913, it had been decided that Lady Alice Stanley, Countess of Derby, was a perfect candidate for the traditional launching ceremony; a date of either 22 or 23 April 1913 had been recommended by John Brown for the ceremony, so time was of the essence. None other than Cunard's chairman, Alfred Booth, wrote to Lady Derby, conveying the company's wish. After she accepted the offer, Cunard's Board of Directors asked the John Brown shipyard to send a formal invitation to her Ladyship to christen the steamer. Before the month was out, it was reported:

The precise date of the launch has not yet been officially announced, but it is understood that provisional arrangements are being made for floating the vessel either on Tuesday, April 22nd, or Wednesday, April 23rd.

The hull of the vessel is fully plated from stem to stern, but a few openings have been left along the bilges for the admission of workmen and materials, which will ... be plated up shortly before the vessel is launched. All the decks ... bulkheads, and other structural essentials have been completed. The standing ways are almost wholly laid, and the work of temporarily fitting the sliding ways and launching cradle is very shortly to be proceeded with ...

The propeller [*sic*: propellers] and the rudder have yet to be fitted to the hull, but this, like the ... other necessary work pertaining to the launching make-up will not overtax the working forces at Clydebank in order to have all in readiness for the 22nd or 23rd April. It is expected that some time before the *Aquitania* is launched the battle-cruiser *Australia* and the Union Steamship Company of New Zealand's large steamer *Niagara*, which at present occupy the builders' fitting-out basin, will have left, so that there will be plenty of room left for the large Cunarder.[25]

Ultimately, the launch was set for Monday, 21 April 1913. A crowd which eventually numbered over 100,000 gathered well in advance so that they could witness the momentous event. It was reported:

The occasion was made one for general holiday on the Clyde, and this furnished opportunity for tens of thousands to attend the launching, while excursions from all parts of Scotland, express steamers from Belfast and special trains from London and the Midlands added to the excitement and beauty of the dramatic scene. At least a thousand persons waited all night in the meadows in order to get a good point of vantage for the moment when the giant, clifflike ship moved into the water.[26]

As the morning waned, anticipation began to build in the packed shipyard, and along the banks of the river where the crowds had gathered to watch:

It was just half an hour after noon that the Countess of Derby sent the beribboned bottle of wine

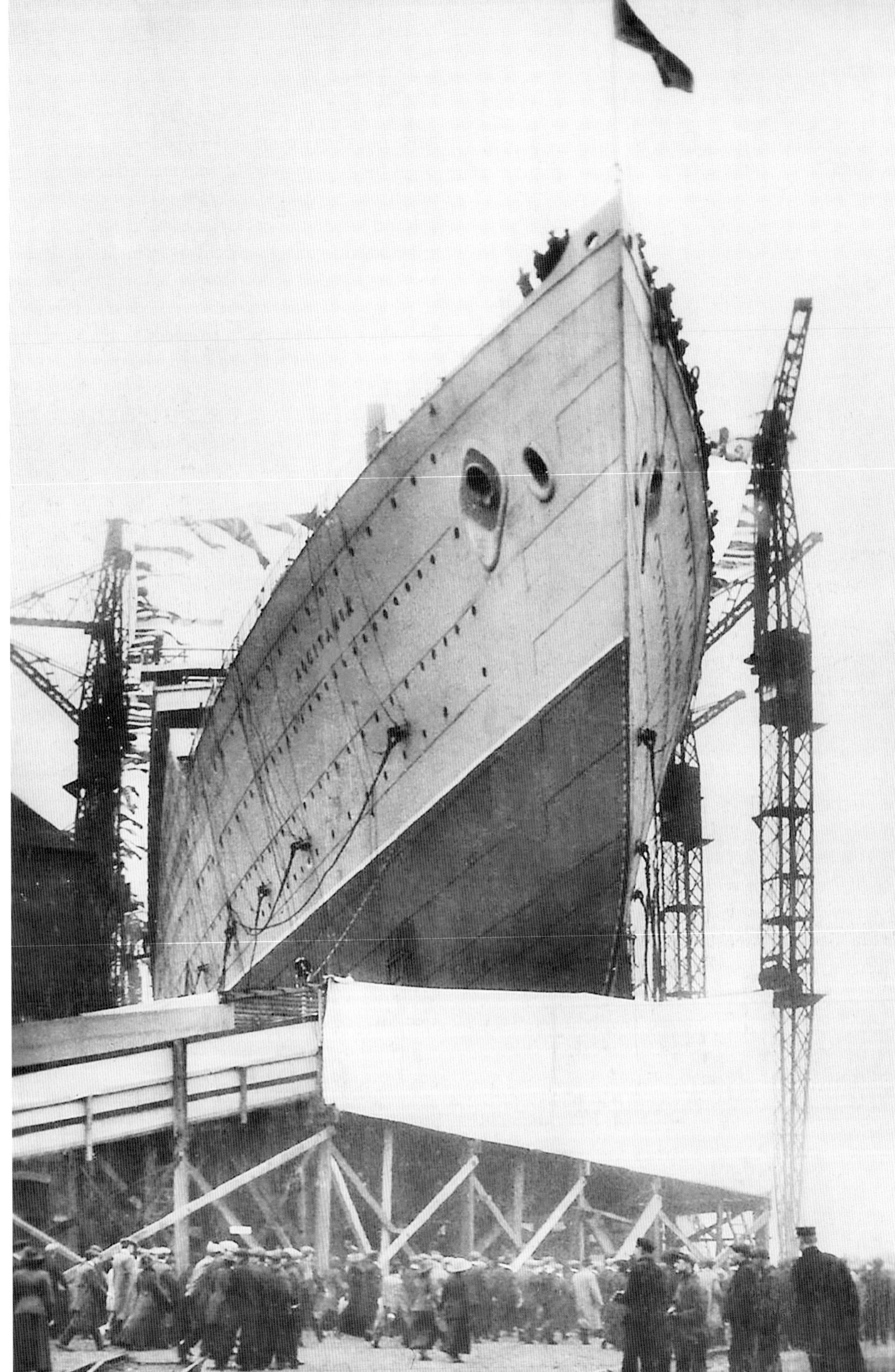

This photograph was taken shortly before the ship was christened, on 21 April 1913. Workmen can be seen atop the liner's forecastle. More shipyard personnel are visible in the lower right foreground of the photo, while both men and women, obviously guests for the event, can be seen just in front of the grandstands. (Clyde George Collection)

crashing against the starboard bow of the ship. It broke at once and spattered the grey plates with its contents.

'Good luck, *Aquitania*!' The godmother's wish was echoed by the crowd of onlookers. A shrill shrieking of syrens on every hand drowned the cheers, and then a sudden hush fell.

Few minutes in life are so tense as that in which one waits for a new vessel to start on her way down the slip. How much more so to-day, when the thick massed crowds in the shipyard, on roofs, on cranes and gantries, and in the fields on the further shore, waited to see if the Clydebank colossus would behave like a lady.

An inch or two at a time the *Aquitania* moved. There was a creaking and a groaning in the blocks beneath her keel.

Success was assured, and she seemed to swim down the ways to the Clyde, disturbing the water so little that one would say it was a ripple and not a wave that she made.[27]

A technical journal reported:

As a spectacle [the launch] was magnificent ... The technical details involved in the floating of a mass of 22,000 tons of steel, the declivity of the ways, the speed attained ... would, indeed, have been of great interest, but the builders, John Brown and Co., Limited, are keeping them a secret for the present. ... Thus very little is left to say that will appeal to technical readers, but there were just two small points which struck us in connection with the launch itself. One was the considerable amount of smoke developed by the burning of the

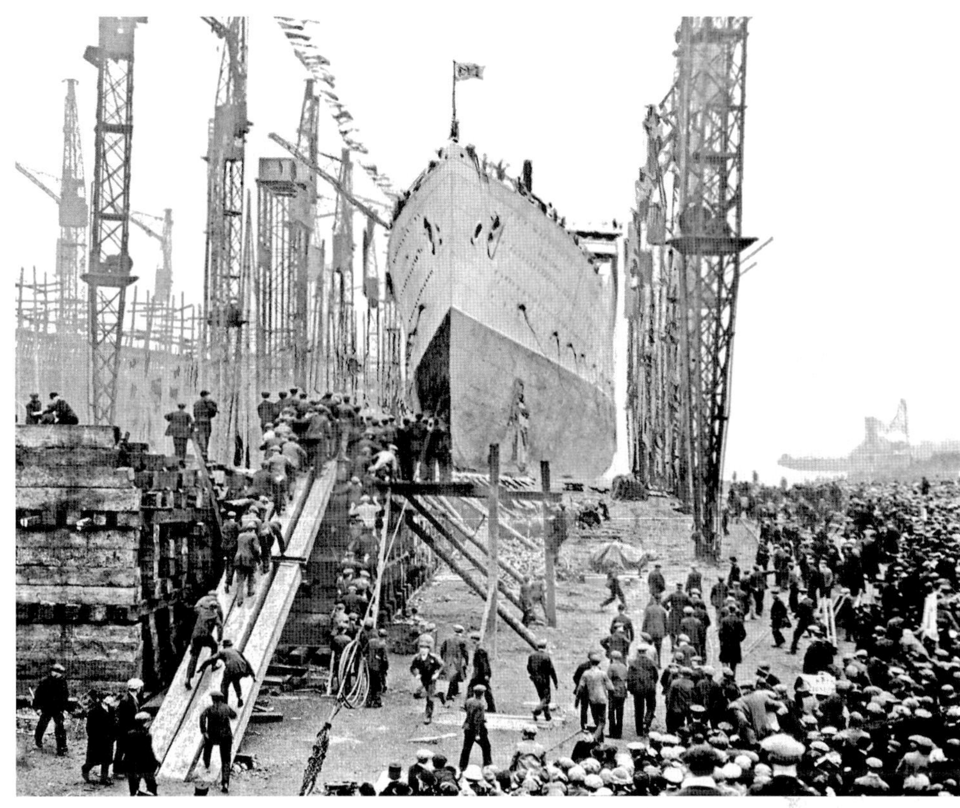

This fantastic view shows the *Aquitania* receding down the ways into the water, while excited spectators begin to break line and rush after her. (Clyde George Collection)

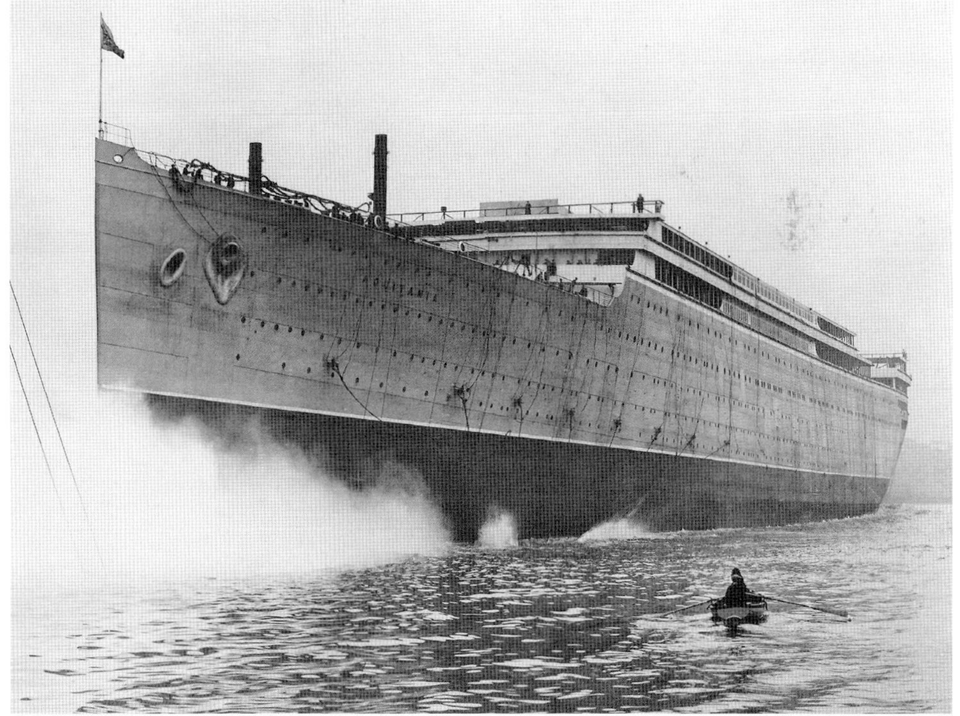

The enormous hull of the *Aquitania* slowly comes to a stop in the middle of the Clyde River. The small rowboat in the foreground gives some sense of scale. Interestingly, the windows along A and B Decks are apparently not windows at all. Along B Deck, their shape is far too narrow to match their final appearance; on A Deck, the 'windows' begin at a point too low above the deck, and one 'window' and the bulwark below it are actually tipped open to allow for a cable to pass through. It would appear that these were clever facades to give the impression the superstructure was more complete than it actually was. (Clyde George Collection)

grease on the ways; this was, in our opinion, greater than in the case of the slightly larger *Vaterland* launched recently in Germany. An examination of the after end of the ways after the launch seemed to indicate that the pressure had been somewhat near the limit, as the surface of the wood had the appearance of being broken as if roughed up with a file. Probably the grease gets thin after the passage of half the length of the cradle, and it is just at this time that the weight on the forward cradle is so enormously increased – that is, just when the stern is water-borne and the rest of the weight is all taken on the forward poppets.

The other point is the remarkably small amount of headway on the ship after she had been brought to rest by the 1400 tons of bundles of chains dragged over the shore by wire cables attached to eye plates along the sides of the ship. The fact that one of these cables, even with its movable resistance, snapped shows how futile the old plan of dropping an anchor to take the way off would be when such huge weights are involved. The gradual braking effect of the bundles ... certainly seems to be the soundest method imaginable.[28]

These last comments may have been a broad swipe at the launching of the *Imperator* in Germany the year before, when anchors were employed during the launch and someone had forgotten to attach one of the anchors' chains to the ship, with the resulting loss of both the anchor and its chain. The actual length of the *Aquitania* had by then been made public knowledge, meaning that she was the longest British-built ship in history; yet an odd sort of saturation point had been reached with these new large liners, it seemed, for the same journal openly said:

Tugs begin to move the incomplete hull of the vessel; the ones at the stern pull to starboard, and those at the bow pull to port, rotating the liner for the short tow to the fitting-out basin. (Clyde George Collection)

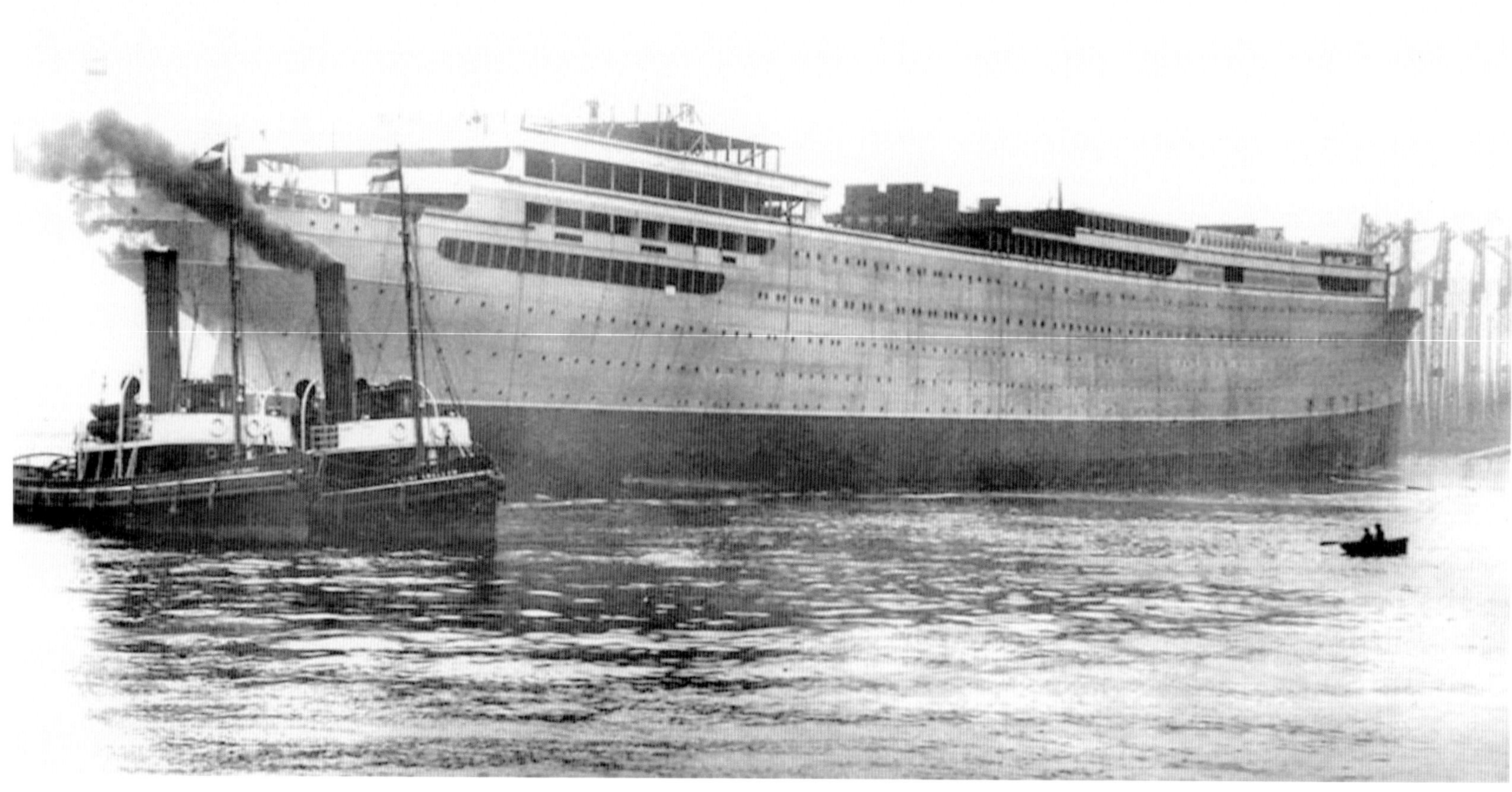

Our supply of superlatives was, we are afraid, exhausted in dealing with the *Olympic* and *Titanic*, so that it is almost a relief to be able to say that the *Aquitania* is not the largest ship in the world, much as we should have liked to be able to say so for the sake of the credit of the country. Nor are superlatives needed yet with regard to the accommodation – the comforts provided, the palm courts, racket courts, and other luxuries, for about them the officials of the Cunard company are at the present moment silent. Doubtless from their attitude they have some surprises in store which they desire to show in actual being before they are staled by verbal repetition or copied by rivals in a ship which comes into service a few weeks earlier [*Vaterland*].

With the launch complete, a bevy of tugs approached the hull, and brought her over to the newly redredged fitting-out basin. It was here that her engines and equipment would be installed, her superstructure, funnels and masts erected and her interior appointments turned into the wonders that they eventually proved to be.

By August, it was reported that 'every effort is being made to hurry on the work'.[29] By November, the massive turbine engines were being lifted into their final positions inside the ship, and 'to enable them to be lowered into the hull one of the four great funnels has not yet been placed in position'.[30] With that work completed, the final funnel could take its place. On 10 January 1914, it was reported:

> The last of the four funnels of the new Cunarder *Aquitania*, Britain's largest liner, was placed in position yesterday in the Clydebank Dock. Several hundred men were employed in the operation.[31]

Unfortunately, a series of labour disputes helped to ensure that completion of the ship fell badly behind schedule. The originally specified delivery date of 30 November 1913 had come and gone by the time her final funnel was placed atop the superstructure. On 17 February 1914, the press were finally able to give the date on which the new liner would make her maiden voyage: Saturday, 30 May 1914. However, there may have

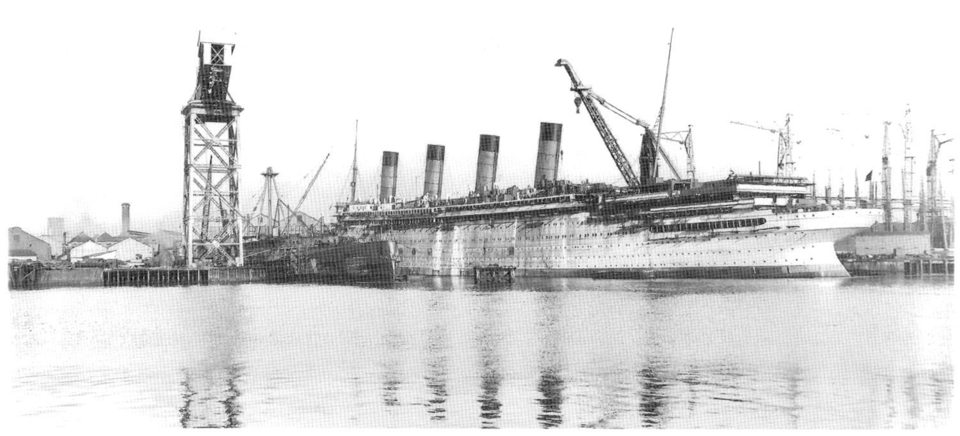

This view shows the *Aquitania* in the fitting-out basin at John Brown's shipyard. All four funnels are in place; from the position of the crane, it would appear that the fourth funnel has only just been set in position, an event which took place on 10 January 1914. Notice here that the areas of A and B Decks which would eventually be covered with windows, and which had facades of windows and bulwarks in place through the launch and most of the fitting out, are now open again, and workers' scaffolds show that the permanent enclosures would shortly be installed. (Clyde George Collection)

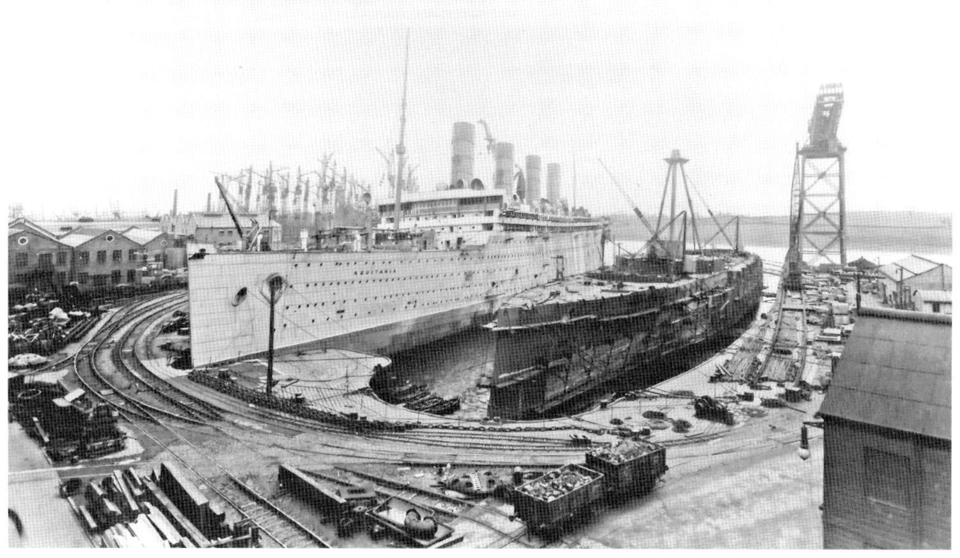

been concerns in some quarters about whether the ship would actually be finished in time. This is because, in early 1914, the joiners had gone on strike. By the end of March, Thomas Bell reported that there was 'a terrific scramble' to get the work finished in time for the maiden voyage.[32]

Back then, a ship's first chief engineer was typically selected quite early, and he would then spend time

A general view of the fitting-out basin as work continues on the *Aquitania* in early 1914. In this photograph, work on setting the windows along A Deck, where the First-Class Garden Lounge would eventually be placed, has moved forward over their state in the preceding view. An incomplete warship fills the other side of the basin. (Clyde George Collection)

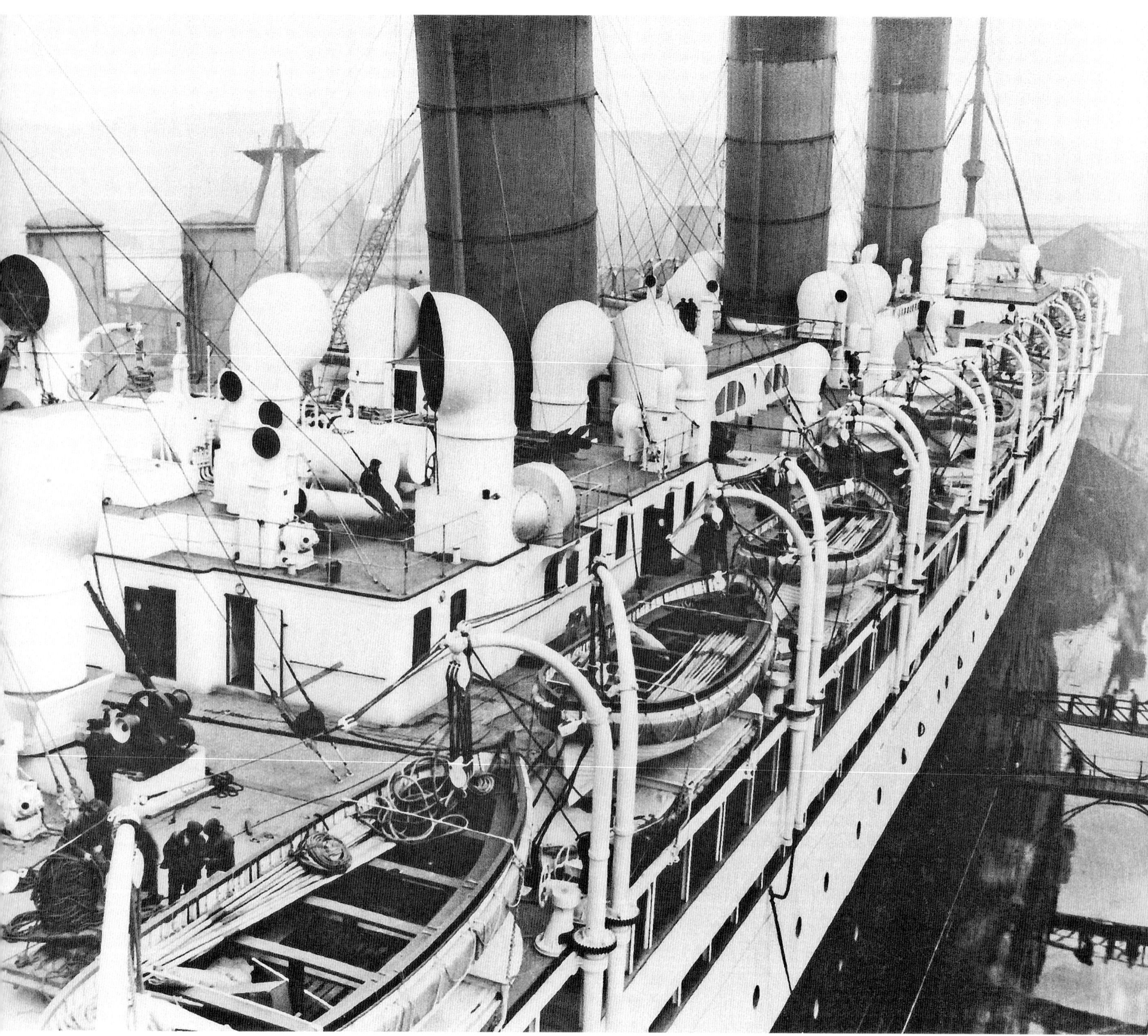

A splendid photograph, likely taken from atop a fitting-out crane, shows the freshly installed lifeboats along the Boat Deck, as well as the forward three funnels. (Clyde George Collection)

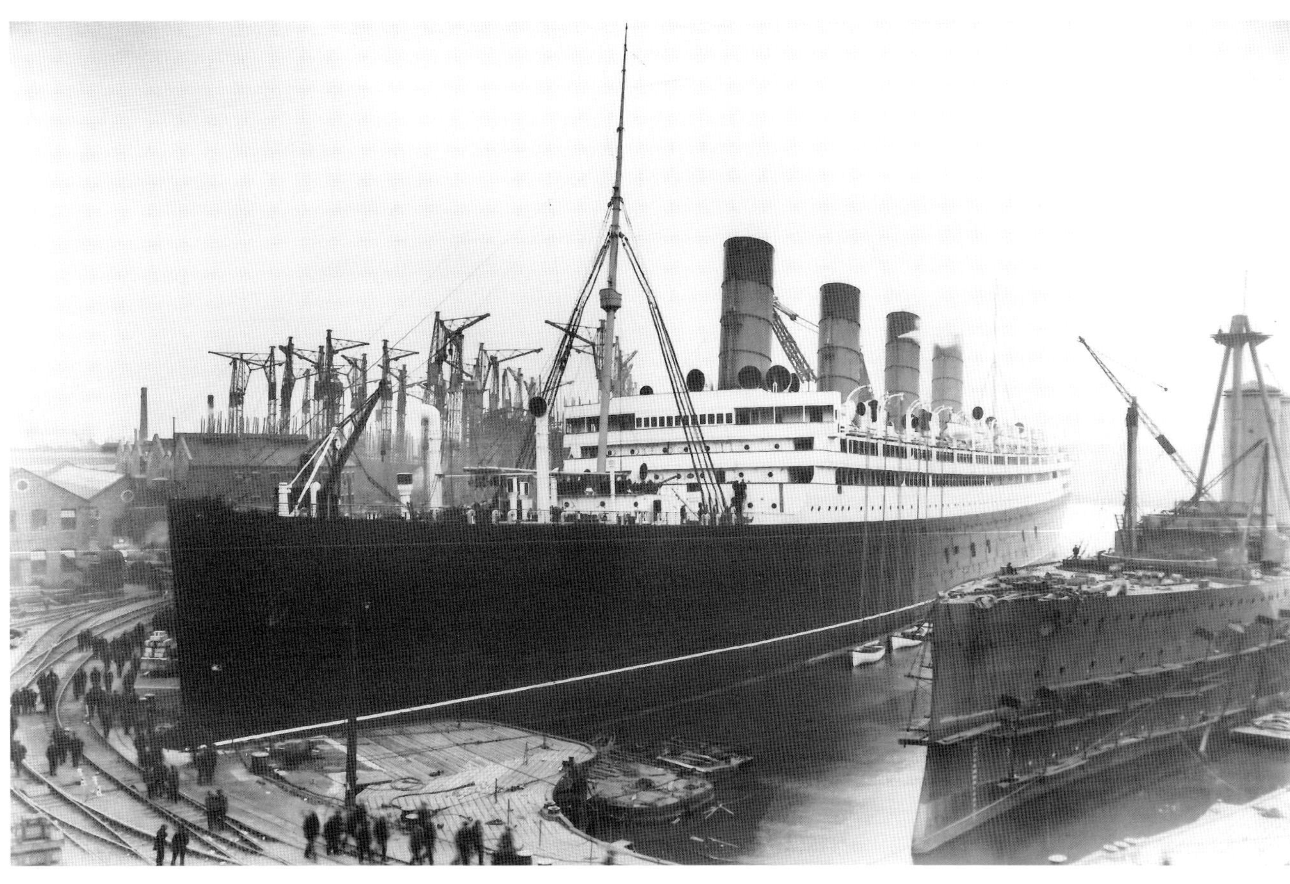

This photograph, taken from a very similar location as the one on page 27, shows that the *Aquitania* is nearing completion. Her hull has received its final livery, her steam powerplant is active, as evidenced by the steam emanating from the third funnel, and it would appear that the operation of some of the lifeboat davits is being tested. (Clyde George Collection)

on-site during the fitting out of the ship. This would ensure that they were thoroughly familiar with the new equipment, and that work on the machinery was being performed satisfactorily. Cunard had announced that the new ship's first chief engineer would be Alexander Dunbar, who had been with the company since May 1883, and who had been the chief engineer of the *Lusitania* from February 1911; they sent him up to the yard to oversee the fitting out of the machinery. Sadly, Dunbar came down with a 'severe attack of pneumonia' as the ship was nearing completion. Before he had fully recovered, Dunbar – who was 'anxious to be out during the completion of the *Aquitania*' – had 'resumed his overseeing duties on that vessel'. The result was predictable: he relapsed, and passed away on 4 April.[33] Although it was a sad moment for the company and the new liner, work had to proceed. Archibald Bryce was named *Aquitania*'s new chief; he had been with Cunard for over thirty years, and had also served as chief engineer on the *Lusitania*.

William Thomas Turner, RNR, was appointed as the *Aquitania*'s first commander; he was then the commodore of the Cunard fleet, and had served as master of both the *Lusitania* and *Mauretania*. Turner was a distinguished officer with a penchant for getting the most out of his ships and his men; it was while he had commanded the two speedsters that many of their speed records had been set. Just the previous year, Turner had the honour of receiving the king and queen aboard the *Mauretania* during the ceremonies surrounding the opening of the new Gladstone Dry Dock in Liverpool, and the accompanying merchant shipping review. Cunard had created the post of staff captain following the *Titanic* disaster, to help ease some of the burdens of responsibility for the skippers of their crack liners; but for the *Aquitania*, they saw fit to appoint two staff captains to aid Captain Turner. In this way, either Turner or one of his two staff captains could always be on the bridge; beneath these three captains, seven additional officers were appointed, almost all of whom had masters' certificates.

AQUITANIA: AN INTERIOR TOUR
FIRST-CLASS

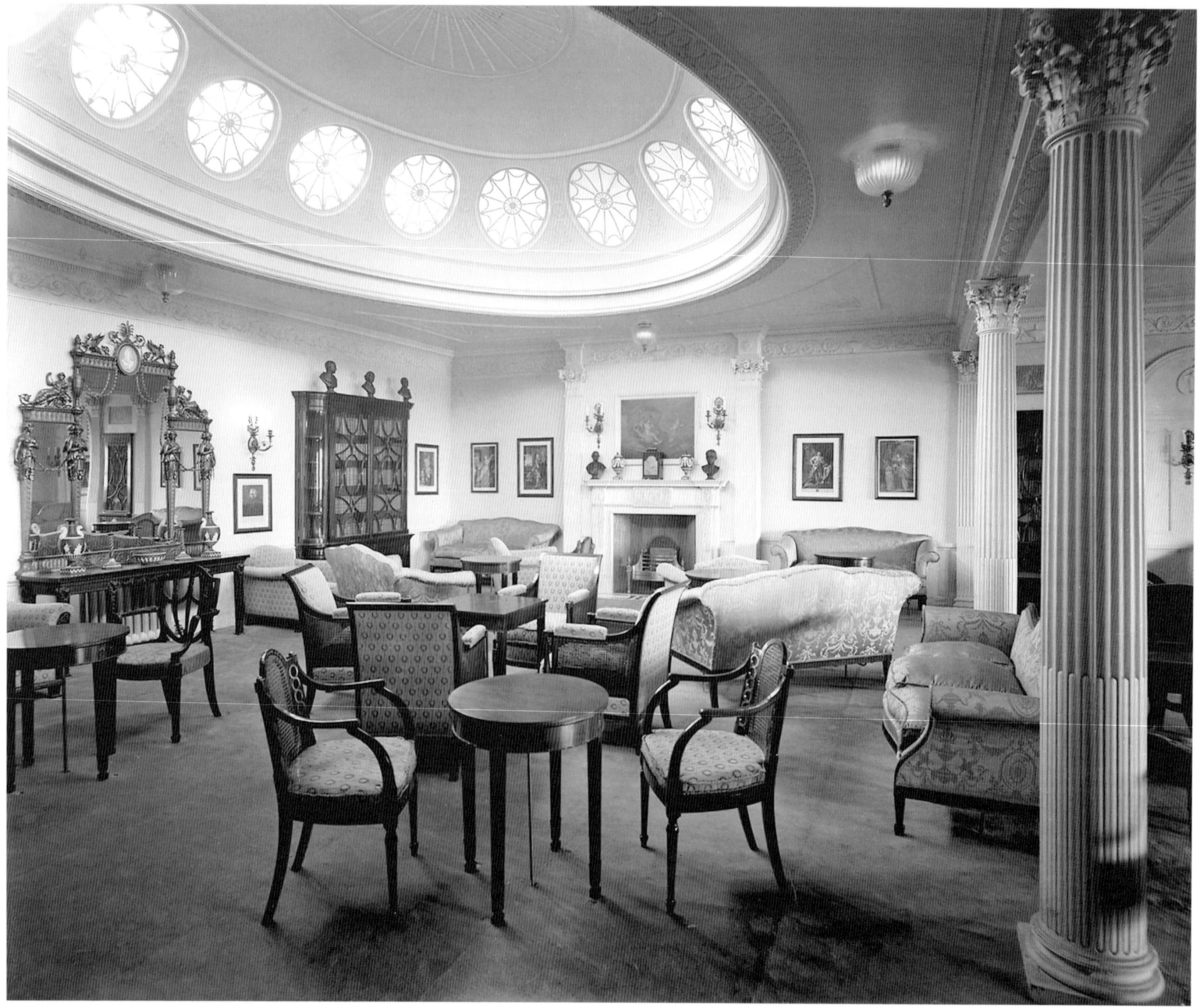

The First-Class Drawing Room was located on A Deck, tucked just behind the uptake for the forward funnel and forward of the Entrance. The room was decorated in the Brothers Adam style, *c.* 1780, and bore a strong resemblance to the First-Class Drawing Room on the *Lusitania*. (Clyde George Collection)

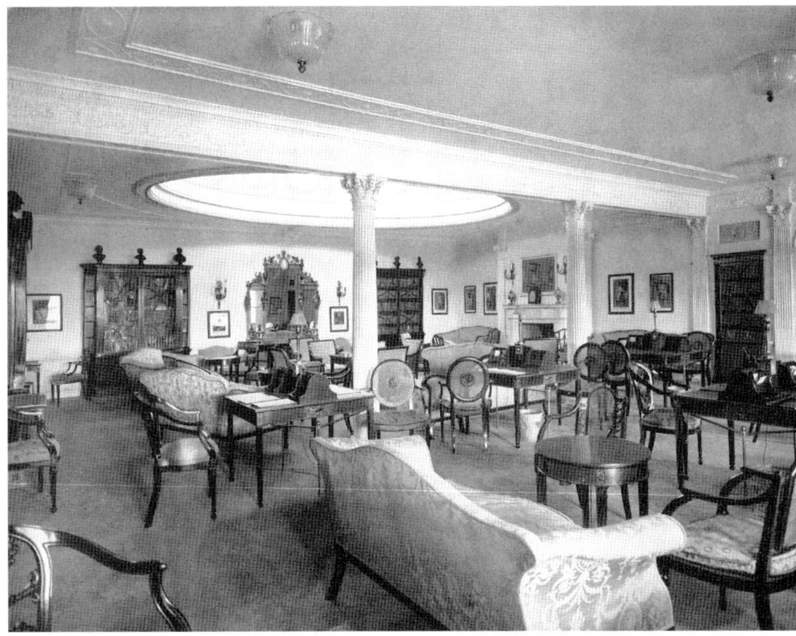

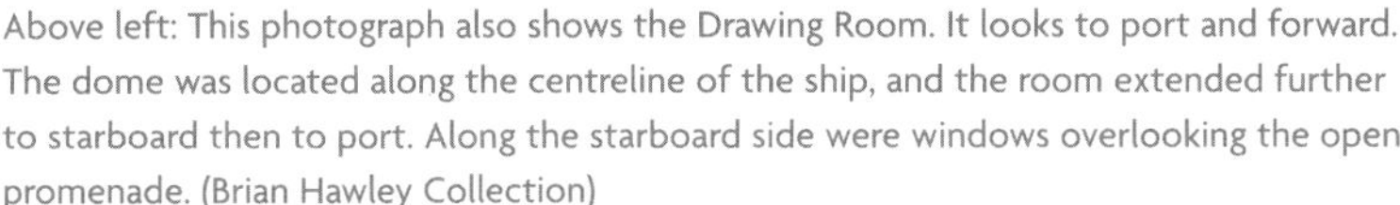

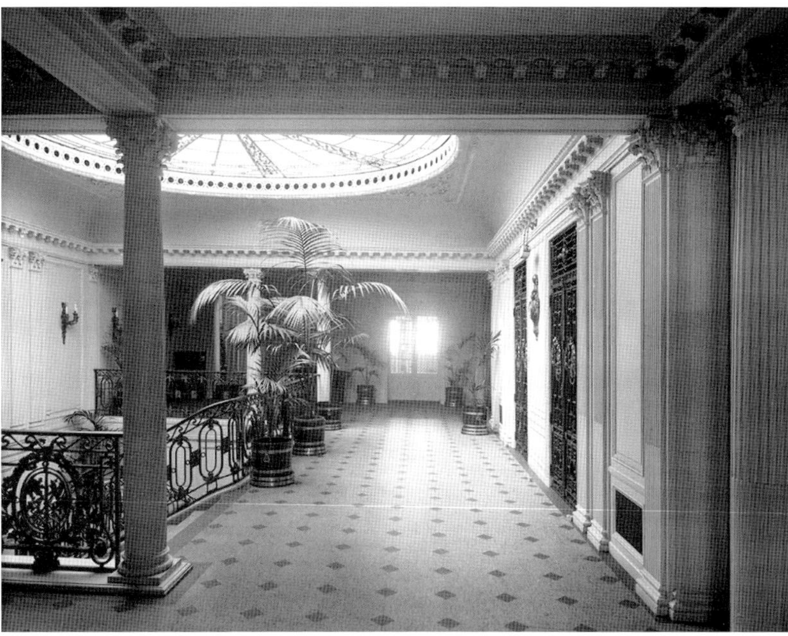

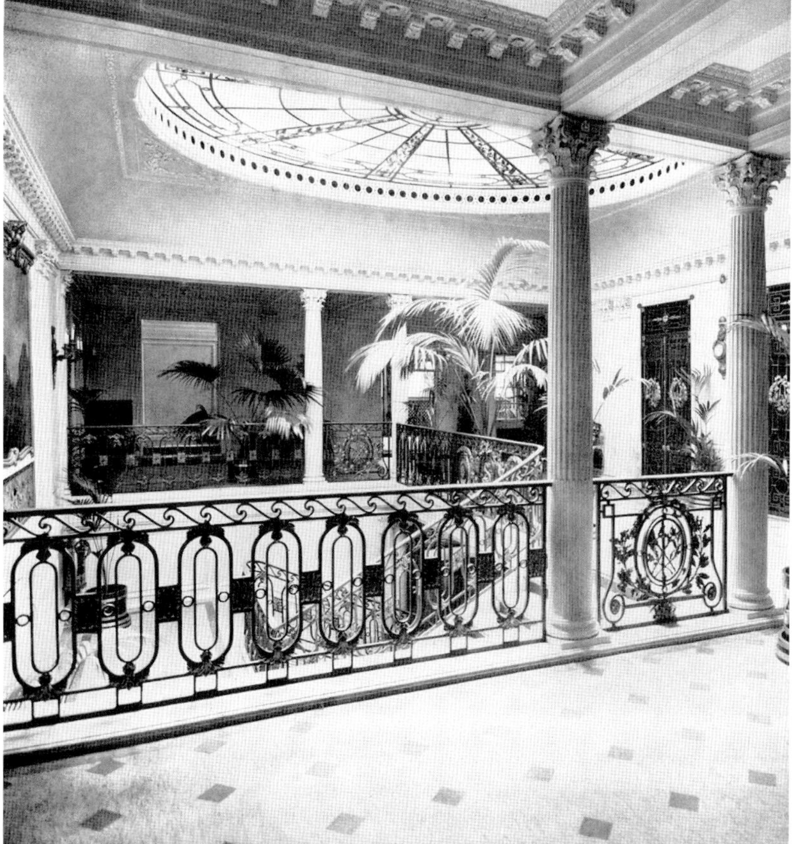

Above left: This photograph also shows the Drawing Room. It looks to port and forward. The dome was located along the centreline of the ship, and the room extended further to starboard then to port. Along the starboard side were windows overlooking the open promenade. (Brian Hawley Collection)

Above right: Just aft of the Drawing Room was the First-Class Entrance and Main Staircase, decorated in the Louis XVI style. This photo looks to starboard from the port side. On the left, or forward end, are the stairs; on the right are the pair of lifts designed to assist weary passengers reach the deck they needed. (Clyde George Collection)

Right: This photo, also looking from port to starboard, gives a better view of the balustrade around the stairs and how the staircase wrapped its way around as it descended. (Eric Sauder Collection)

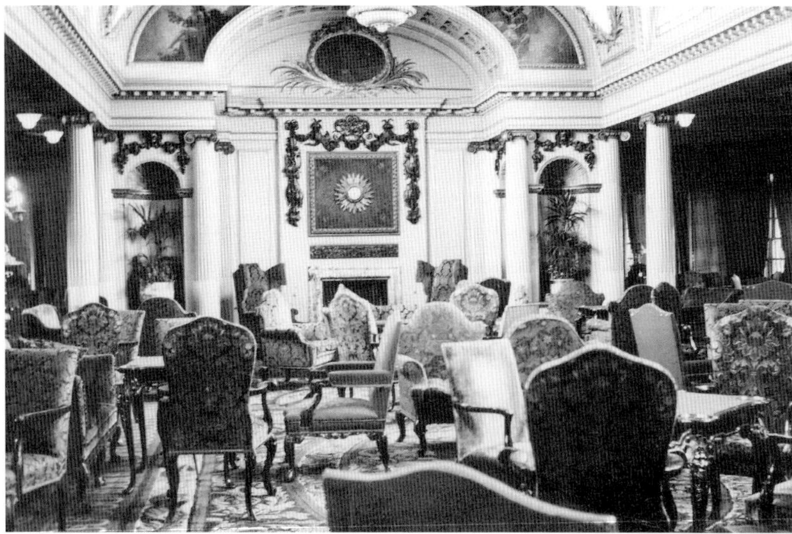

Moving aft from the Entrance on A Deck past the uptake for the second funnel, which was ensconced on either side by a salon sitting area, one entered the Lounge. This large room was intended as the social centre for passengers of both sexes during a crossing. Primarily George I in style, it also was referred to as the 'Palladian' Lounge, as Georgian architecture was heavily influenced by Palladianism, a philosophy of design based on the work of sixteenth-century Italian architect Andreas Palladio. The oak floor was designed for dancing. At the aft end was a semi-circular stage, while at the forward end, seen here, was a fireplace. (Ioannis Georgiou Collection)

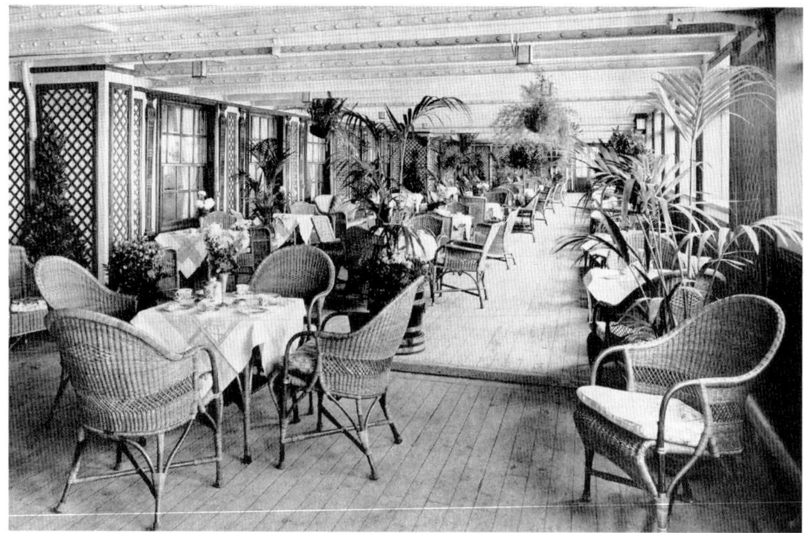

One of the Garden Lounges, which flanked the First-Class Lounge and also the Entrance. Each of these rooms was roughly 150ft long. The walls were covered in teak trellis, and the rooms were furnished with small tables and wicker chairs. The Garden Lounge on one side was designated for smoking, and the other as non-smoking. However, the smoking side was used so heavily during the maiden voyage that Cunard thought it prudent to allow smoking on both sides. Captain E.G. Diggle recalled that these rooms were 'filled with the fragrance of lovely flowers tended by the gardener'. (Eric Sauder Collection)

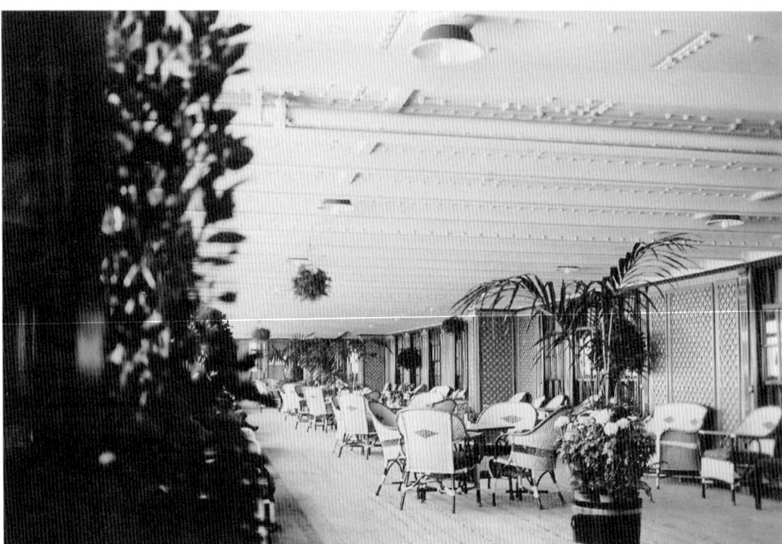

Another view of the Garden Lounges, very artistic in its flavour, showing how the rooms looked between the wars. The wicker chairs have been painted in a light colour, with a darker diamond pattern on the backs. (Steven B. Anderson Collection)

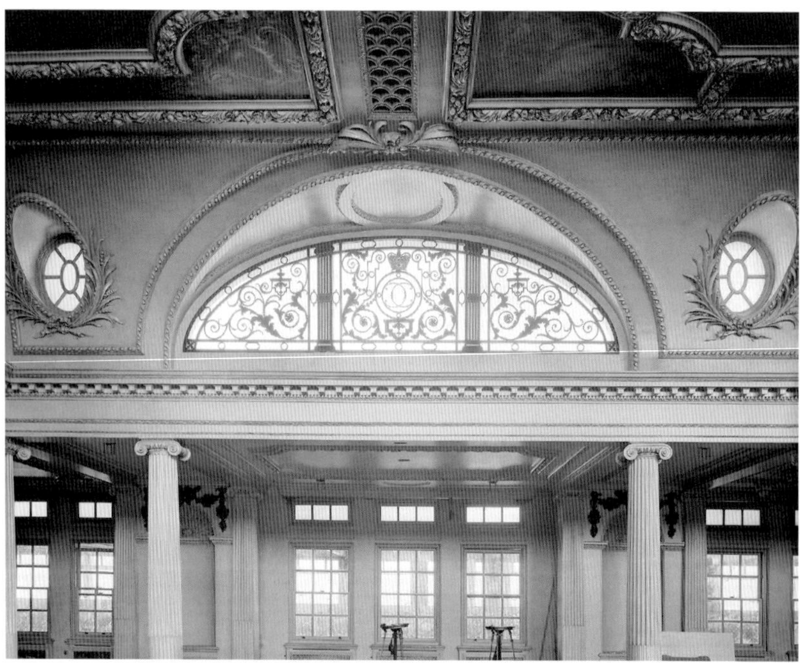

The Lounge was located amidship, couched on either side by the Garden Lounges, which overlooked the sea through large windows. Windows from the Lounge, seen here, gave a view onto the Garden Lounges and thence to the sea beyond. Above, the higher central portion of the Lounge ceiling, which was painted with exquisite murals, had side windows which overlooked the Boat Deck. This view shows the Lounge, and the Garden Lounges outside, in an unfurnished state. The chandeliers are also on the deck. (Eric Sauder Collection)

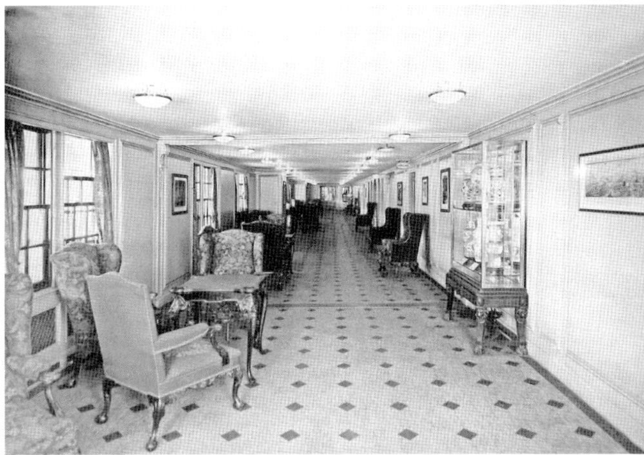

Above: This view shows the Long Gallery, which was located on the port side of amidship, inboard of the open First-Class Promenade, and astern of the Palladian Lounge. It later came to be known as 'High Street', and by the close of the 1920s featured a variety of shops from which customers could purchase goods during the trip: a men's clothier; a fancy goods store where women could purchase handbags, perfume and the like; a chocolate and sweet shop – a favourite haunt for 'flappers', it was said; a tobacconist; and conveniently enough, a jewellery store that offered engagement rings with which to formalise blossoming shipboard romances. (Author's collection)

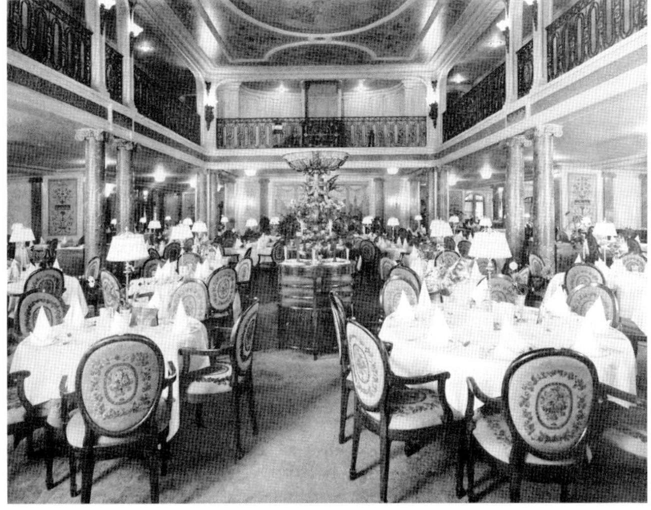

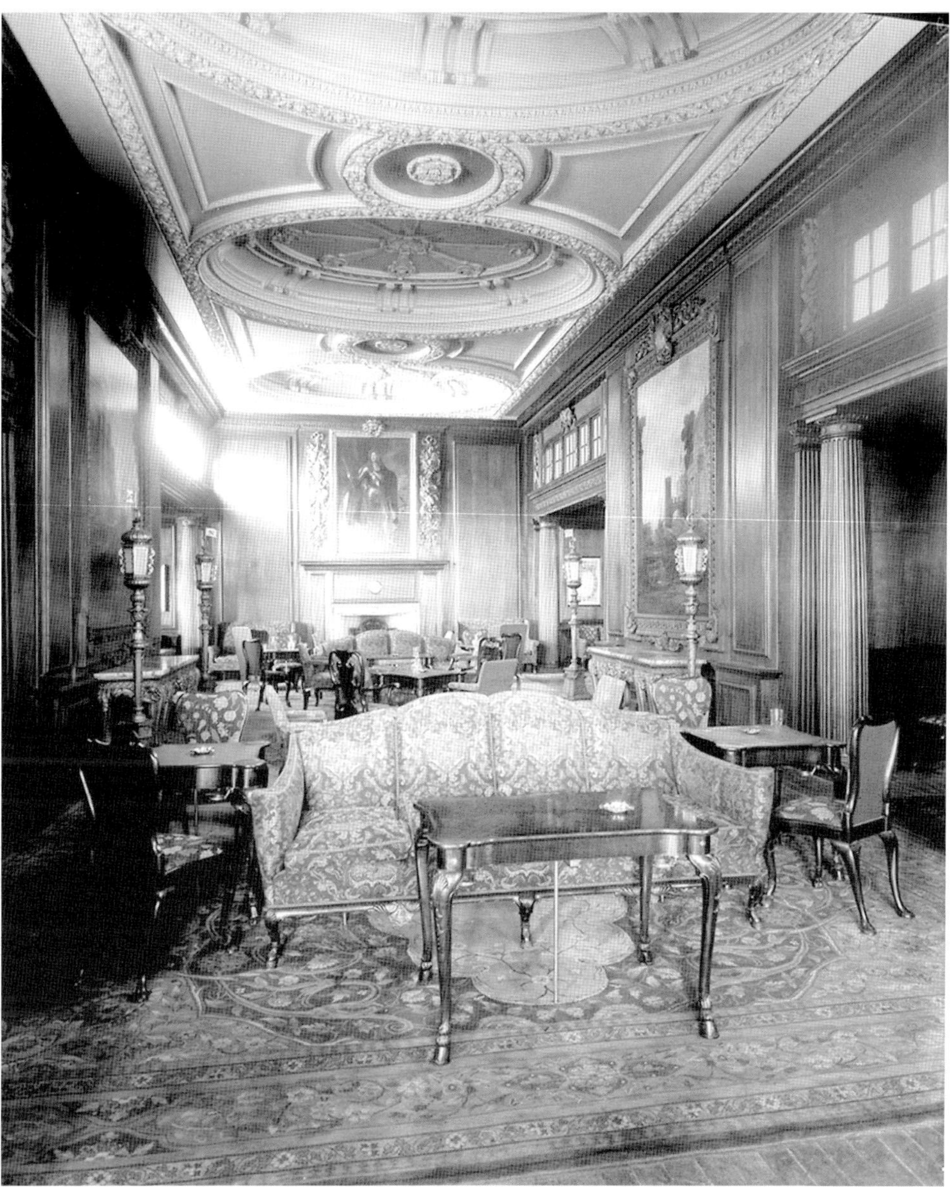

Above: The Charles II-style Smoking Room was located aft of the Long Gallery on A Deck. It was adapted from the architecture of the Greenwich Hospital. From the central square of the space there branched out a series of smaller alcoves, featuring oak-beamed ceilings. The central hall of the Smoking Room, amidships, had a higher ceiling with side windows admitting light into the room from the Boat Deck above. (Clyde George Collection)

Above: Despite the dangers of seasickness, among the favourite daily activities was a meal in the Dining Saloon. *Aquitania*'s First-Class Dining Saloon was officially referred to as the Restaurant – a clear attempt by Cunard to break with old shipboard terminology. Rendered in Louis XVI style, but with the carved mahogany panelling painted grey, tables and seats were arranged only on D Deck. However, Cunard's architects came up with a clever way to escape the dreaded 'single-deck Saloon' idea which some of them felt was such a mistake on the *Olympic*-class ships: the central portion of the room, seen in this photo, featured a well opening on to C Deck above, the ceiling of which bore glorious murals. One potential drawback was the fact that on both outboard sides of the well on C Deck were passenger cabins. Although none opened directly onto the walkway over the Saloon, noise and aromas were sure to drift into the passengers' staterooms where, hopefully, it was appreciated rather than unwelcome. (Brian Hawley Collection)

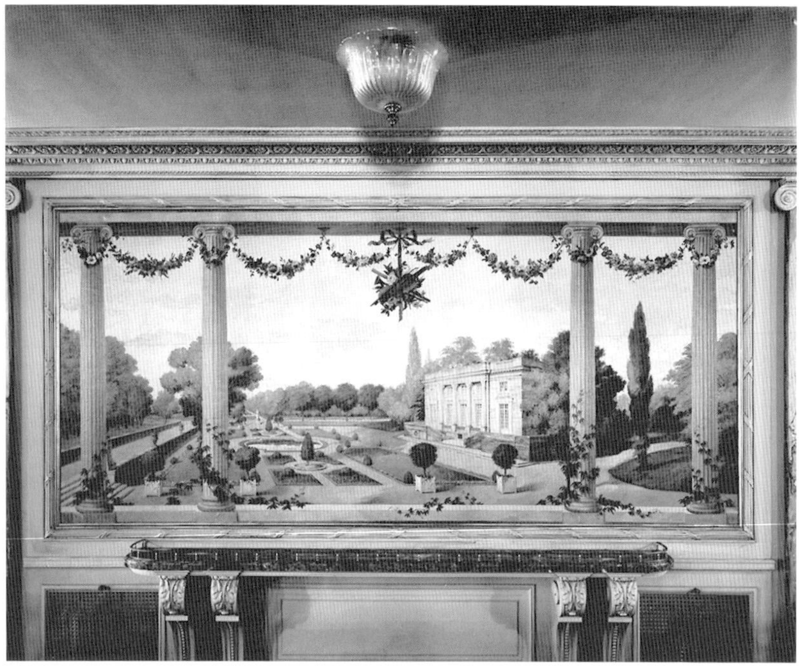

The Dining Saloon, or Restaurant, was filled with beautiful art. This masterpiece was titled *Parc du Grand Trianon*, and depicted the gardens at the Palace of Versailles. (Clyde George Collection)

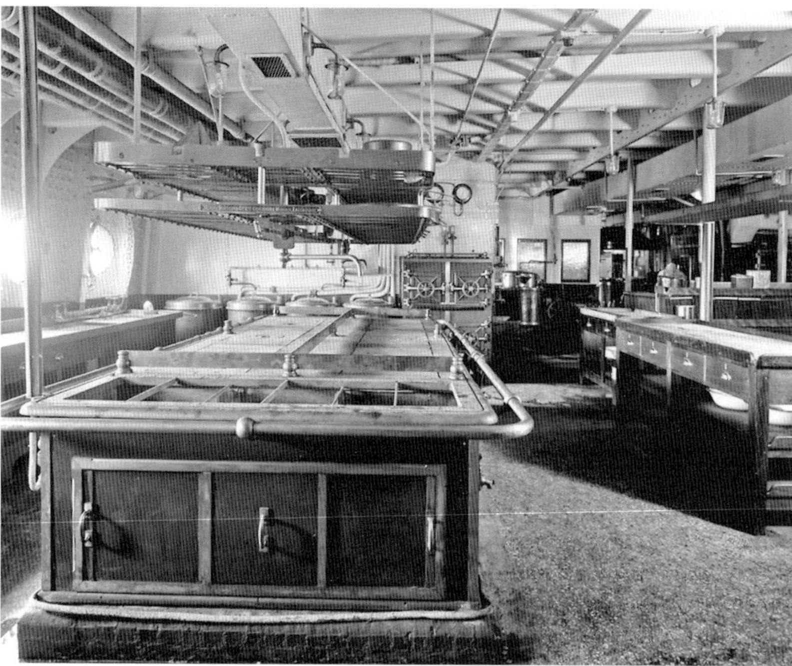

The fine cuisine which First-Class passengers enjoyed was prepared in the First-Class Galley. It was located aft of the Dining Saloon on the starboard side of the ship. (Clyde George Collection)

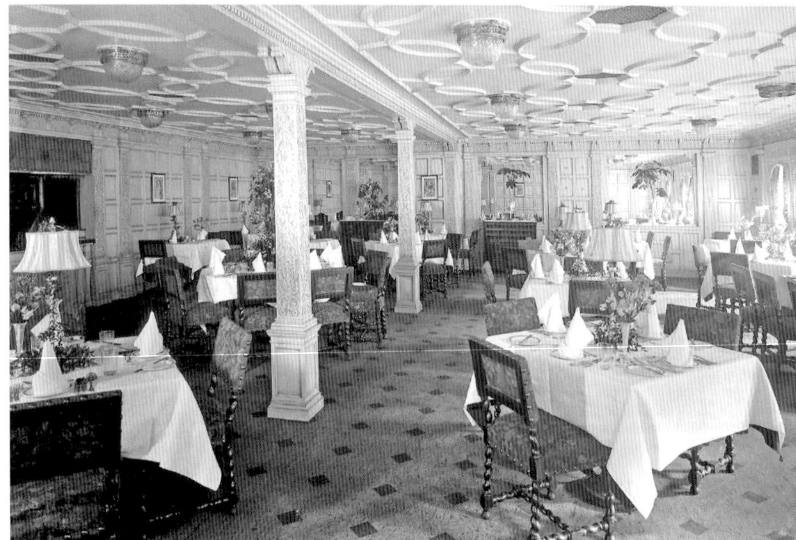

On the port side, opposite the First-Class Galley, and just astern of the Dining Saloon, was a separate dining space: the Grill Room. Jacobean in décor, it strongly resembled the First-Class Dining Saloon on the *Olympic* and *Titanic*. Contrary to popular belief, this Grill Room was not an extra tariff or *á la carte* restaurant, for the Cunard company were staunchly opposed to the concept, despite its enthusiastic adoption by the Hamburg-Amerika and White Star Lines. It seems Cunard felt that this sort of facility created a super-elite sort of First-Class passenger, whereas they preferred to give peak service to all of their passengers. The Grill Room was eventually abolished in 1936, and the space was converted into new passenger staterooms to keep up with the high demand for capacity on board the liner. (Eric Sauder Collection)

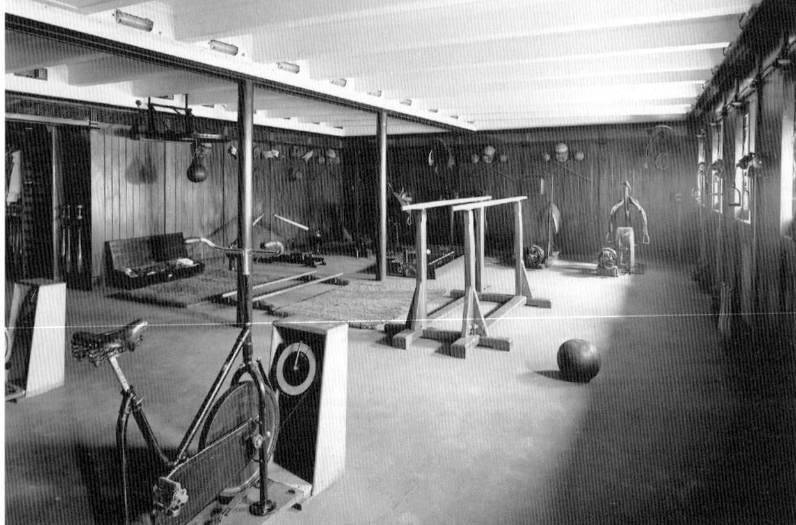

Adjacent to the Swimming Bath on E Deck was the Gymnasium. The space itself was functional, but its décor was no match for that of the Gymnasia aboard the *Olympic*-class ships. Although the location of this Gymnasium was an advantage in that it created a single area for physical exercise and refreshment, the large Palladian windows overlooking the Boat Deck on *Olympic* and *Titanic*'s gyms allowed for finer views. (Eric Sauder Collection)

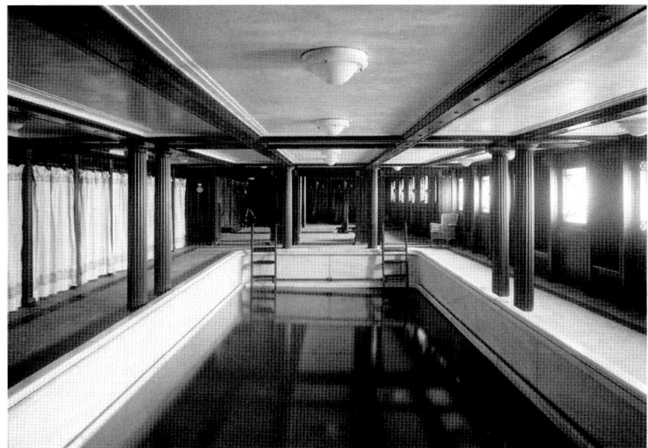

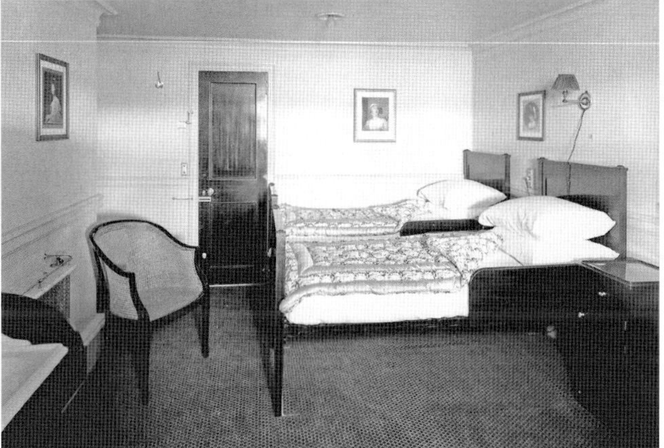

A First-Class stateroom. (Brian Hawley Collection)

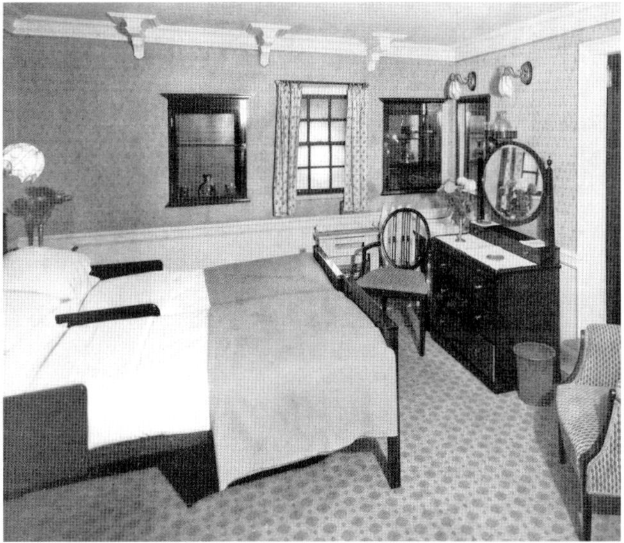

Aquitania's cabins and staterooms underwent many modifications over the years. This is typical of the later style of offerings available to passengers. (Eric Sauder Collection)

The First-Class Swimming Pool was located on the starboard side of E Deck, between the uptakes for the forward two funnels. The novelty of swimming baths aboard liners was still quite new, and a feature not even seen on the *Lusitania* and *Mauretania*; the first liner to sport one was White Star's *Adriatic* of 1907. Although this space is better decorated than the swimming baths on the *Olympic*-class ships, where bare steel bulkheads and undressed portholes were everywhere, it pales by comparison to the grand Swimming Baths on the German *Imperator*-class liners.

Although *Aquitania*'s Swimming Bath was widely used by passengers, candid photographs of the space in use – indeed, any photographs of the space other than a handful of carefully composed shots such as this one are nearly impossible to find. Additionally, this space does not seem to have met all needs encountered during the ship's career. At least one temporary swimming pool was erected on B Deck aft, in the crook between the First-Class and Tourist- (ex-Second) Class segments of superstructure in the vicinity of the electric cranes, during the ship's later tropical cruises. (Clyde George Collection)

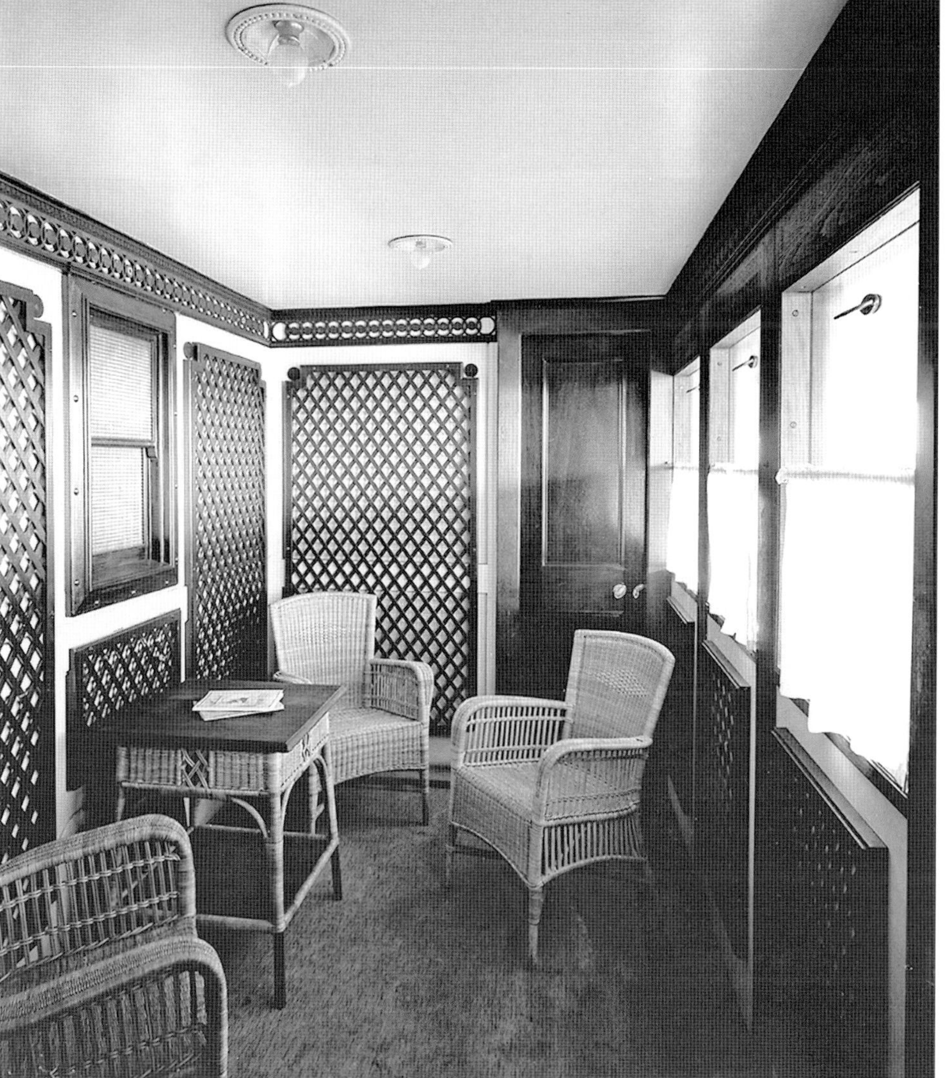

The private verandah of the Gainsborough Suite, which was located on the starboard side, just off the First-Class Entrance. This and the Reynolds Suite, on the opposite side, were the *Aquitania*'s premiere accommodations. (Clyde George Collection)

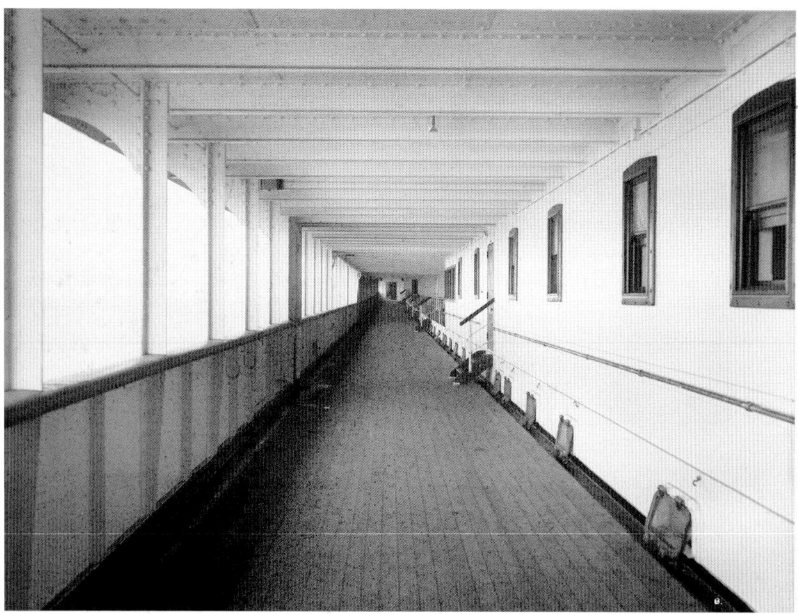

A view looking forward along the port side of the B Deck Promenade. The windows on the right belong to the Holbein Suite (foreground) and Romney Suite (beyond the steps). (Brian Hawley Collection)

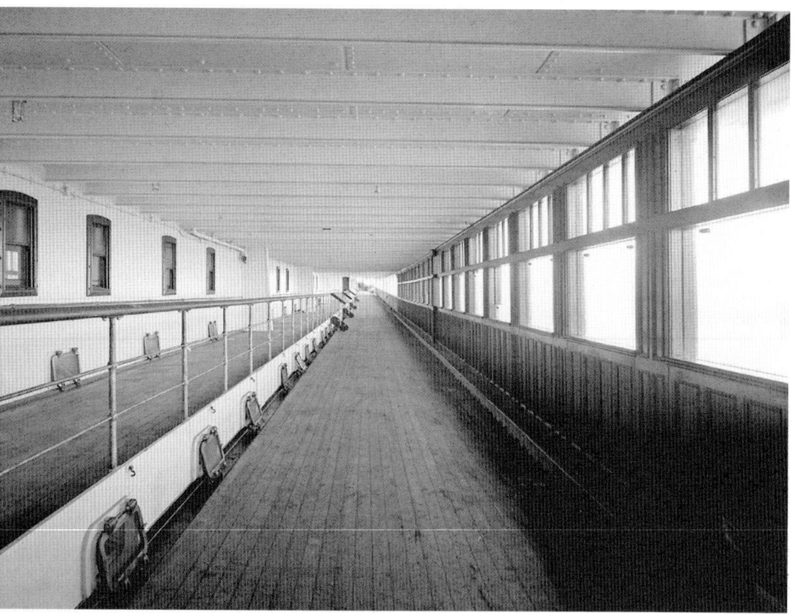

Looking aft along the enclosed segment of the port side B Deck Promenade. The raised platform inboard gave a higher vantage point for obtaining sea views when passengers relaxed in deck chairs. It was dispensed with in the 1926 refit, when the inner suites were enlarged. (Clyde George Collection)

SECOND CLASS

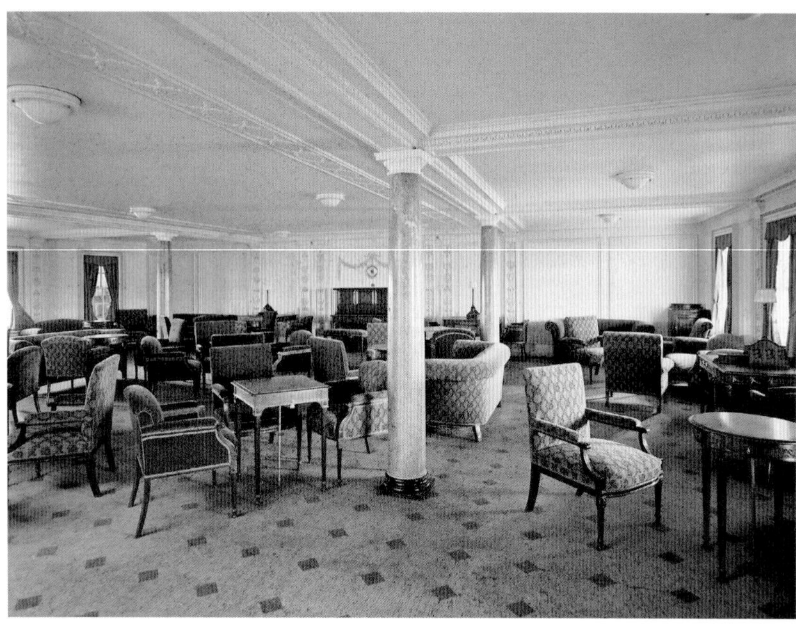

The Second-Class Lounge, which was situated aft on A Deck. This view looks astern. The space might have passed for First-Class décor not too many years before. (Eric Sauder Collection)

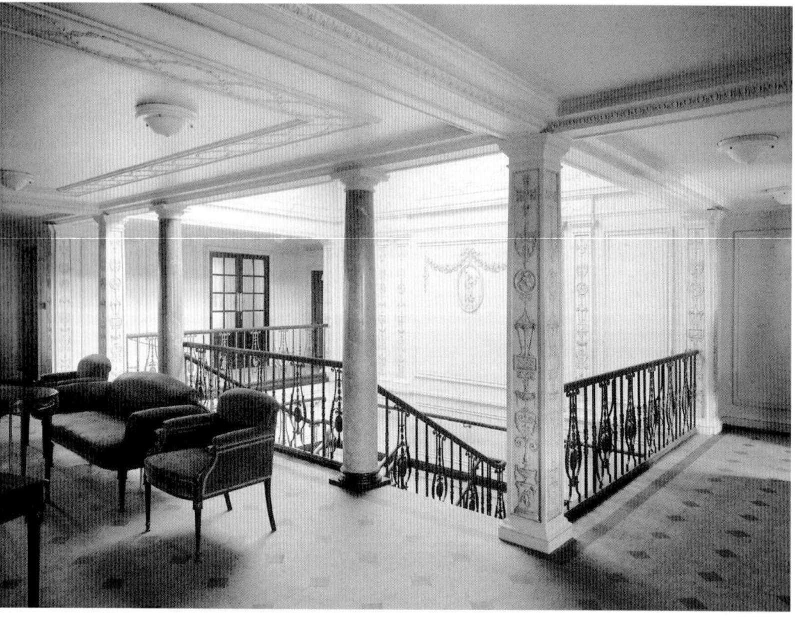

At the forward end of the Second-Class Lounge were located the main stairs, capped off by a skylight. This view looks forward and to port. The doors and vestibule on the opposite side of the staircase led outside onto the open deck. (Eric Sauder Collection)

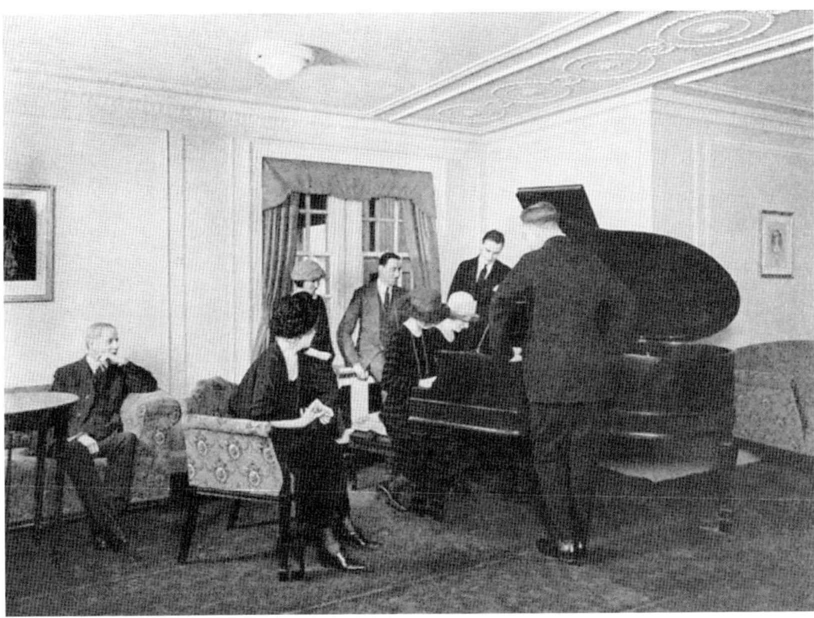

Directly below and forward of the Lounge was the Second-Class Drawing Room, seen in this photograph. It featured a grand piano which in this view, staged for Cunard publicity in the 1920s, is being used by passengers for entertainment. (Ioannis Georgiou Collection)

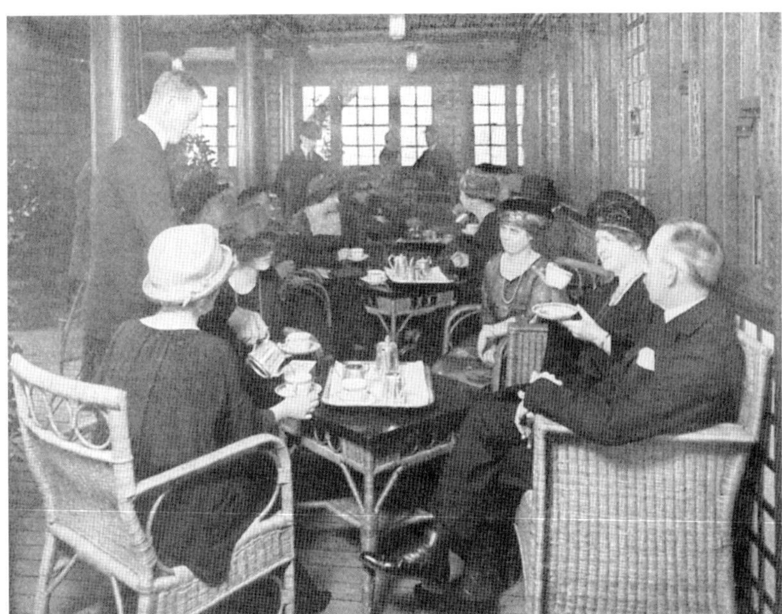

Second-Class passengers enjoyed a Verandah Cafe that was more than a match for that installed for First-Class passengers on the *Lusitania* and *Mauretania*. It was located directly astern of the Smoking Room on B Deck. This view looks to port; it was taken in the 1920s, and was used in Cunard marketing. (Ioannis Georgiou Collection)

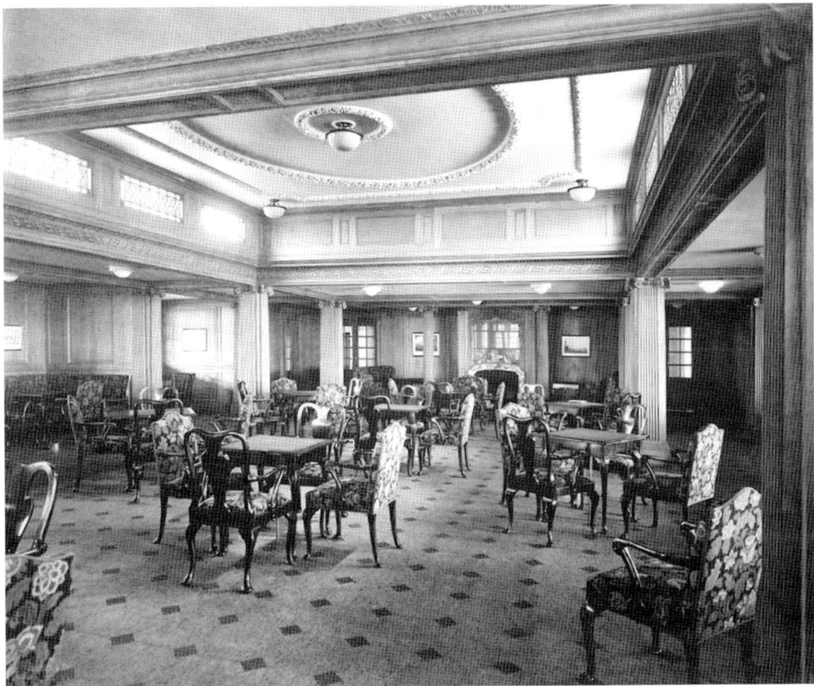

Astern of the Drawing Room on B Deck was the Second-Class Entrance, where the stairs came down from the Lounge above. Aft of the Entrance was the Second-Class Smoking Room. This view looks forward and to port, and shows the doors leading forward to the Entrance. (Clyde George Collection)

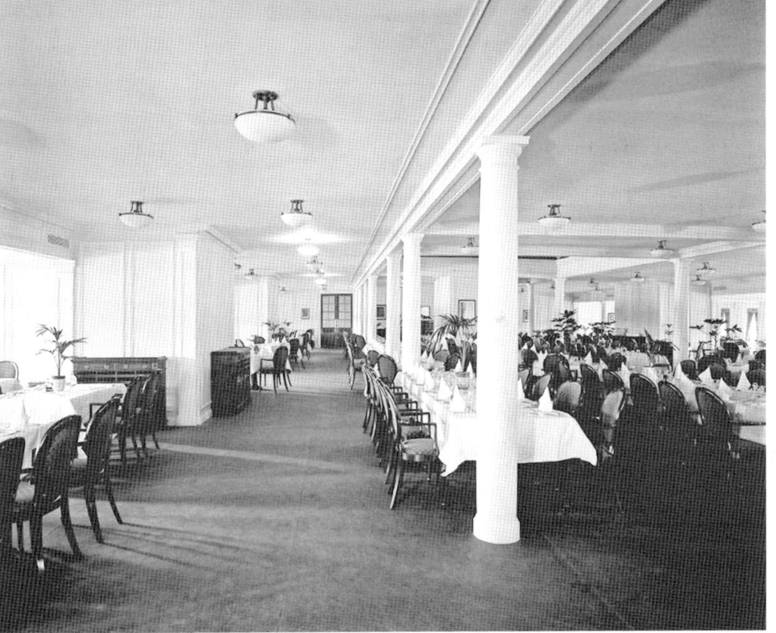

Down on D Deck, aft of the First- and Second-Class Galleys, was the Second-Class Dining Saloon. Although less ornate than its First-Class counterpart, and featuring larger tables instead of more intimate seating arrangements, it was a pleasant and comfortable space. In the centre of the room, a well opened up to C Deck above. This view is looking aft along the starboard side. The area in the centre of the room, between the rows of columns, all the way aft to the central well, was the scene of significant alterations in the 1932–33 refit, at which time a cinema was installed. (Clyde George Collection)

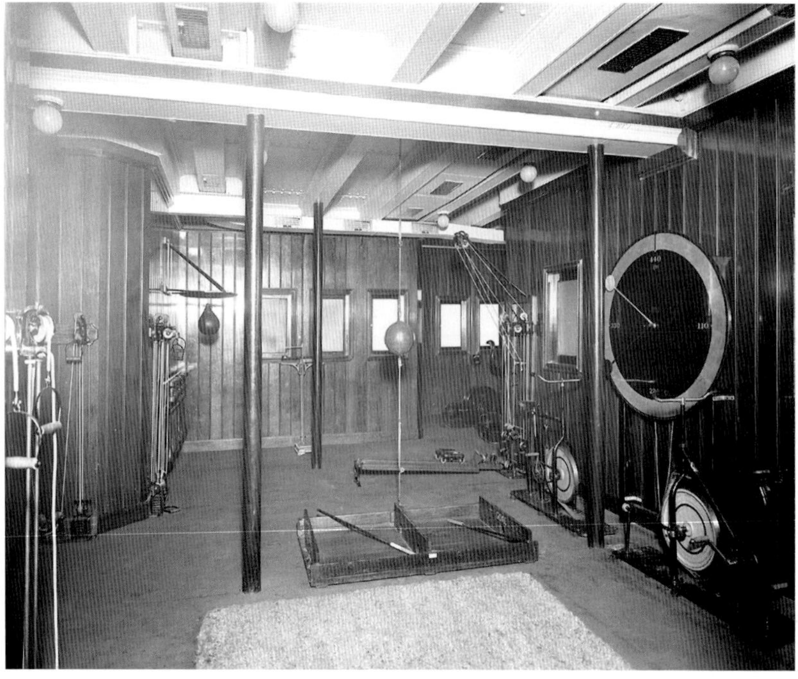

The Gymnasium, located astern of the Dining Saloon and Entrance, was a unique offering for Second-Class passengers in 1914. It was removed and replaced by a Winter Garden during the 1929 refit. (Clyde George Collection)

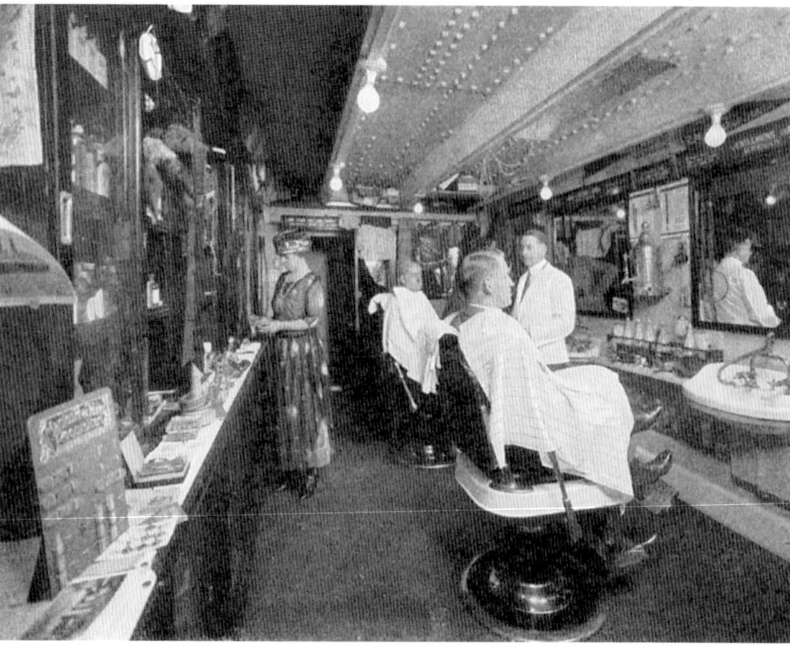

The *Aquitania* offered a barber shop for each class of her passengers. It was here that souvenirs of the trip were offered for sale, and some of these are visible on the left. While a woman peruses the offerings, a pair of gentlemen await the barber's services. (Ioannis Georgiou Collection)

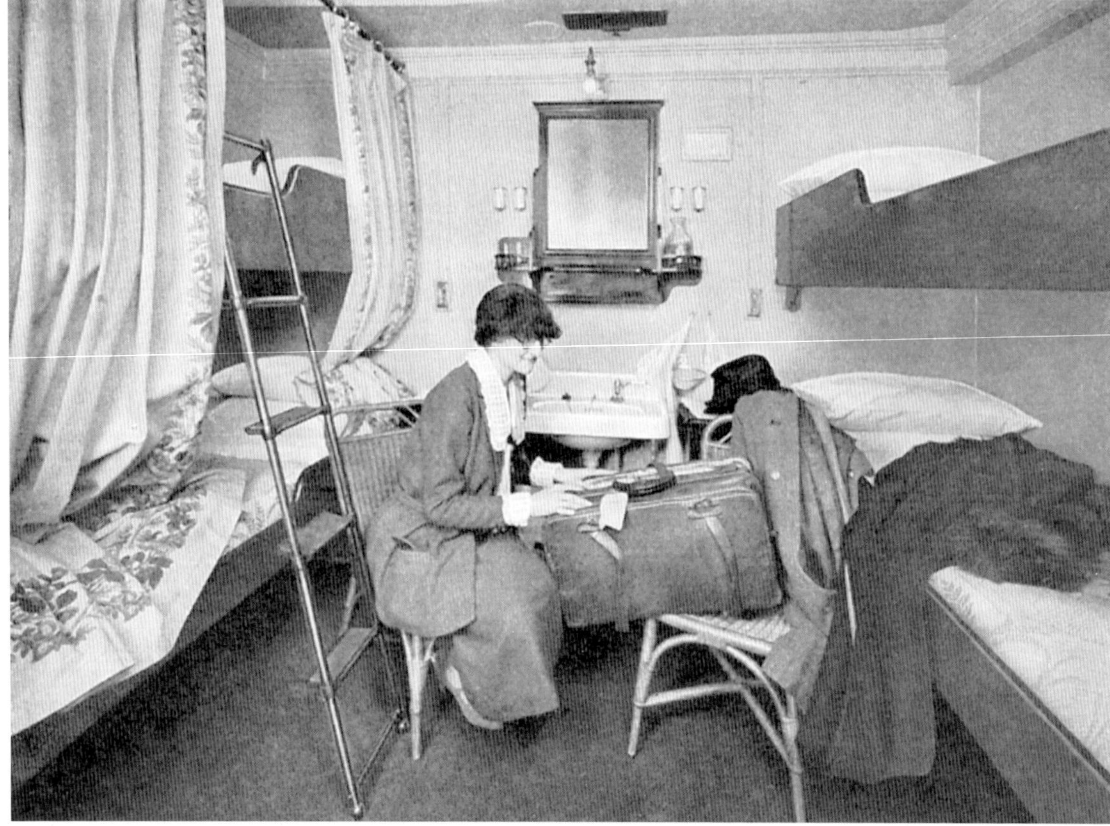

A woman makes herself comfortable in a four-berth Second-Class cabin. (Ioannis Georgiou Collection)

THIRD CLASS

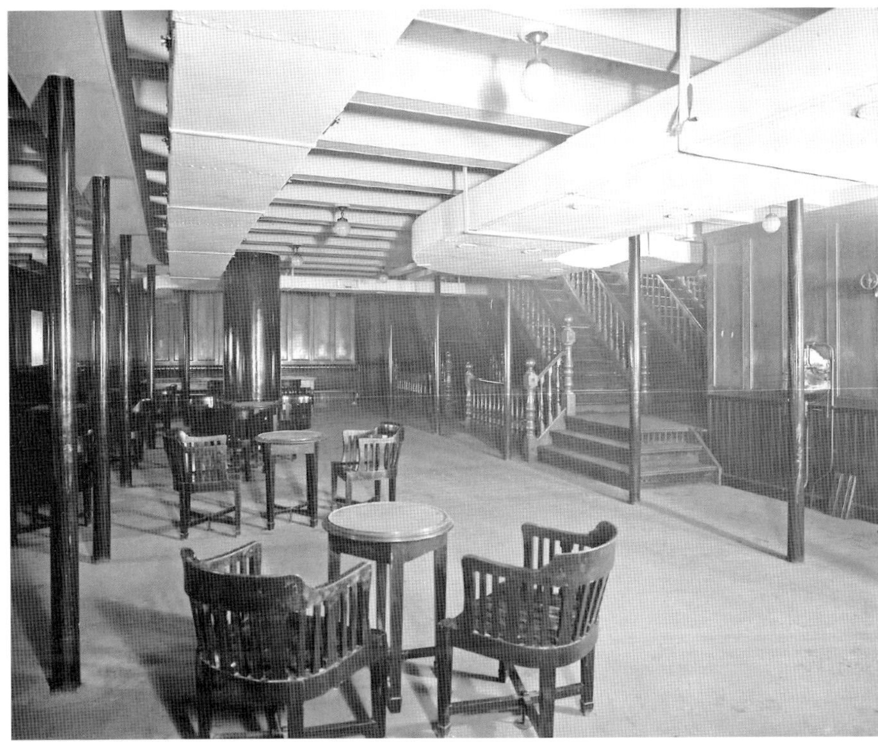

The Third-Class General Room had none of the frills of public rooms in First or Second Classes. However, it was outfitted with sturdy furniture designed to hold up under harsh use by large numbers of passengers. It was also clean very much an upgrade from the lifestyle or comforts that many of its intended passengers were accustomed to. (Eric Sauder Collection)

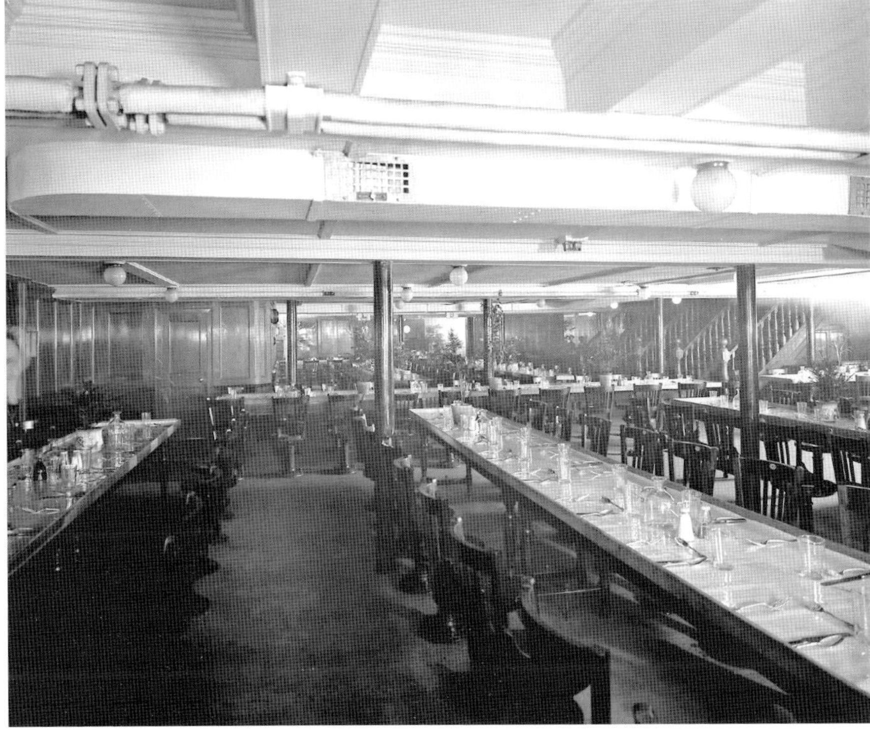

A segment of the Third-Class Dining Saloon shows the same simplicity and rugged design, but here passengers were able to partake of basic, if high-quality meals, and were tended to by waiters. Such a luxury might have been their first experience at being served by someone else, rather than serving others. (Eric Sauder Collection)

The *Aquitania*'s Engine Room Starting Platform. Looking like something from a Jules Verne novel run amok, this was the location from where her engineers could control the enormous powerplant she had been endowed with. (Eric Sauder Collection)

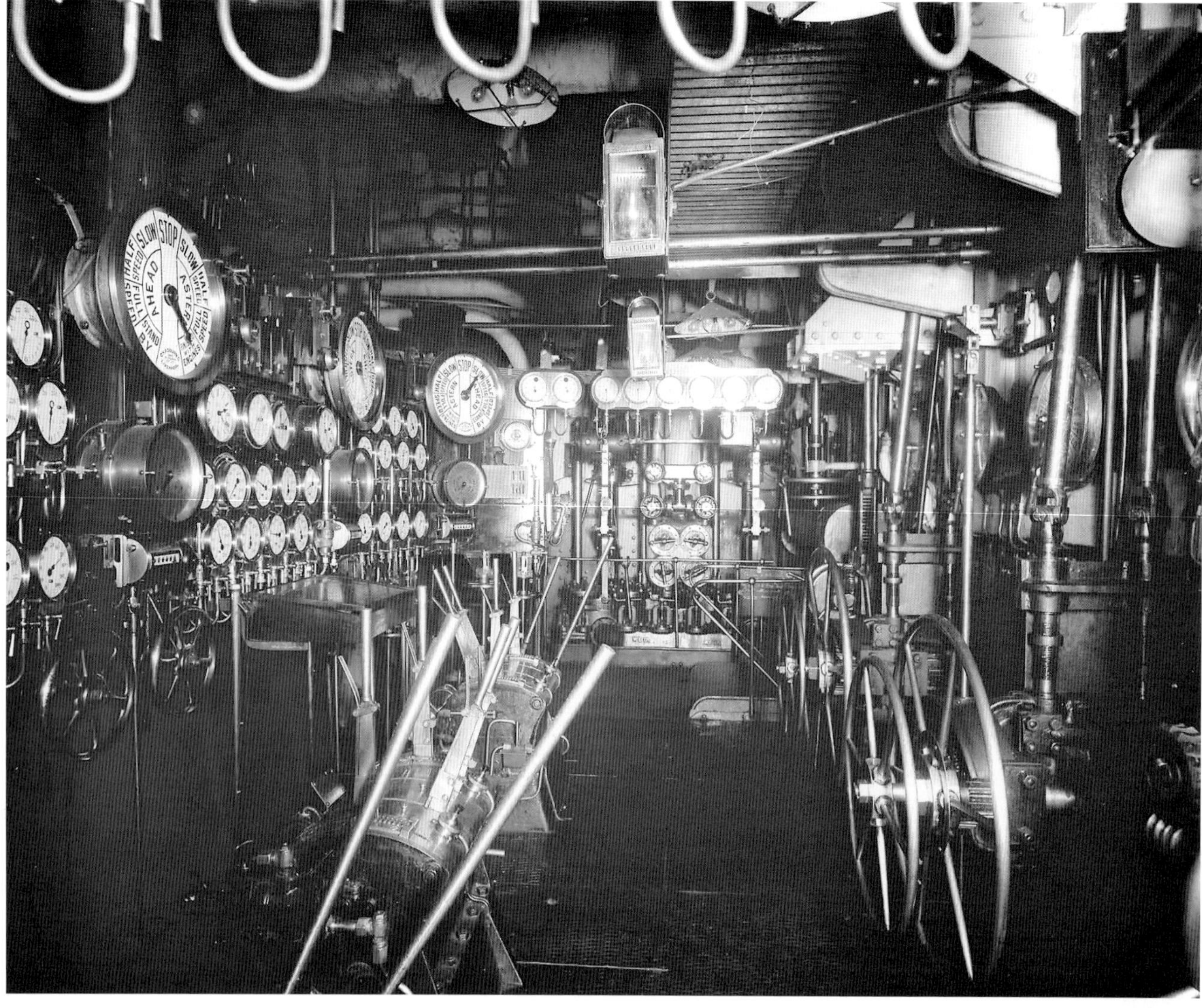

In early May 1914 the *Aquitania* was still incomplete. Ernest Fahrenheim, Cunard's victualling superintendent, later recalled that the First-Class Smoking Room remained unfinished, and that 'it was only by having an army of men at work night and day that it was got ready in time'.[34] Other portions of the ship, particularly in Second Class, were so hopelessly behind schedule that there was no way they could be ready in time for the maiden voyage. Delivery was already more than five months past due, and the 30 May maiden voyage was rapidly approaching. It was said that 'but for the indomitable determination of the Cunard company', the strikes 'would have delayed her sailing far beyond the appointed day'.[35]

At last, on Sunday 10 May the *Aquitania* left the John Brown shipyard for the first time, bound for Greenock. So great was the interest in her first passage to open water that local churches postponed their services from noon until 12.30 p.m., so that parishioners could watch her go.[36] The press reported that the crowds which gathered numbered about 100,000, but *Engineering* magazine mentioned a total of 500,000–600,000. The weather was calm but misty; the damp, however, did little to decrease enthusiasm:

Loud cheers broke out as the great liner, towering sixty feet above the water, began to move down the Clyde, towed by six tugs, and a chorus of cheering accompanied her throughout the two hours' journey to Greenock.[37]

One of the liner's later captains, E.G. Diggle, wrote:

[If] the drizzling rain that fell was a discomfort to the ... people gathered along the banks of the river, the breeze added a touch of the mysterious to the slow but continuously steady progress of the majestic ship with her attendant tugs, preceded by the Clyde Trust commodore steamer, and a convoy of pleasure steamers. She came out of the mist in her great bulk, passed by the crowds gathered at each vantage point, and receded again in impressive, stately silence. When she eventually reached the Firth, however, the weather cleared and the sun shone. She then made a short spin

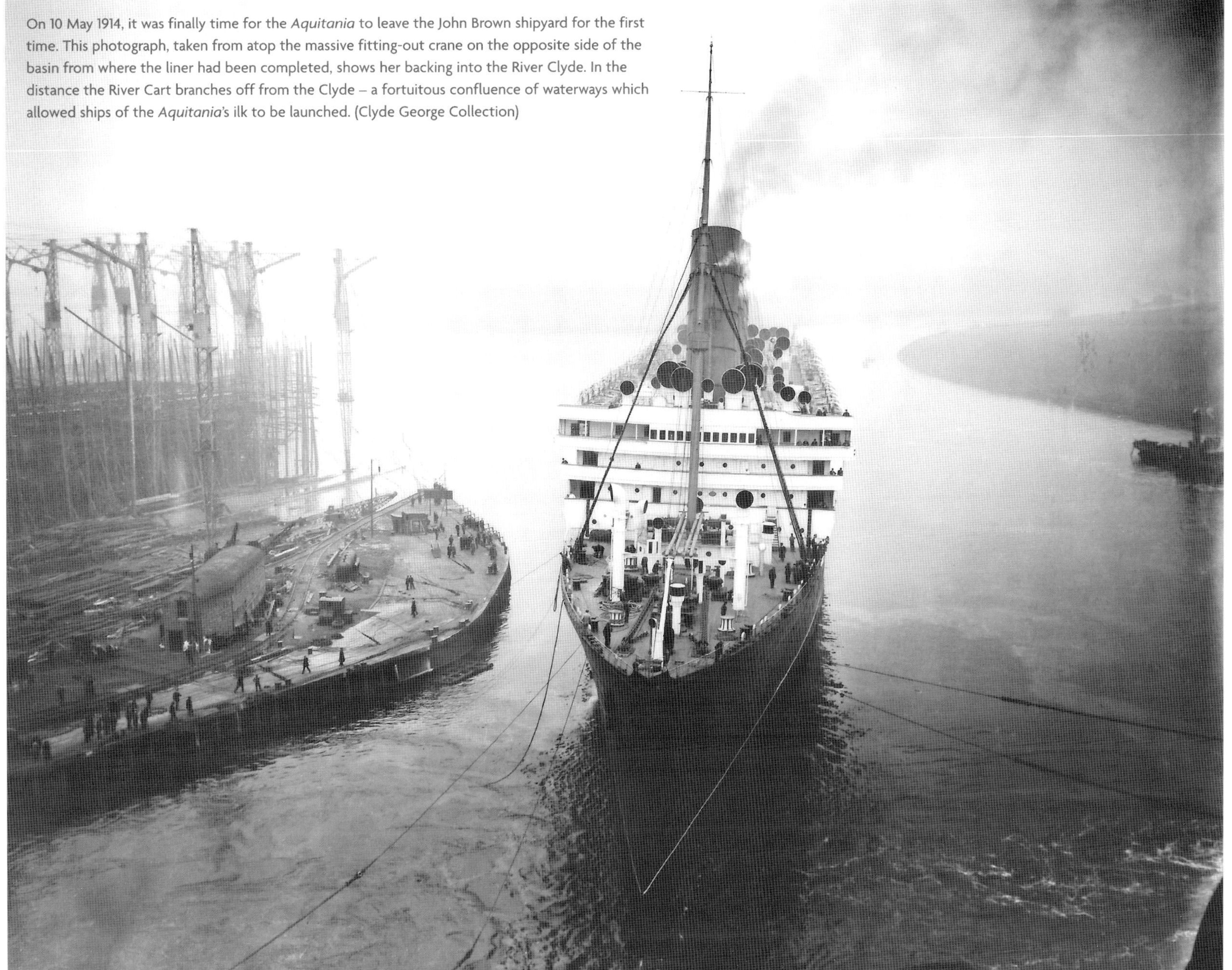

On 10 May 1914, it was finally time for the *Aquitania* to leave the John Brown shipyard for the first time. This photograph, taken from atop the massive fitting-out crane on the opposite side of the basin from where the liner had been completed, shows her backing into the River Clyde. In the distance the River Cart branches off from the Clyde – a fortuitous confluence of waterways which allowed ships of the *Aquitania*'s ilk to be launched. (Clyde George Collection)

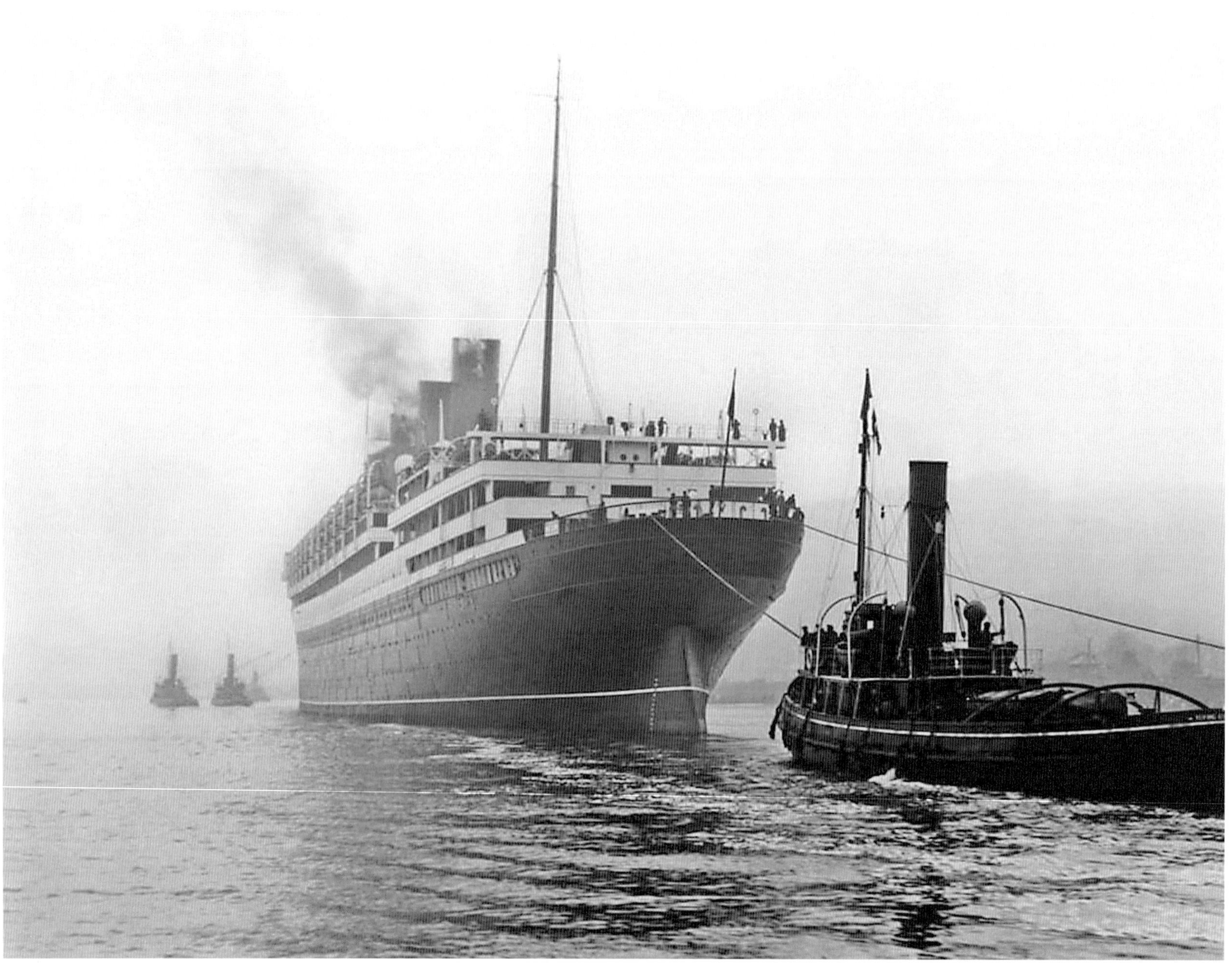

Seen from a boat on the Clyde, the *Aquitania* is shepherded down the river towards open water. (Clyde George Collection)

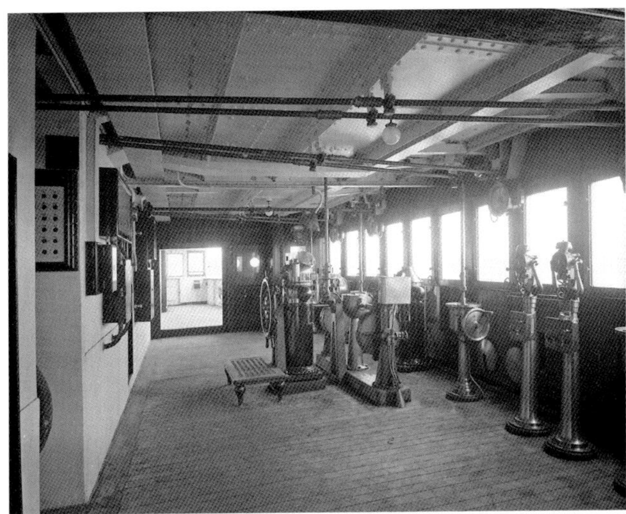

A fine view of the Wheelhouse. This view looks from the after starboard corner over to port. A dozen rectangular sliding windows provided a view forward. The ship's wheel is clearly visible amidship. The bulkhead on the left was the forward wall of the Chart Room. Unlike *Lusitania*, *Mauretania*, *Olympic* and *Titanic*, *Aquitania* did not have a separate Wheelhouse enclosed within the main bridge. The recess at the extreme left was for fire-detecting equipment. On the right, in the foreground, stand two navy phones for communication with other areas of the ship. Just beyond them is a telltale for the ship's starboard inboard turbine. (Clyde George Collection)

A splendid view from the port bridge wing, overlooking the extraordinarily long forecastle. The view was taken while the liner was at anchor before the trials began. The photograph demonstrates the trouble of trying to keep track of anything close to the bow from the original bridge's height. (Clyde George Collection)

at 8 or 9 knots, twelve boilers being alight. Finally, she anchored to coal up and prepare for her trials.[38]

On Monday 11 May, she took on board over 2,000 tons of coal and water ballast; when this was finished, she was loaded to approximate her draught during a typical crossing. The trials began the following morning, Tuesday 12 May, and ran through the following day. They were of relatively short duration because the maiden voyage was slated to commence only eighteen days later, and not because of the line's 'confidence' in their satisfactorily result, as periodicals claimed. Despite the shortened length of the tests, the ship was put through her paces.

She was gradually worked up to full speed, and her manoeuvrability was thoroughly tested. Runs of successively increasing speeds were made over the famous Skelmorlie measured mile, and the operation of the changeover valves of the turbines was tested. Then she embarked on a series of full-power runs between Holy Isle and Ailsa Craig. Despite a Force 5 sea and the dirty condition of her hull, she easily managed 24 knots. She remained at sea overnight, and further tests were made the following day. Based on their knowledge of vibration troubles with the two earlier speedsters, John Brown had taken extra care to stiffen the ship's hull right from the outset, and the liner thus proved remarkably free of vibration.

After this, *Aquitania* steamed down to her new home port of Liverpool. She arrived there on Thursday 14 May, and was placed in the new Gladstone Dry Dock

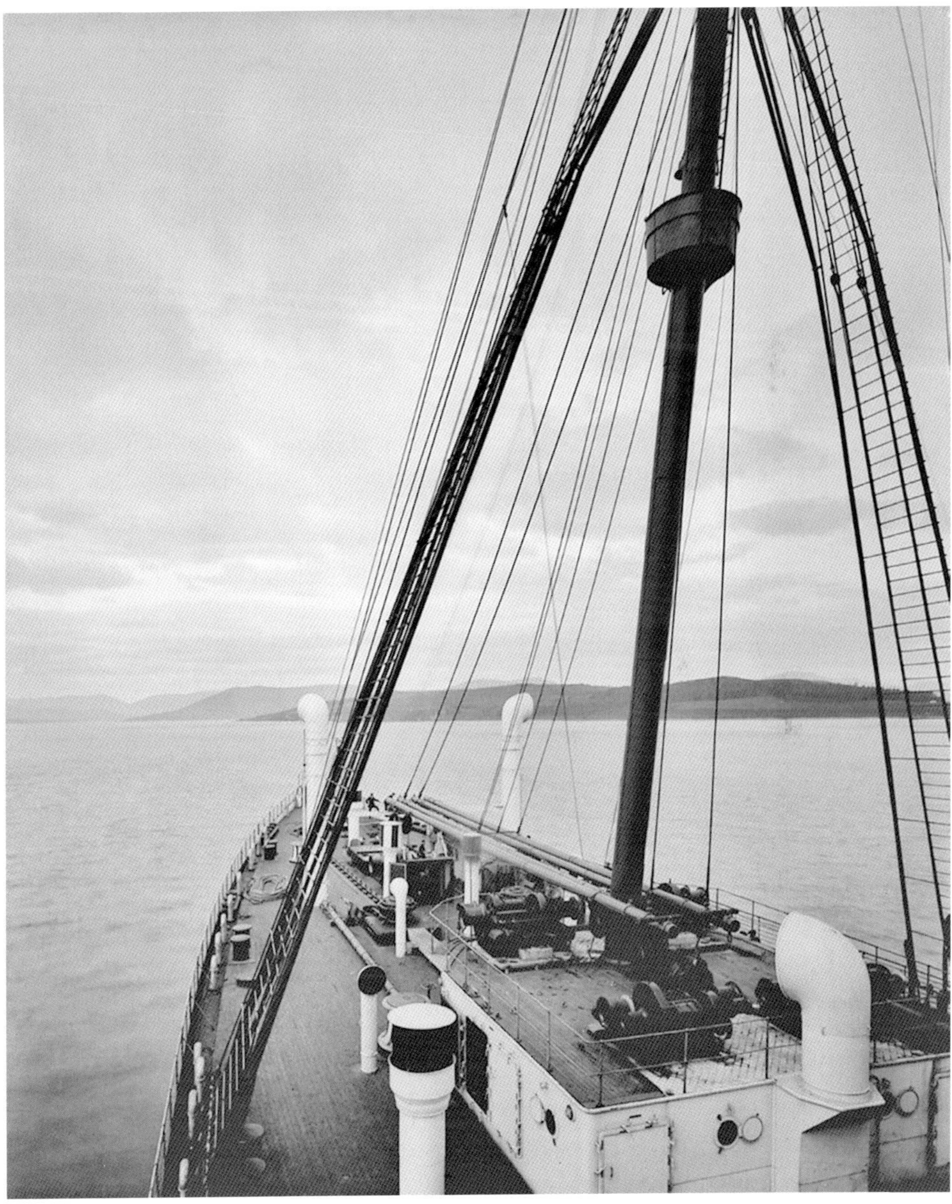

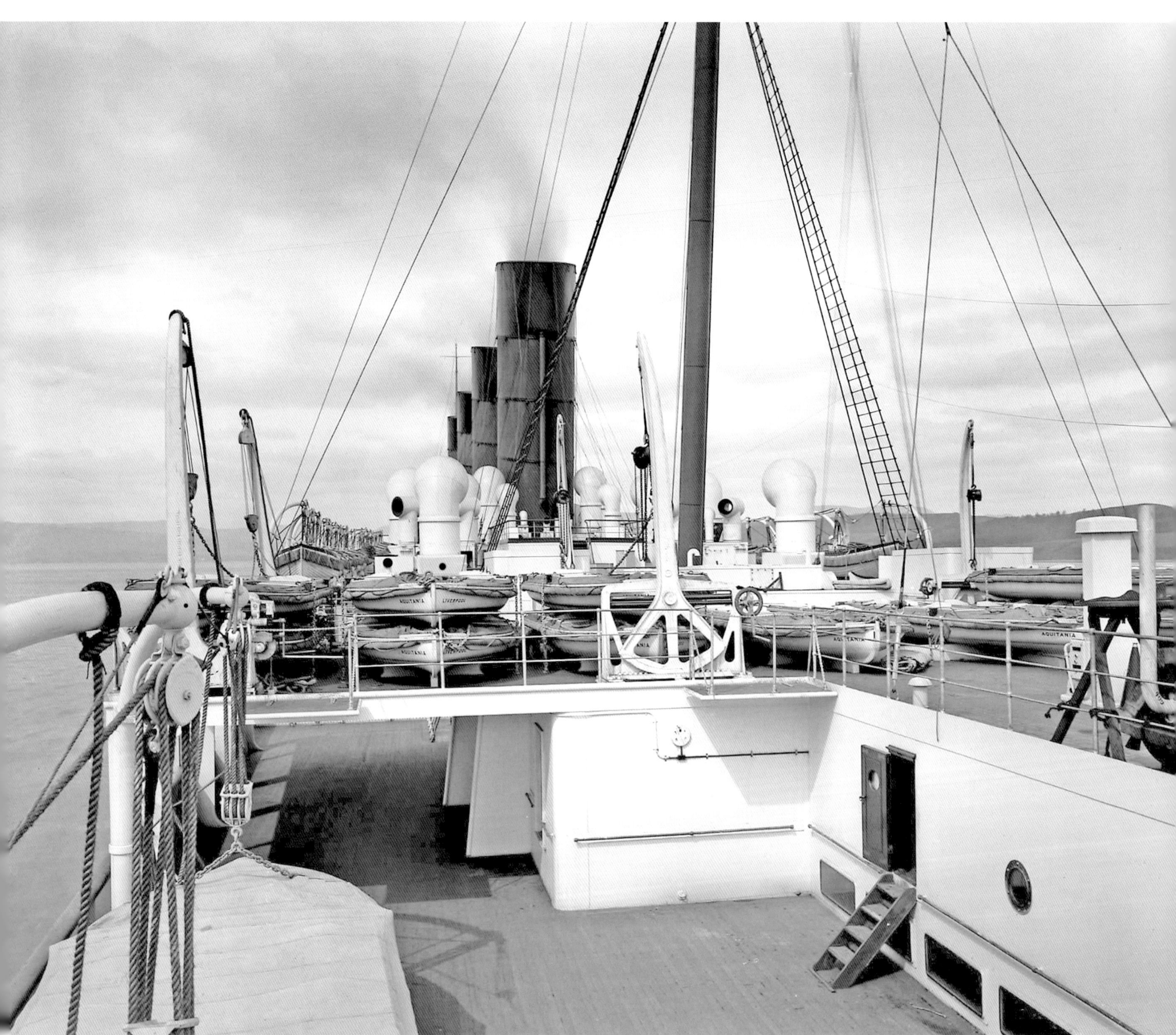

A view taken from the port side of the Stern Docking Bridge while the ship was anchored before her trials. A long line of lifeboats marches toward the bow, while her quartet of 'Cunard red'-and-black funnels tower overhead. (Eric Sauder Collection)

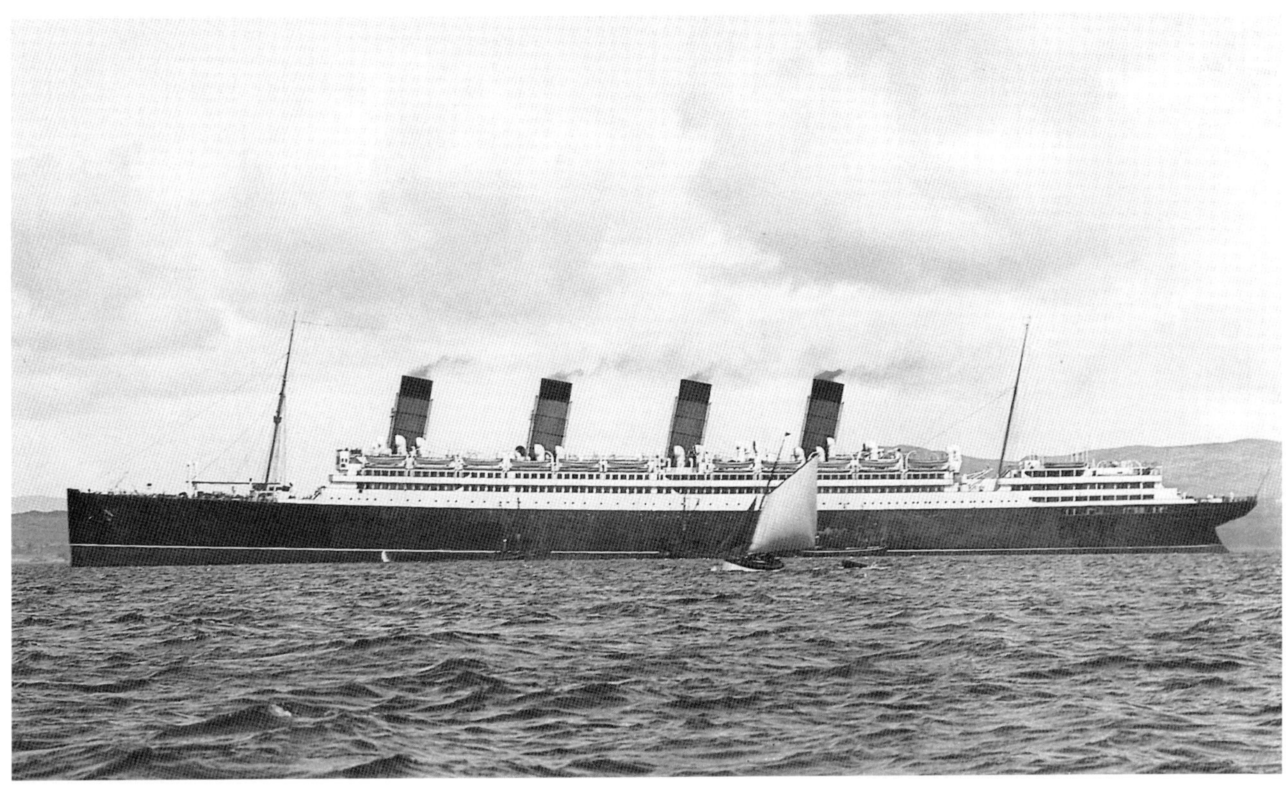

A perfect port profile of the *Aquitania* before her trials. She rides high in the water, particularly at the bow. A couple of smaller craft lay along her flank, while another sails toward the photographer. (Clyde George Collection)

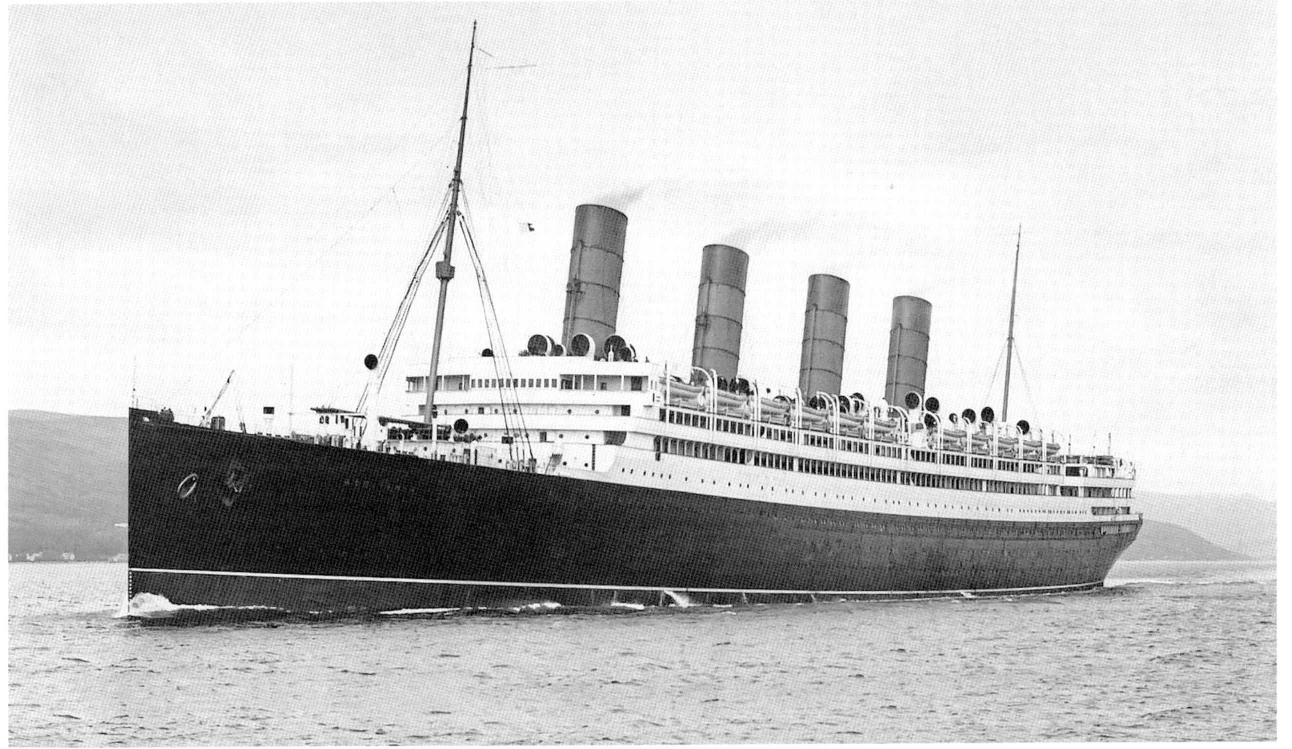

This photograph was likely taken on 12 May as the *Aquitania* began her abbreviated sea trials. (Clyde George Collection)

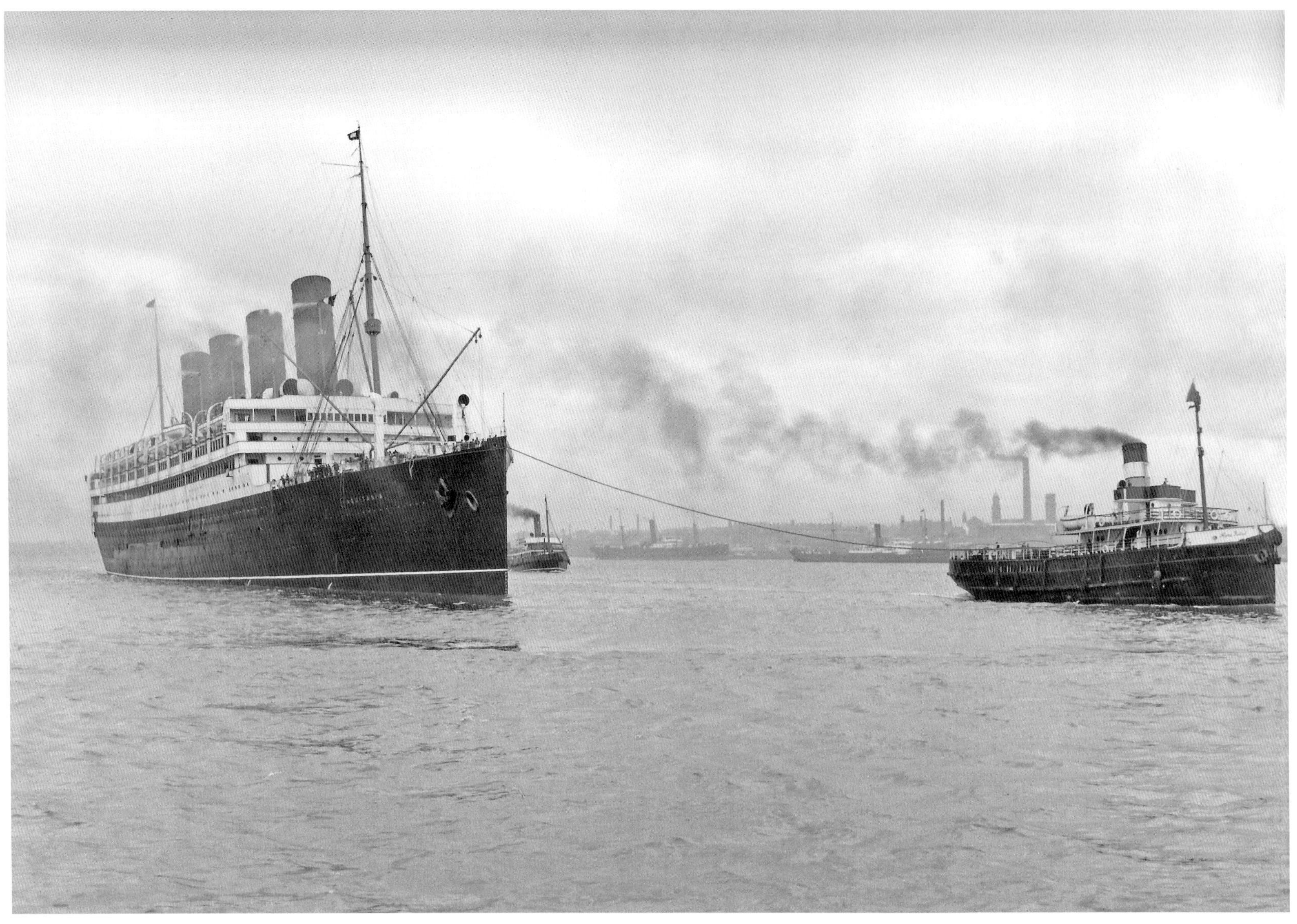

The *Aquitania* enters the Mersey in Liverpool for the first time on 14 May 1914. (Clyde George Collection)

on Friday. The work of preparing this dock to accept the new ship had begun several days in advance, and was carried out by Messrs H. & C. Grayson, Ltd. Once the ship was in place and the water began to drain away, it only took a couple of hours to place the side shores. Once the water had been completely removed, work on cleaning and painting her underwater surfaces could commence.

On Monday 18 May, the *Aquitania* was formally accepted by the Cunard company. Whereas the initial contract price for the ship had been set at £1,414,222, Cunard in the end paid a total of £1,684,683 for the liner

to completely prepare her for the maiden voyage.[39] Best of all, Cunard had so completely turned their financial 'ship' around since 1902 that they were able to pay for the cost of the new liner themselves, without any financial assistance from the government.

On the morning of Friday, 29 May, the first public inspection of the ship was held. Nearly a thousand guests boarded her while she lay in the dry dock. Among these was the Lord Mayor of Liverpool, H.R. Rathbone, and his wife, as well as various Cunard officials, representatives of the press and others. Cunard's general manager, A.D. Mearns, and their assistant general manager, S.J. Lister,

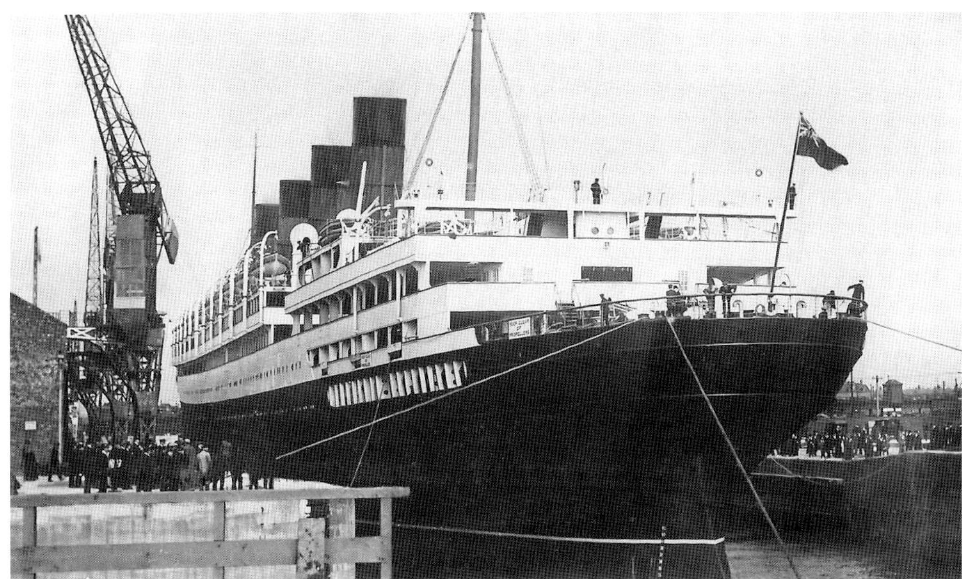

The day after arriving, the *Aquitania* enters the Gladstone Dry dock for the first time. (Clyde George Collection)

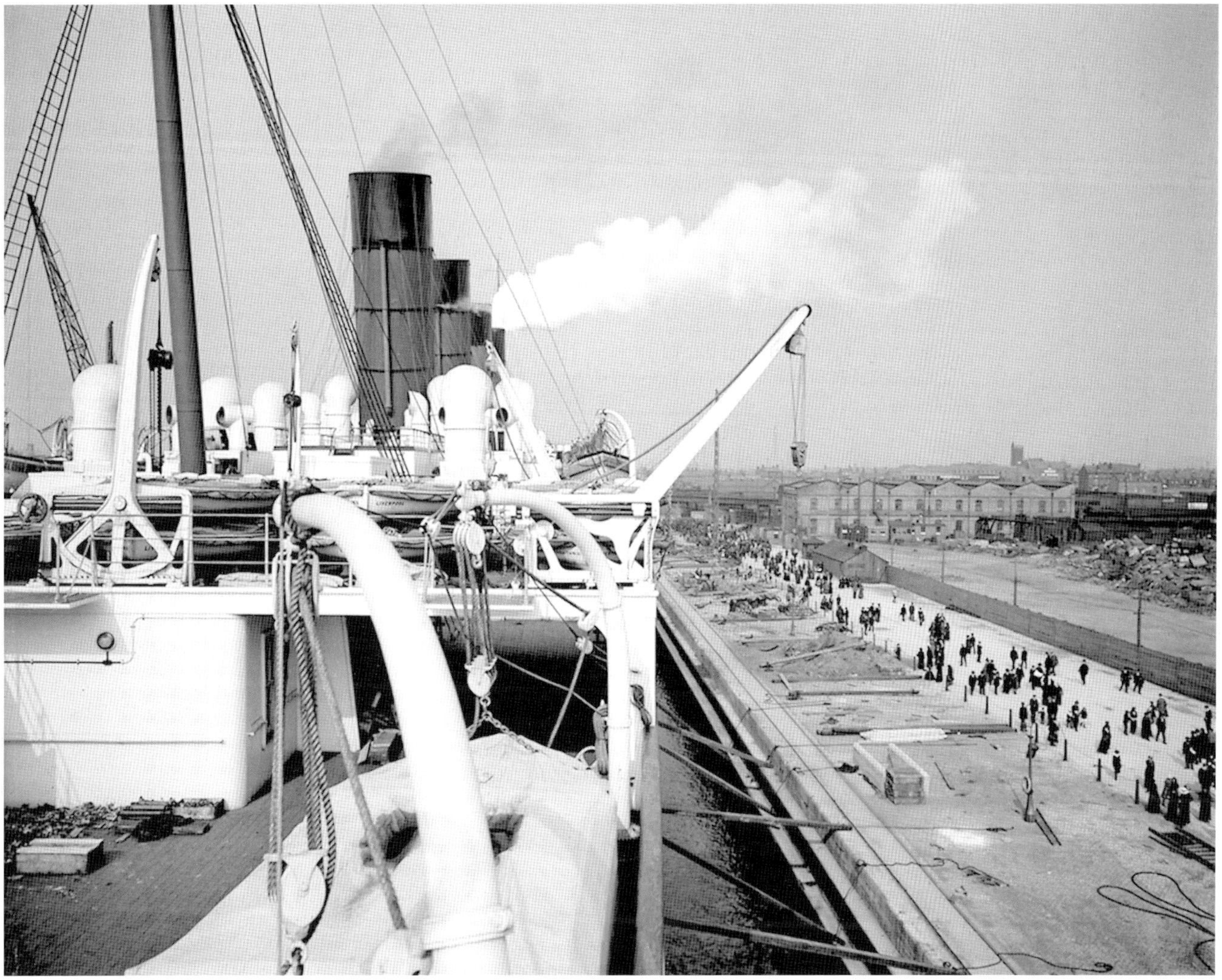

Having been positioned in the dry dock, shoring timbers have been placed from the dock wall to the hull of the ship to prevent her from toppling over once the water had been drained. As the water begins to recede, people on the ground below stare up at the enormous ship. The liner towers over every building visible in this photograph. (Clyde George Collection)

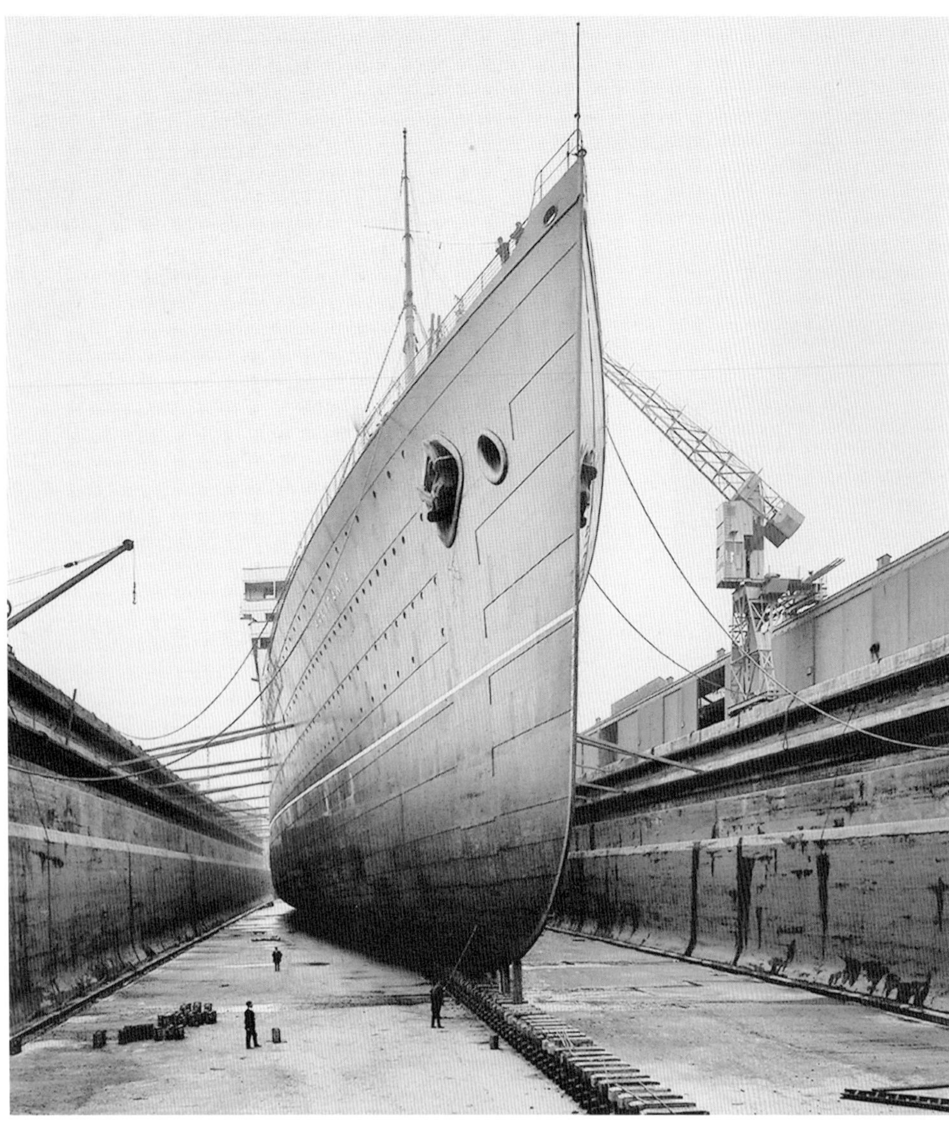

This photograph gives a sense of scale that very few other views can match. The workmen on the floor of the dry dock look like Lilliputians beside the new liner. (Clyde George Collection)

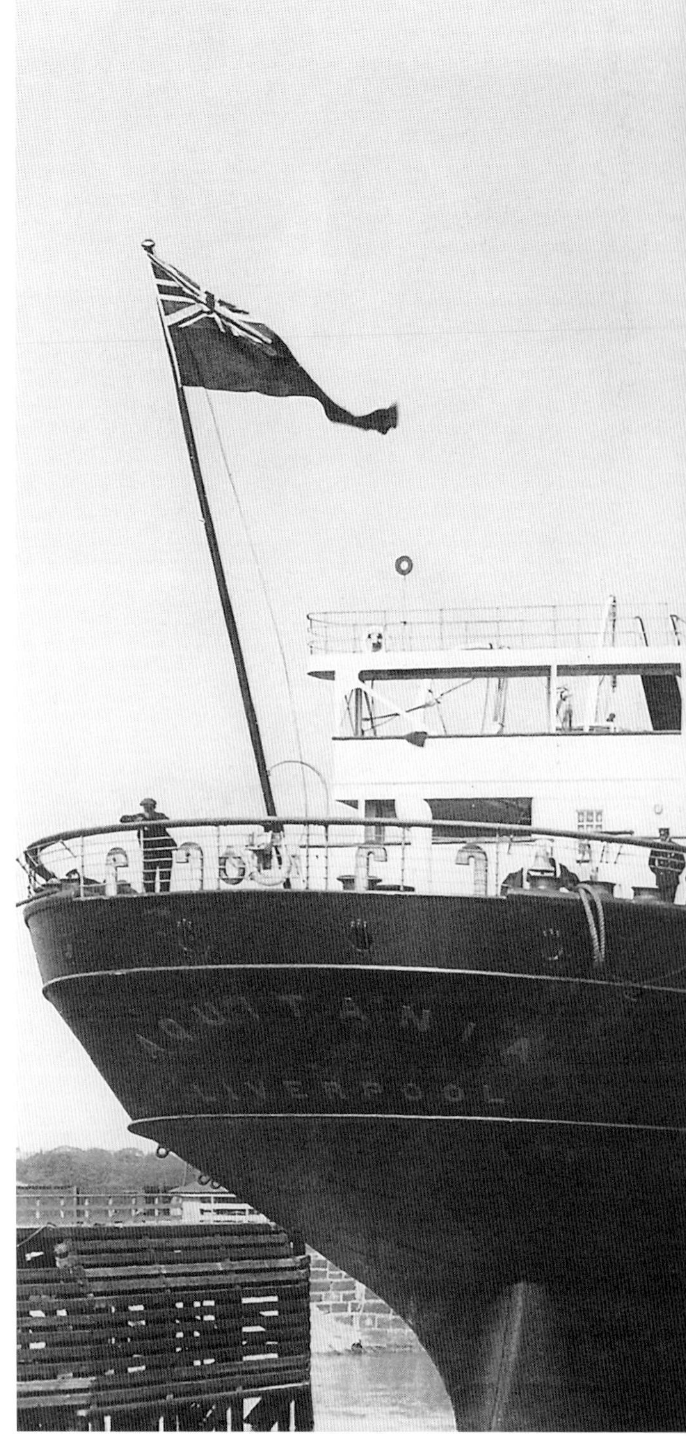

received the guests. In the afternoon, the ship was removed from dry dock and moored in the river.

Unfortunately, as the day progressed, abysmal news began to stream in from Canada: the Canadian Pacific Steamship Company's *Empress of Ireland*, a 14,191-ton liner, had sunk in the early hours of that morning. She had been steaming through fog, having just left Quebec City bound for Liverpool, when she was struck by a collier named the *Storstad*. The *Empress* had sunk in a stunning fourteen minutes, claiming the lives of 1,012; all of this had transpired in the placid waters of the St Lawrence River, and even though both ships were within sight of land. It was the worst Canadian maritime disaster, and the loss of life was

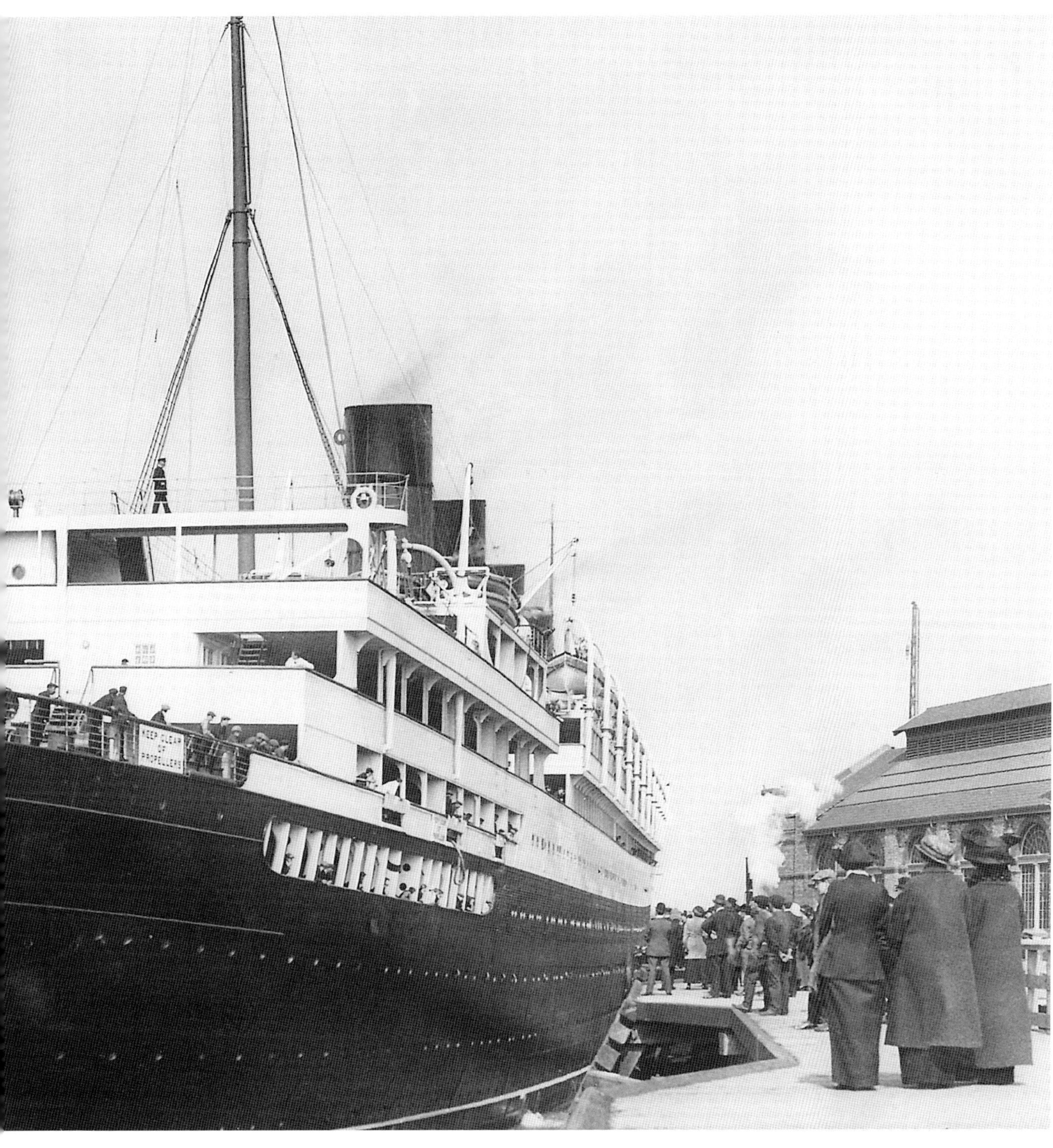

This photo shows the newest Cunard liner – for she had only been formally accepted during her stay in the dry dock – being backed carefully out of dock in the afternoon of Friday 29 May. A first public inspection had been held on the ship that morning and a large crowd has turned out to see the event. (Clyde George Collection)

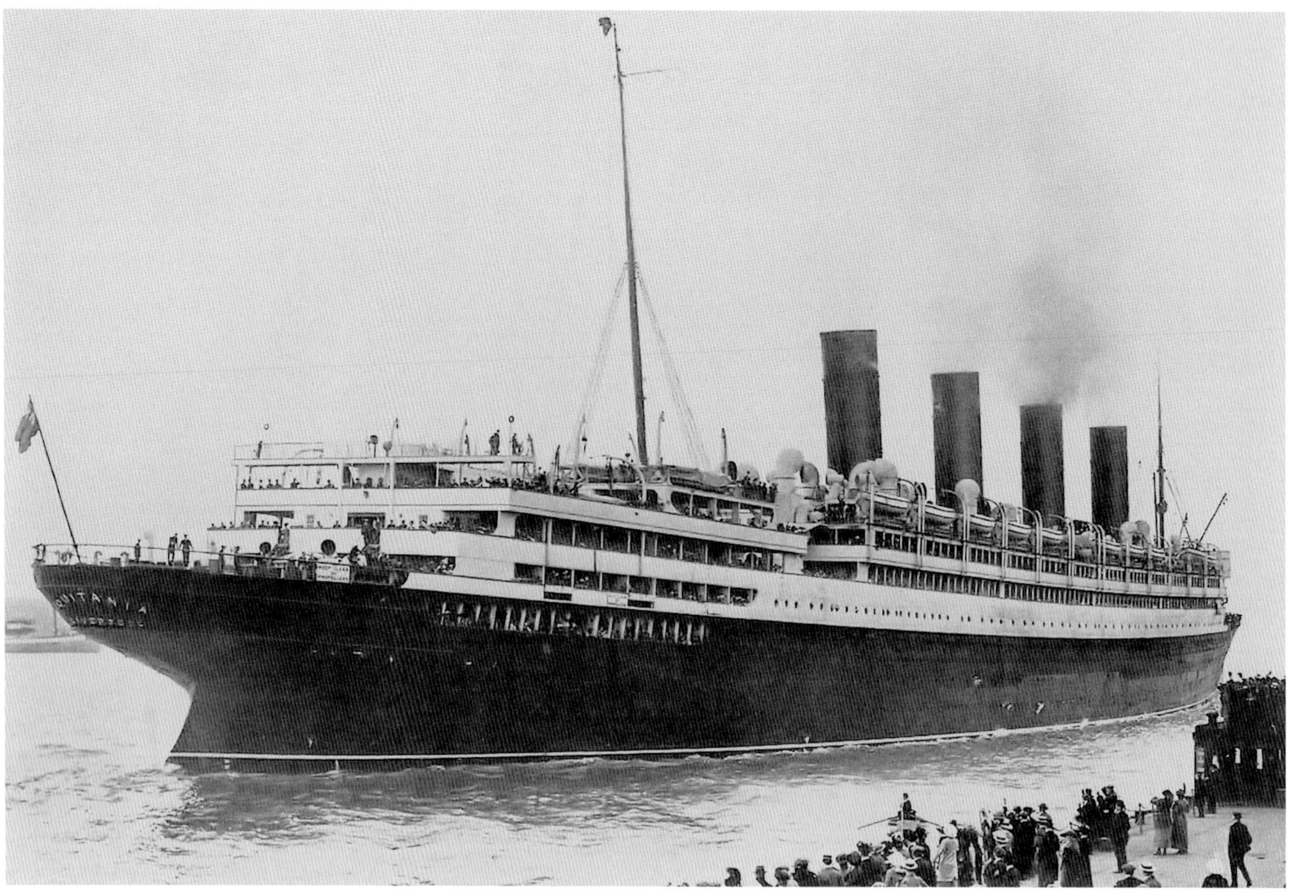

The career of one of the longest-lived and most successful Cunard liners in history officially begins: *Aquitania* casts off from the Prince's Landing Stage in Liverpool at 2.30 p.m. on 30 May 1914. (Clyde George Collection)

over two-thirds that suffered by the *Titanic* just over two years before.

The *Empress of Ireland* affair cast a pall over the maiden voyage of the *Aquitania*. On the morning of Saturday 30 May, the day the ship was scheduled to sail, there was no small amount of anxiety expressed:

> There was considerable apprehension among the passengers on the *Aquitania* boat train which left Euston station [London] to-day as a result of the Canadian disaster. The company says that no passages were cancelled. One woman, weeping, pleaded with American relatives not to sail, and many others were tearful at their departure. All were eager for the latest Canadian news.[40]

The *Aquitania* tied up with her starboard side against the Prince's Landing Stage, which was abuzz with activity. Passengers began to board the ship; some 1,055 were booked for the crossing.[41] It took time to get them aboard, and to bring their luggage on board and send it to the right cabins. Passengers included Lady Norah Brassey; Herbert Casson; Charles Cherry, the actor; J. Cheever Cowdin; marine engineer, novelist and dramatist Walter S. Cramp; Mr Sheldon Crosby, the American *chargé d'affaires* in Siam, and his wife; Lord Eversley, also known as the Right Hon. George John Shaw Lefevre, who was also the oldest passenger for the trip, at nearly 83 years of age; W.A.M. Goode, secretary of the British committee of the Panama-Pacific Exposition; Mrs John H. McFadden, wife of the cotton king; William N. McMillen, a friend of Col. Roosevelt who was described as the 'biggest man physically' aboard for the voyage; George C. Riggs and his wife, who was known to literature circles as Kate Douglas Wiggin; Charles Scribner, of the Western Electric Company; Mrs Smedley, the former Countess Vivier; and F.W. Whitridge, president of the Third Avenue Railway. Also booked as passengers were a number of workmen who, during the voyage, would work to finish portions of the ship's interior spaces that remained stubbornly incomplete.

Promptly at 2.30 p.m., the ship cast off and began to move down the Mersey towards open water. A crowd, estimated to number in the tens of thousands, had gathered to wish her a fond farewell:

> The banks of the Mersey were thronged with cheering crowds, and a great convoy of craft, with sirens shrieking, accompanied the leviathan downstream.[42]

Once the voyage had begun, it was clear that some passengers were still in a rather gloomy mood and anxious because of the Canadian disaster. However, as time passed the great liner's amenities, comforts and fine performance began to turn the general mood around. *Aquitania*'s passengers were pleasantly surprised by her lack of vibration and steadiness at sea. Her daily averages were surprisingly high: to noon on Sunday, she steamed 475 miles at an average of 23.19 knots; to noon on Monday, she added another 576 miles, averaging 23.1 knots; to noon on Tuesday, she steamed 602 further miles at an average of 24.3 knots; to noon on Wednesday she made another 527 miles at 21.57 knots; to noon on Thursday, she again made 602 miles, this time at an average of 24.19 knots; she covered the final 399 miles to Ambrose light at an average speed of 22.43 knots. For the entire crossing, she had averaged 23.17 knots.[43] This performance was in spite of the fact that on Wednesday morning, she was forced to proceed at 'Dead Slow' for three hours because of ice and fog. It was said that on the Monday–Tuesday run, she had touched a full 25 knots at times, an incredible feat for a ship designed only for 23 knots.

To put these 602-mile runs in perspective, the world record for a day's westbound steaming set in 1901 by the *Deutschland* had been 601 miles. On her maiden voyage in September 1907, the *Lusitania* had failed to break that record. When she did raise that bar, on her second westbound crossing, she had upped it to 608, and then

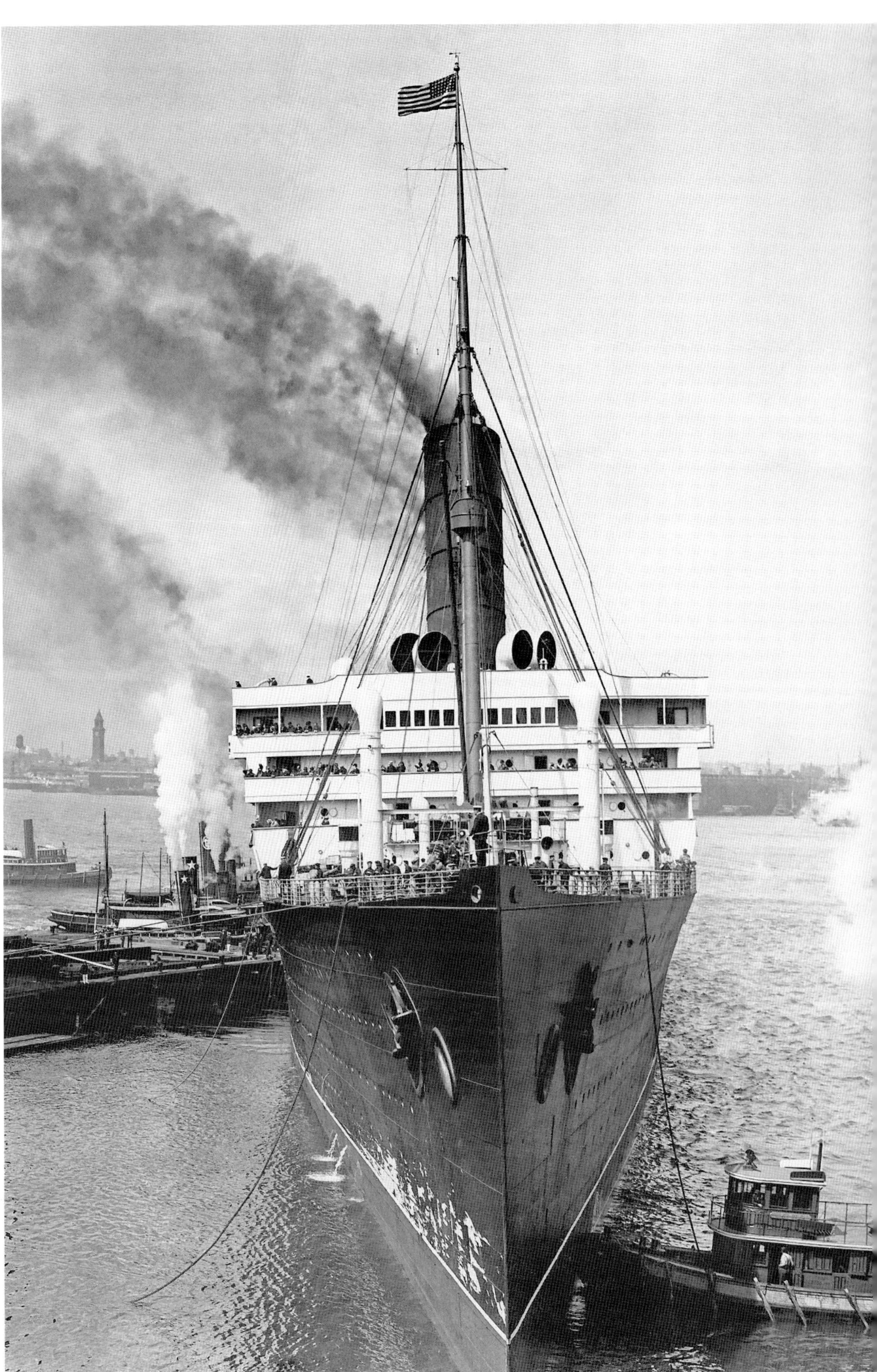

Having completed her maiden crossing successfully, the *Aquitania* docks at Cunard's Lower West Side pier on the island of Manhattan. Notice the paint scoured away from the sides of her prow, just above the waterline. (Clyde George Collection)

617 miles. Just a few years before, *Aquitania* would have held the Blue Riband for such a performance.

As the *Aquitania* steamed up from Quarantine, Captain Turner stood triumphantly on the bridge with the New York Harbor Pilot. Having passed Quarantine Station at 7.45 a.m., she proceeded into the North River.

She was welcomed by a noisy chorus from the vessels in harbour as she steamed up New York Bay. To ensure that she had right of way, two patrol boats moved just ahead of her, warning all other craft to keep clear.[44]

It was said that she 'was greeted vaporously by everything afloat with a steam whistle and by the cheers of a multitude that thronged six piers to the north and south of the slip at the foot of Fourteenth street'.[45] She was alongside the Cunard Pier by 9.15 a.m.

The docking procedure was quick and without incident. Seventeen tugs were involved in the effort, and Captain Turner used the ship's propellers deftly, with the port propeller driving 'Full Ahead' and the starboard one set to 'Full Astern', to warp her in. Port Captain Roberts told the press that only careful planning could have allowed for her to nestle alongside her pier 'much like a duck might

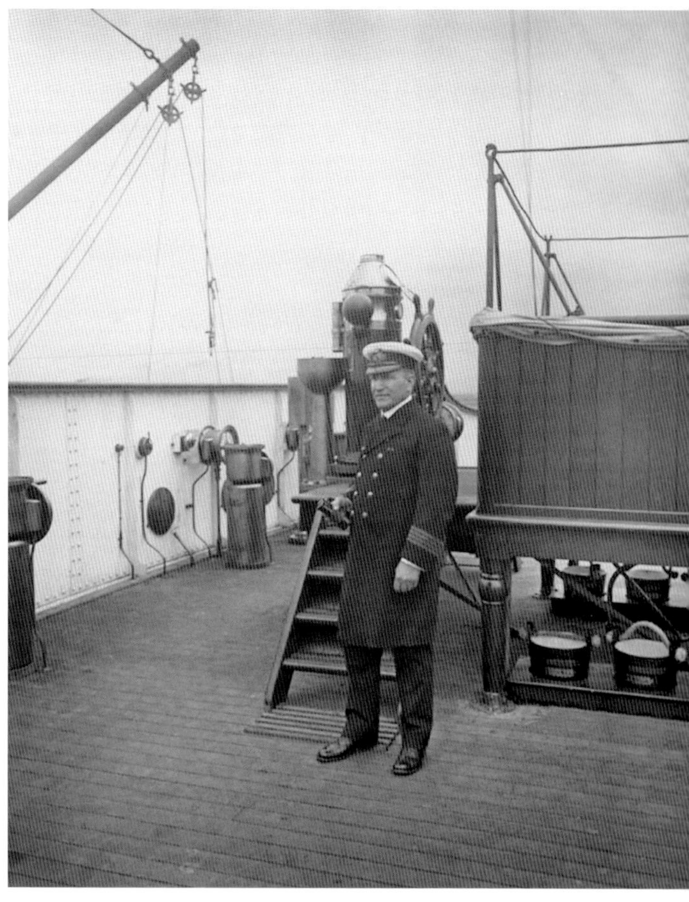

paddle up to a flat in a placid lake'. Another reason it had all gone so smoothly, he pointed out, was that 'the boss of all the British skippers, Capt. W.T. Turner, was at the helm'. She tied up at 9.40 a.m., a bugle sounded, and in short order passengers were disembarking. The passengers were all very enthusiastic about the *Aquitania*. New York newspapermen called her 'trim and graceful', and said that she 'followed the racy lines of the *Lusitania* and *Mauretania*'.[46] Kate Douglas Wiggin wrote a 'rhapsody' about the *Aquitania*, and gave it to Chief Purser James McCubbin. It read, in part:

The *Aquitania* is not only the last word of comfort, convenience, luxury and splendor but there is something marvellous about her adaptability to her great task. Far from exhibiting any of the uncertainties and timidities that might accompany a maiden effort, her ease and freedom and swiftness of motion seem to bespeak the practised

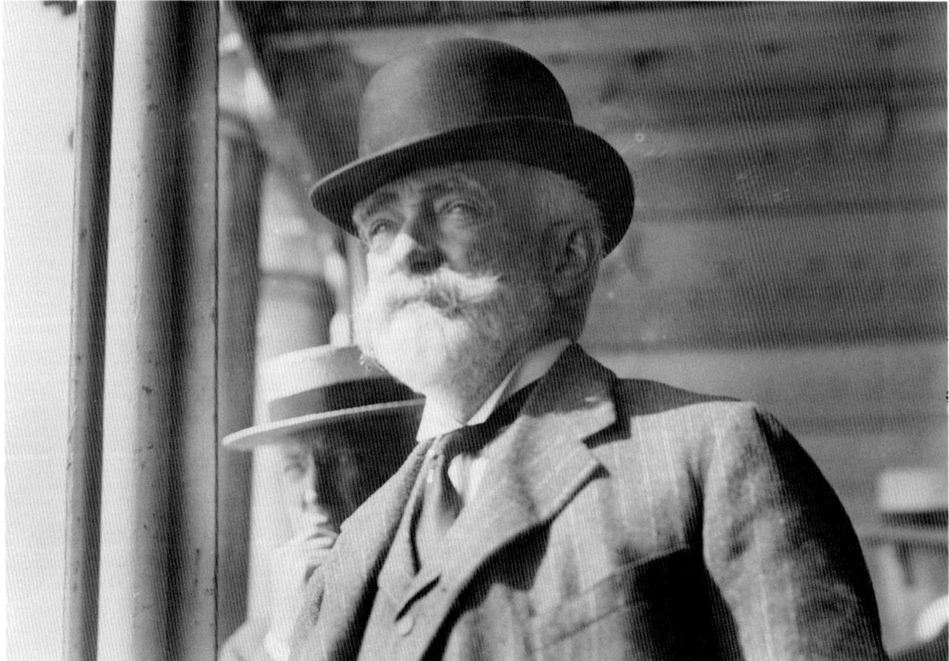

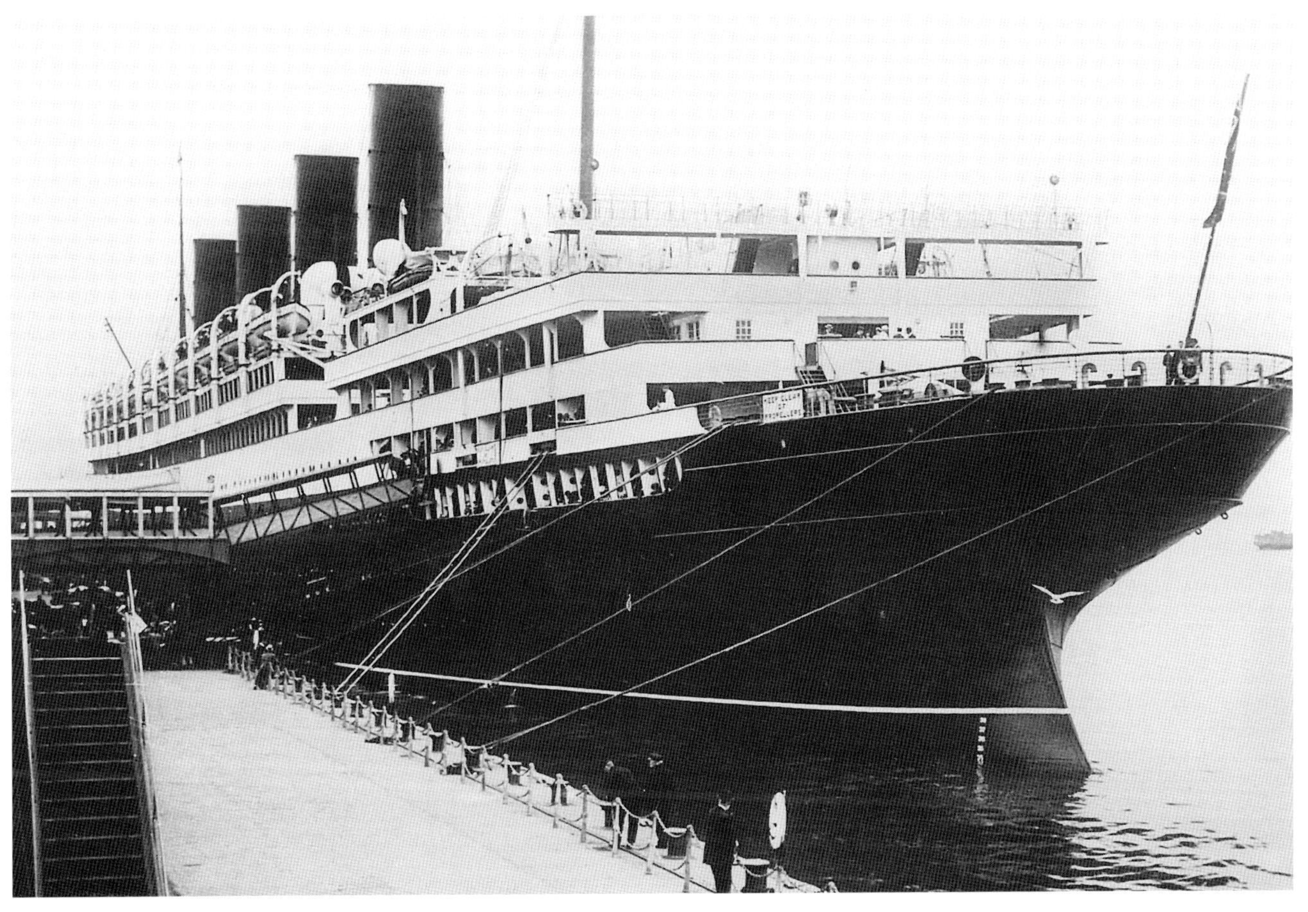

The *Aquitania* returns to the Prince's Landing Stage in Liverpool, having successfully completed her first round-trip voyage on the Atlantic. What very few could have imagined was that in just a few short weeks, the world would be engulfed in a conflict that it would never recover from, and one that would forever alter the new liner's career. (Clyde George Collection)

perfection of a veteran. 'Ease is the lovely result of forgotten toil.' The *Aquitania* herself is the 'lovely result'; the men of constructive genius who toiled unceasingly for months to achieve her must have been proud this morning when she glided serenely into port.[47]

Indeed, only one person seemed out of sorts that day: Lott Gadd, the *Aquitania*'s barber. He explained that it was because, for the first time in his twenty-five-year career as a seagoing barber, he had cut one of his customers. The reason was not because the ship was vibrating or the sea was rough; it was instead because his now-wounded passenger had refused to take off his monocle during his shave. 'I was 'hypnotised,' he explained in his thick accent. 'I must 'ave been 'ypnotised. 'E wouldn't take off his monocle. 'E shut 'is other eye, but 'e stared at me with the monocle eye. 'E followed every movement. And so I cut 'm. It was 'orrible! 'orrible!'[48]

The *Aquitania* remained in New York for five days. Crowds of fascinated New Yorkers buzzed about the Cunard pier and begged to be allowed on board to inspect the ship, but Cunard officials declined, saying that she would instead be opened for inspection on her second stay there. On 8 June, company officials did host a luncheon on the ship for 700 guests, mostly city and federal officials, along with various railway and steamship men. After the luncheon, the guests were given a guided tour of the ship.[49]

Someone who visited the *Aquitania* during one of her earliest stays in New York was named Harry Grattidge. He was a new officer on the Cunarder *Carpathia*, and stopped in to see the *Aquitania* while his ship was there. He later recalled:

[About] this time I fell in love. Not with a girl, for that was always happening, but with a ship, the mightiest ship that Cunard, or indeed the British Merchant Navy, had ever known. It was not easy then to imagine

44,786 tons [*sic*: 45,647 in 1914] of ship, but that was the measure of the beloved *Aquitania*. The fine clean sweep of her lines prompted even hard-boiled seamen to christen her 'The Ship Beautiful'. And when I went over her in New York, still young and stuffed with dreams, and marveled at her Adam Library and the fine plasterwork in her Palladian Lounge, there and then I swore a vow: *Because we have begun our service in the same year, one day I will be your Master*. I was as certain of that as a youngster can be, for no one had confided in me the great truth that life gets in the way.[50]

It would be many years, but Grattidge — who eventually became a captain and thence commodore of the Cunard Line — would eventually make good on his promise.

On 10 June, *Aquitania* was to return to England. Captain Turner backed her deftly into the river and turned her nose to move down-river just as easily as he had brought her in. She carried a fine load of 2,649 passengers — said to have been a record for the number of passengers embarked at an American port on an English ship — and averaged 23.45 knots on the eastbound crossing.

Cunard had officially become the first major steamship company to successfully implement a well-balanced three-ship service. White Star's trio was in shambles after the *Titanic* disaster, and Hamburg-Amerika only had two of their new ships in service at the time. Even better, the *Aquitania*, when compared to her competitors, was frequently coming up as the declared favourite.

It was all a fine start to what looked to be a busy season; so many people were then travelling to London for the summer that they overwhelmed the hotel capacities of that city. Some were said to have slept on cots or in bathrooms overnight, space was at such a premium.[51] The *Aquitania* made two more round trips between England and the United States, completing each in perfect safety and showing that she performed wonderfully in rough seas.

Cunard found that the new liner's facilities were very much appreciated. The Long Gallery was well populated by passengers, who used it both as a lounge and for card games; the Garden Lounges were extremely popular, being quite full at all times of the day; the Grill Room was more popular than expected; the Swimming Bath was also well used. There were a few teething problems: the majority of these centred around ventilation issues. It was also found that the lifts did not work perfectly, and that the dance floor in the First-Class Lounge needed to be enlarged because it was proving so popular.[52] But these were, in the grand scheme of things, minor details. *Aquitania* was a fine ship, and Cunard had every reason to expect a profitable summer season for her.

It was not to be.

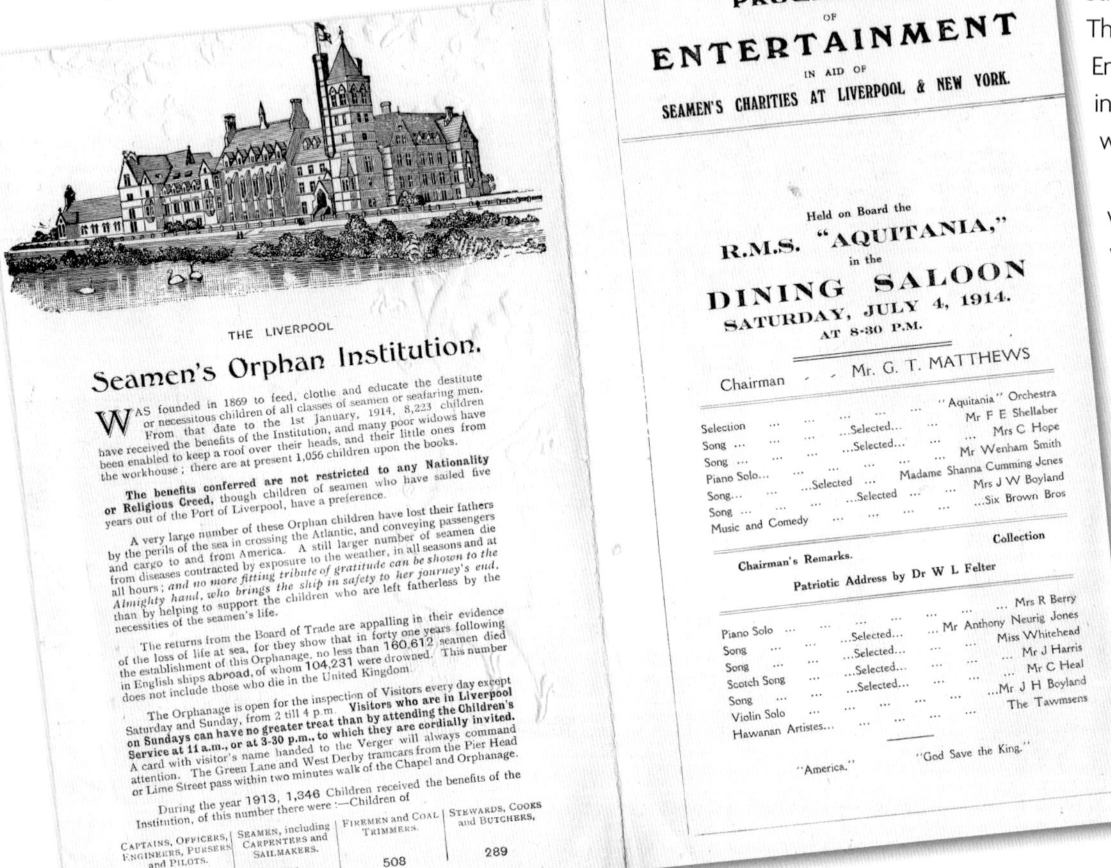

This programme is for the charity concert held to benefit the Seamen's Orphan Institution on Saturday 4 July 1914. It was held in the Dining Saloon, and featured music both from the 'Aquitania Orchestra' and a variety of solos from passengers. (Mike Poirier Collection)

CAPTAINS, OFFICERS, ENGINEERS, PURSERS and PILOTS.	SEAMEN, including CARPENTERS and SAILMAKERS.	FIREMEN and COAL TRIMMERS.	STEWARDS, COOKS and BUTCHERS,
155	394	508	289

Selection	…Selected…	"Aquitania" Orchestra
Song	…Selected…	Mr F E Shellaber
Song		Mrs C Hope
Piano Solo	…Selected	Mr Wenham Smith
Song	…Selected	Madame Shanna Cumming Jones
Song		Mrs J W Boyland
Music and Comedy		…Six Brown Bros

Piano Solo	…Selected…	Mrs R Berry
Song	…Selected…	Mr Anthony Neurig Jones
Song	…Selected…	Miss Whitehead
Scotch Song	…Selected…	Mr J Harris
Song		Mr C Heal
Violin Solo		…Mr J H Boyland
Hawanan Artistes…		The Tawmsens

CONFLICT

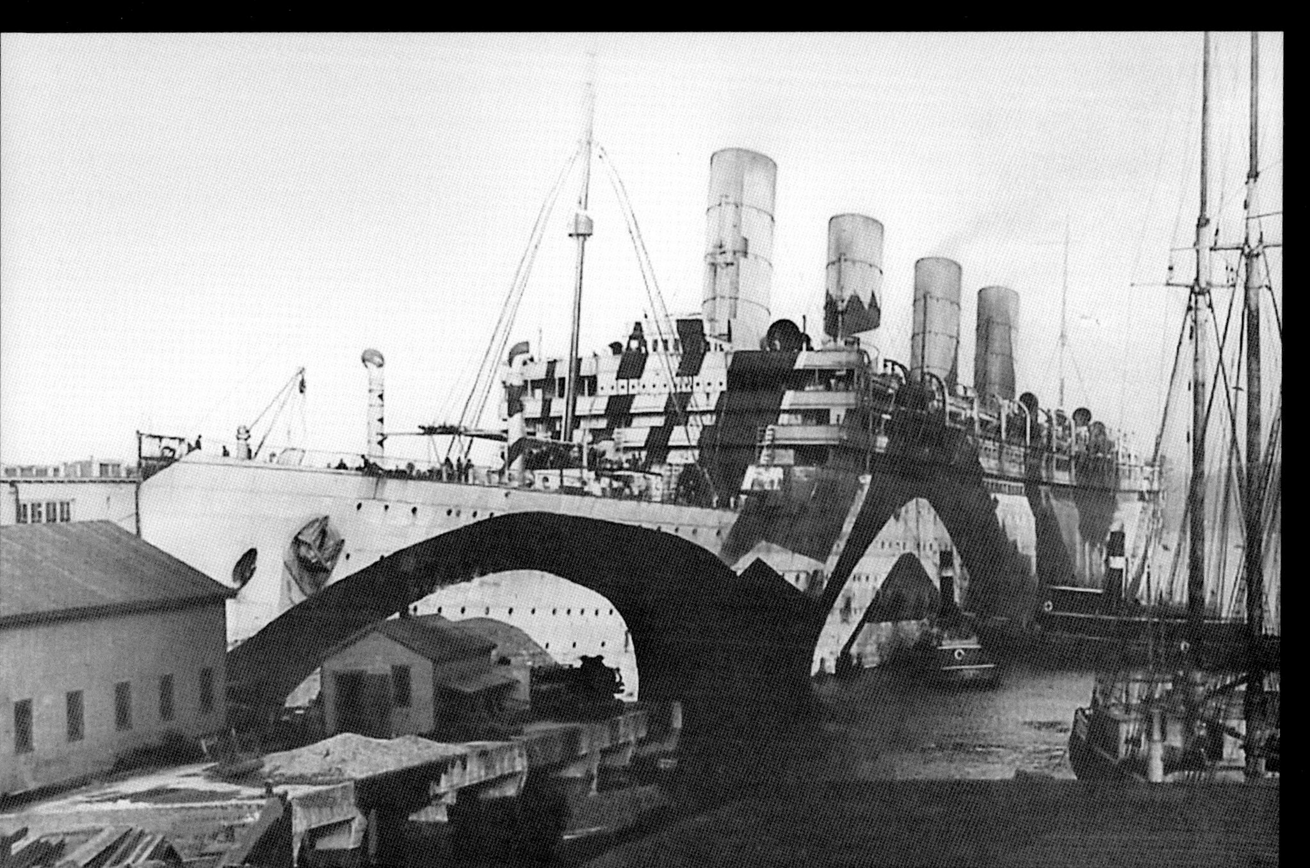

This post-war image was taken in Halifax, Nova Scotia, in January 1919, while *Aquitania* was employed in the repatriation of Canadian and American troops. Evident is the ship's dazzle camouflage, as well as the wear and tear of wartime service. (Clyde George Collection)

On 28 June 1914, the Austro-Hungarian Empire's Archduke Franz Ferdinand and his wife Sophie were brutally assassinated in the streets of Sarajevo. The murders were committed by a Serbian Nationalist. Tensions in Europe were already high, and this spark set off a chain reaction that would lead to the First World War. On 28 July, Austria–Hungary declared war on Serbia, and in the following days, feelings of nationalism, as well as a complicated web of international alliances and treaties, caused a cascade of war declarations between nations. After declaring war on Germany on 4 August, Great Britain also found itself swept up in the conflict, before most realised the full ramifications of what was happening. By the end of the war, 16 million lives would be extinguished.

On 7 August, the British Admiralty requisitioned the *Aquitania* to serve as an armed merchant cruiser. The *Mauretania* had likewise been called into service. The Admiralty requisitioned the vessels in accord with the terms of the agreement between the Admiralty and Cunard, signed 30 July 1903, which had provided a loan to fund construction of the *Lusitania* and *Mauretania*. That agreement essentially placed the whole Cunard fleet at the Admiralty's disposal to 'command the Atlantic' in time of war.[53] While in Liverpool, the *Aquitania* was quickly converted for its new role. Its luxurious interiors and fittings were stripped out and placed in storage. Sir Archibald Hurd, a journalist who would go on to write the official history of the Merchant Navy, described the *Aquitania*'s conversion:

> ... the *Aquitania* and *Caronia* ... were fully dismantled and fitted out as armed cruisers in the first days of August. ... This, of course, meant the ruthless stripping out of all their luxurious fittings and those splendid appointments ... and for all these articles storage had to be found on shore at the shortest notice. Some idea of the work involved in this conversion can best be gathered perhaps, by realising that no less than 5,000 men were employed upon this herculean task, and that more than 2,000 waggon [*sic*] loads of fittings were taken ashore from these two liners.[54]

The vessel was also fitted out with 6in deck guns. As soon as these modifications were made, the newly christened HMS *Aquitania* was assigned to patrol the Western Approaches. She set sail on her first patrol on 8 August, returning to the River Mersey on 16 August. During her second patrol, *Aquitania*, which by then was less than 3 months old, suffered a mishap. On 22 August, the vessel was sailing through a heavy fog bank and collided with the Leyland liner *Canadian*. The *Aquitania*'s stem was crushed in the accident, and while it was never in danger of sinking, the ship had to limp back to Liverpool. The damage was extensive enough that repairs would take months.

The accident with the *Canadian* had further highlighted an unforeseen flaw in *Aquitania*'s design. During the vessel's brief pre-war life, her officers found that the ship's distinctively long forecastle, while giving the ship a unique look, reduced visibility looking forward from the bridge. In order to correct this problem, a small viewing platform was hastily installed on the bridge roof, which was nothing more than a temporary and haphazard fix. A sturdier, smaller pilothouse later replaced this thrown-together one. While this was a more permanent solution, many thought that this unexpected add-on to the vessel detracted from *Aquitania*'s overall appearance, and looked every bit like the tacked-on addition that it was. Later in her career, the square windows on the original bridge's forward bulkhead would be replaced with a row of portholes.

Prior to this accident, on 11 August, the *Lusitania* and *Mauretania* had been released from government service, because they were expensive to operate and the Admiralty could not find a proper use for them. Ultimately, it was concluded that the two speedsters, as well as *Aquitania*, were simply too large to serve as armed merchant cruisers. Not only did they burn through coal supplies quickly, but they were soon revealed to be unsuitable for this role in other ways. Passenger liners simply lacked the defensive armour needed to survive sustained battles with other vessels. This concern was perhaps best illustrated by the clash between the Hamburg South American Line's *Cap Trafalgar* and the Cunard Line's HMS *Carmania*, both of which had been converted for military service by their respective nations. This battle, which took place on 14 September 1914, became known as the Battle of Trindade. This was the first time in history that two passenger liners were engaged in armed combat, and it resulted in the *Cap Trafalgar* sinking. The *Carmania*, whilst emerging as victor, sustained extensive damage to its bridge and superstructure, and had to be dry-docked for lengthy repairs in Gibraltar.

The Admiralty decommissioned the *Aquitania*, laying her up through the winter and the following spring. While the *Aquitania* languished in port, the Cunard Line was desperate to find another vessel to fulfill its role in the transatlantic passenger route, even though the war had greatly diminished the number of passengers willing to brave the war zone that the Atlantic was becoming. Cunard selected the *Campania*, an ageing liner that had recently been chartered to the Anchor Line to complete crossings from Glasgow to New York. The *Campania* filled *Aquitania*'s role in the passenger service from August to October 1914, before proving too old and run-down for the duty.

As early as September 1914 — within days of the collision with the *Canadian*, and far earlier than is generally accepted — documents prove that the Admiralty was considering returning the *Aquitania* to Cunard for commercial service. Indeed, a surviving original carbon copy of an estimate of cost, drafted on 9 September, shows that the Admiralty was actively engaged in the process of returning the great vessel to the Cunard Line. This document placed the cost of restoring the vessel to commercial service at a total of £175,310. Amongst others, these costs included £55,000 alone for joinery work 'in connection with all cabins, materials & wages', £26,500 for 'work in connection with 15 public rooms' and £15,000 for 'electric lighting installation, removals, including casings', £11,000 for painters and £1,500 for 're-shipping and renewals to replace breakages and losses' of crockery, glass, silver and cooking gear. This was a steep cost for a vessel which had served just two voyages in military service. Ultimately, they decided to leave the vessel laid up, pending a final decision regarding the role it could best fill for the Royal Navy, if any.

In February 1915, there was again discussion of returning *Aquitania* to civilian service. On 5 May, Cunard publicly announced that they were returning the *Mauretania* to the North Atlantic, and further said that they were leaving open the option to make the *Aquitania* available, 'if business demanded' her use. Indeed, demand for passenger travel across the Atlantic was experiencing a tremendous uptick in the late winter and spring of 1915. This was in spite of Germany's burgeoning U-boat campaign in the waters around the British Isles, and significant concerns that large passenger liners would make easy targets for the underwater hunters. These fears were validated when the *Lusitania* was sunk by the submarine *U-20* on 7 May. This event quelled discussions about returning both *Aquitania* and *Mauretania* to Cunard for civilian service.

Simultaneously, and despite the demonstrated hazards from U-boats, the British Admiralty saw a need to press both liners into service as troopships. The Gallipoli Campaign in the Mediterranean, meant to wrest control of the Straits of the Dardanelles from the Ottoman Turks, was going poorly for the Allies. The passenger-hauling capacity of the *Mauretania* and *Aquitania* made them perfect candidates for ferrying troops into the combat zone. *Mauretania* was requisitioned for this service on 11 May, and *Aquitania* followed on 18 June.

Aquitania was moved to the Gladstone Drydock in Liverpool, and work commenced on converting her for troop-carrying duties. After these alterations, and a week after being requisitioned, she began service as HMT (His Majesty's Transport) *Aquitania*, setting sail from Liverpool bound for Gallipoli with over 5,000 troops aboard. For two months, the vessel successfully served as a troopship. In August 1915, with the Gallipoli Campaign failing, the need to ship as many troops to the Mediterranean theatre slackened. However, the growing number of casualties from the campaign led to a much greater need for shipping wounded troops home to England. As a result, the Admiralty decided to utilise both the *Mauretania* and *Aquitania* as hospital ships, along with White Star's *Britannic*, which was still sitting unfinished in port. The carrying capacity of the three giant vessels would ultimately save many lives, by getting wounded troops back home quickly.

Once again, the *Aquitania* was sent back to port to undergo another conversion. Work quickly commenced on the vessel. The open public rooms and spaces on the ship, barren and devoid of their luxurious panelling and décor, were turned into sterile operating theatres and recovery wards. Meanwhile, workers also clambered over the exterior of the *Aquitania*, rushing to complete the conversion. She was painted to clearly indicate its role as a hospital ship, thus warning that under international law, she was off-limits as a target to both surface raiders and U-boats. The hull was painted white, encompassed by a green band broken only by red crosses, which clearly indicated her non-combatant status. The funnels were painted yellow, completing the livery.

In short order, the vessel, now dubbed HMHS (His Majesty's Hospital Ship) *Aquitania*, was ready for service in her new role. She now had the capacity for carrying 4,182 wounded soldiers, plus necessary medical staff, nurses and crew. She would see much work transporting wounded troops throughout the remainder of 1915 and into early 1916. Ironically, some of the wounded she transported back to England were likely men she had shipped to the front during her previous role as a troopship.

Following the Allied decision to cut their losses and evacuate Gallipoli in December 1915, the number of wounded needing transportation from that region back to England began falling. *Aquitania* departed port in Southampton on 20 January 1916, bound for Liverpool, where she was due to receive maintenance on her engines. Upon completion of this work, the Admiralty decided that with the lower numbers of wounded coming from the Mediterranean theatre, having the *Aquitania*, *Mauretania* and *Britannic* all in service as hospital ships was overkill. Ironically, all three vessels, intended as rivals in peacetime, but allies in war, were laid up in Cowes, near Southampton. There the giants would wait, as the Admiralty deliberated on whether to release them back to Cunard and White Star, or to find another military use for them.

The Admiralty ultimately released the *Mauretania* back to Cunard on 1 March, after making the announcement of that decision on 24 February.[55] *Aquitania* was similarly released on 10 March. The Admiralty was so confident that the two liners' services would no longer be needed that they paid Cunard to refit them for passenger service. They provided £60,000 for refitting *Mauretania* and £90,000 for *Aquitania*. This later amount was substantially less than the £175,310 estimate of the cost of returning the *Aquitania* to civilian service that had been drawn up in September 1914, when the Admiralty had first contemplated returning her to Cunard. The work was contracted to Harland & Wolff, and over the next several months, commenced at their Southampton facilities.

While the extensive overhaul and refit for civilian service was taking place, the situation in the Mediterranean theatre began to flare up again. As the situation became more active throughout the summer, the Admiralty began to realise, much to their chagrin, that there was still a need for a large number of wounded troops to be transported home from the Mediterranean.

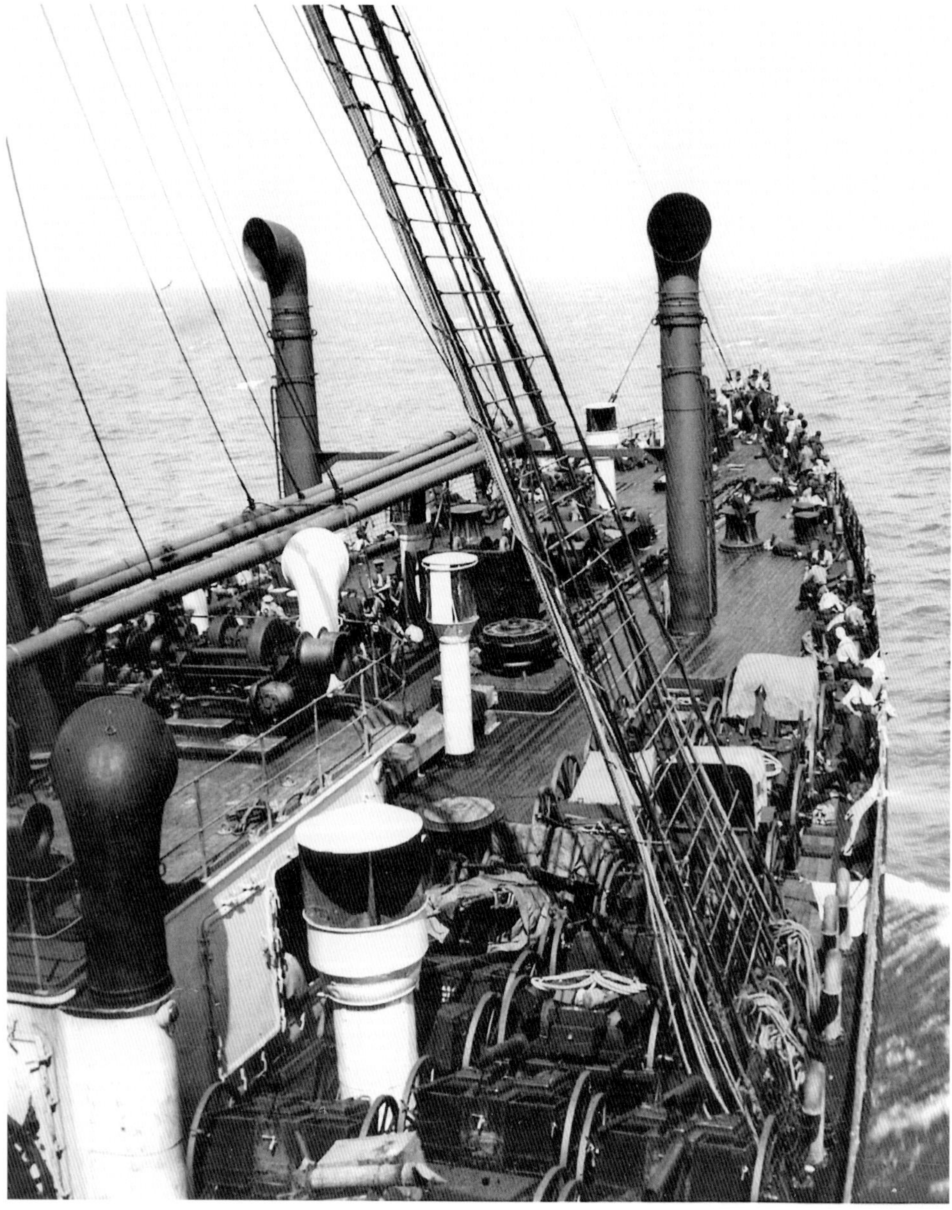

Troops and supplies fill the foredeck of HMT *Aquitania*, as the troop transport heads to Mudros, Greece, in May 1915. (Clyde George Collection)

Announcements.

APRIL FOOLS DAY

Grand Charity Sermon

by the

REV. ROSINGSTON,

of West Kesingdale,

IN AID OF HOSPITAL SHIPS

Admission - - ONE LIBEL.

Proceeds to be devoted to the Fortification
of the Island.

At 8 p.m. **"FERG"**

Presents for the first time on any Stage
his Latest Illusion

"The One-Armed Flautist"

The Thickness of the Hand deceives the Eye!

No R.A.M.C. man should miss this!

The Corps March

"Thy Blight Smile daunts me, Bill"

will be sung by

LIEUT. SLATER,

in a rich Fruity Falsetto, jewelled in two holes,
immediately after the foregoing immense
Illusion.

Please keep your seats.

A COLLECTION WILL BE TAKEN.

Immediately after the Concert CAPT. QUIN
will hold a

Grand Reception

In the Aviary on "E" Deck.

Strait waistcoats provided. All are welcome.

STRETCHERS FOR 11.30 p.m.

Superfluous Hair

A new astonishing epilatory.

The **LINDSAY-JONES SPECIFIC**

The police won't know you after
one application!

"I have no hesitation in recommending it"

A. C. MAJOR.

Obesity.

A rapid remedy for this distressing condition

Absolutely secret and does not
interfere with your usual drinking

Unsolicited testimonial:—"I tried it, What price
buttons now?"—G. DAVIDSON.

Sole Licensee - R. COX.

A page from a rare holiday programme produced aboard the HMHS *Aquitania* in December 1915. The intentionally humorous tone was intended to boost the morale of those aboard. (Authors' Collection)

As nurses observe the scene, wounded troops are carried aboard the *Aquitania* from a smaller vessel. The liner's hospital ship livery, marking it as a non-combatant to enemy vessels, is unmistakable. (Authors' Collection)

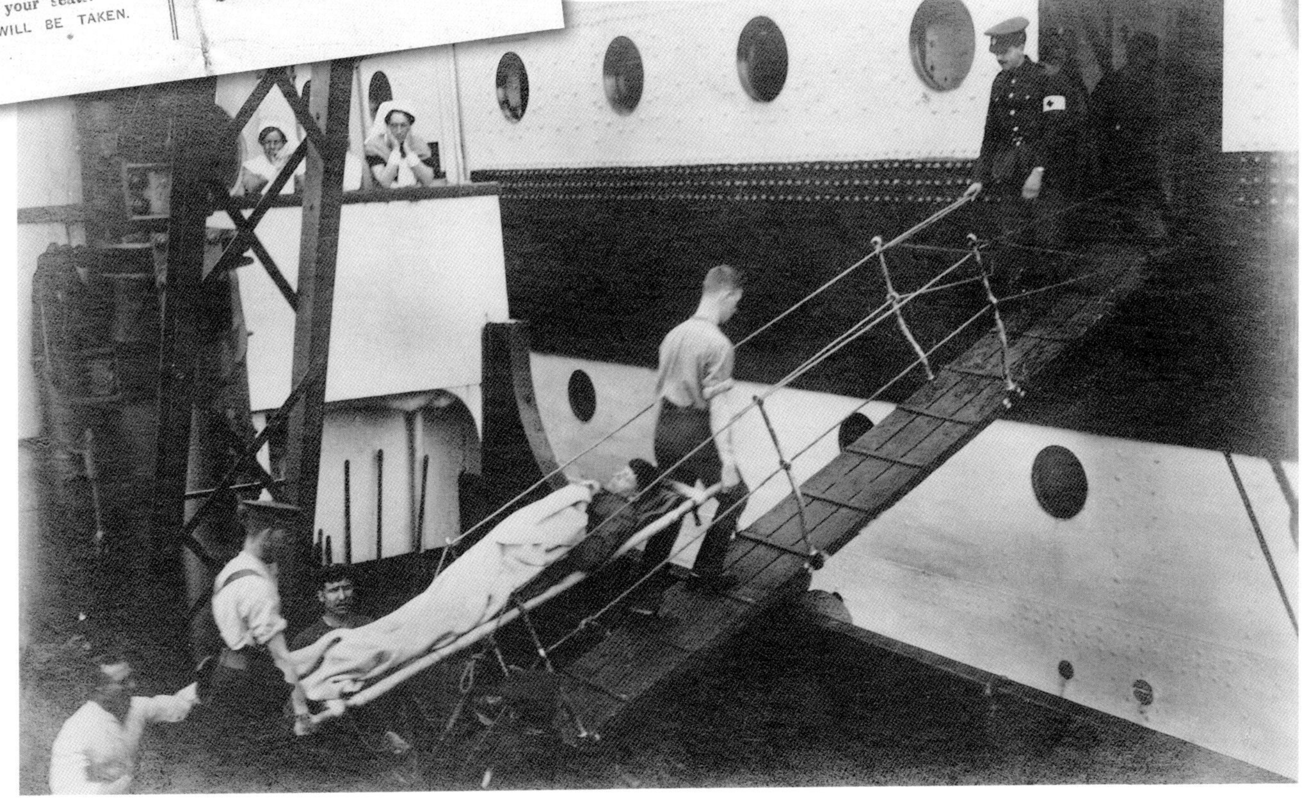

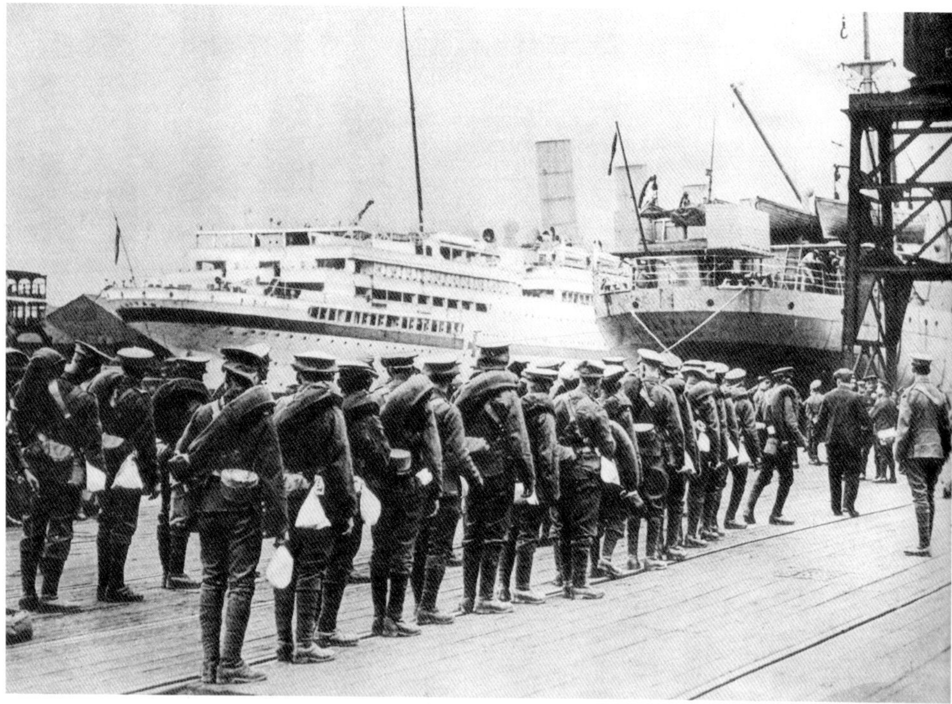

Troops stand on the pier in Southampton, with the HMHS *Aquitania* visible in the background. (Ioannis Georgiou Collection)

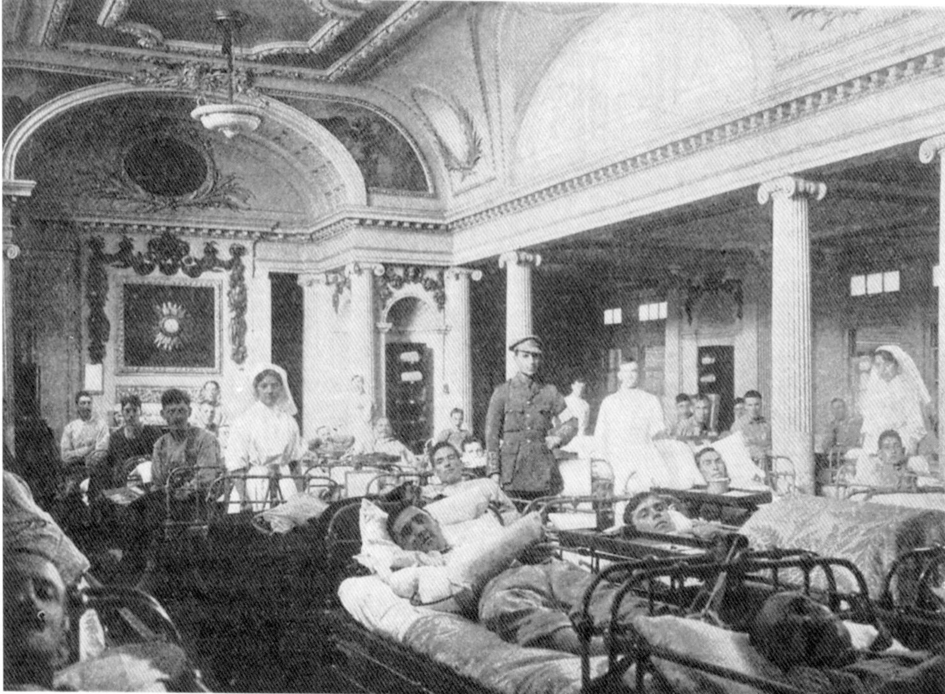

Luxury is juxtaposed with tragedy, as wounded troops convalesce in the *Aquitania*'s Palladian Lounge in 1916. (Ioannis Georgiou Collection)

On 21 July, the Admiralty decided to requisition the *Aquitania* once more. This decision was made, despite the fact that the lengthy process of converting the ship back into a passenger liner was nearly complete. Harland & Wolff's yard workers, who had been labouring away at the restoration work, were suddenly instructed to undo much of what that they had just accomplished, and convert the liner back into a hospital ship.

With much of the ship's panelling and décor already reinstalled, the larger rooms were converted into either operating theatres or patient wards. The Garden Lounges morphed into accommodations, with rows of beds and medical supplies; the Palladian Lounge became a ward for male patients, albeit one with superior décor; the First-Class Smoking Room, with its finely carved panelling, was reserved as a ward for convalescing officers only. Elsewhere, a makeshift chapel was installed; a stage was also placed at one end of the First-Class Lounge, for the presentation of entertainment. The ceiling above the stage was festooned with the flag of the Red Cross, framed by the Union Jack and French Tricolour.

Once this work was done, *Aquitania* served as a hospital ship for the remainder of the year, again partnered with HMHS *Britannic* – which had undergone a similar lay-up, partial restoration and return to hospital ship service.

All of this back-and-forth was seen by many as a needless waste of precious resources. The subject came up on the floor of the House of Commons on 24 July; William W. Ashley, MP, was profoundly critical of what he termed 'by far the worst instance of waste on the part of the Admiralty which has occurred since the beginning of the War'. He recounted the *Aquitania*'s history: her change from a passenger liner to armed merchant cruiser; her conversion back to a passenger ship that January, when he personally visited her in Liverpool; her reconversion to a transport, and thence to a hospital ship; her partial conversion back to a passenger ship that had just been reversed so that she could be a hospital ship again. He said that she, and the similarly treated *Britannic*, were 'the laughing-stock of the South of England', and that the back-and-forth was a 'lavish waste of money' which showed 'vacillation and no direct line of policy'.[56]

While it is true that some of the back-and-forth changes might have been avoided with foresight, the

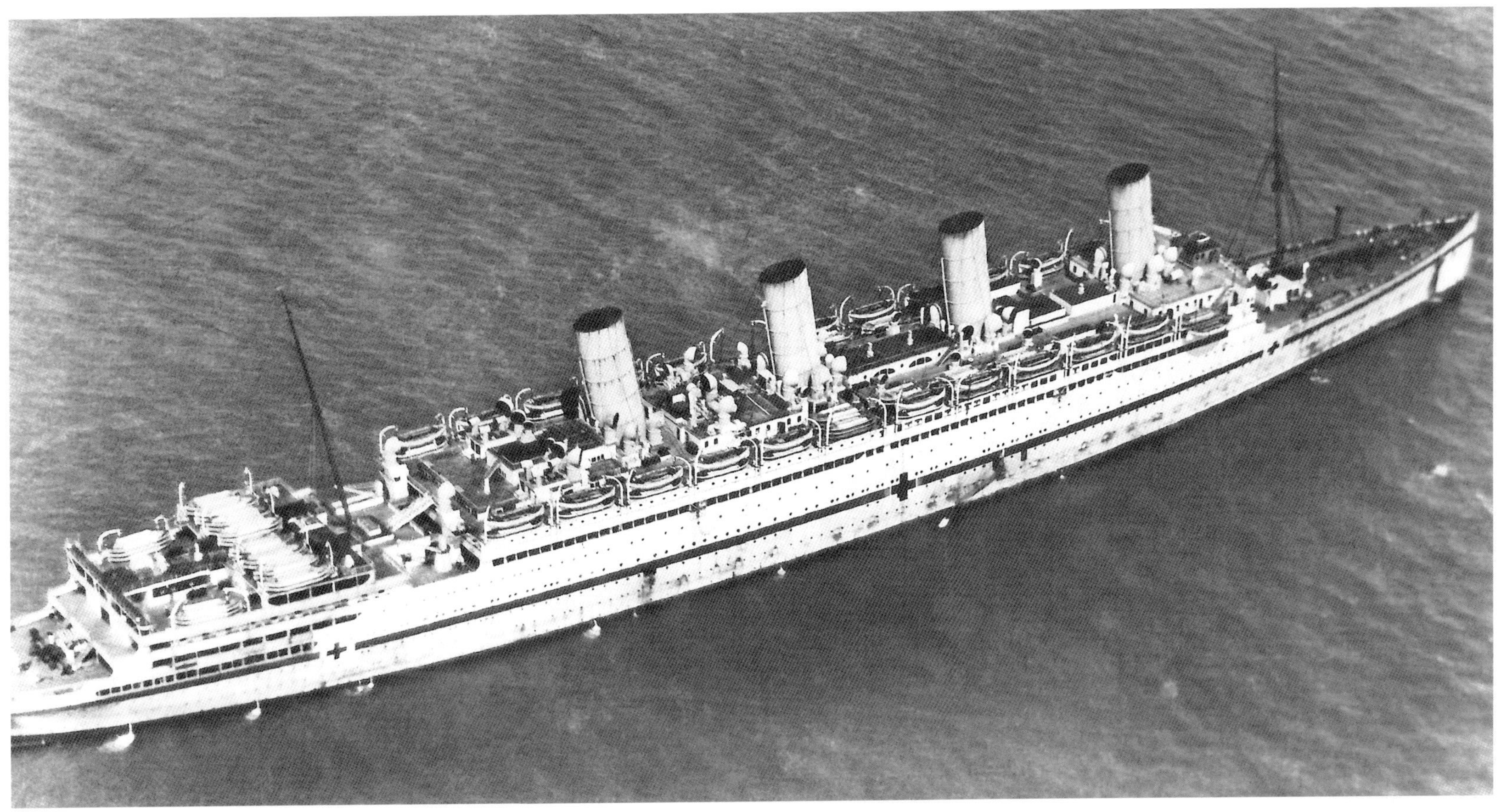

An aerial view of the HMHS *Aquitania* at Mudros Harbour in late 1916. This photograph shows an alteration to the liner's profile: an upper pilothouse atop the original Wheelhouse, forward of the No. 1 funnel. The original open-air upper bridge had been a temporary solution to the unexpected visibility problems encountered from the lower Wheelhouse. The new structure, added in November 1915 by John I. Thorneycroft & Company Ltd, at Southampton, was a more permanent fix to the issue. Interestingly, a careful inspection of photographs taken during the *Aquitania*'s time as a hospital ship give clear indication of when the photo was taken: if there is no pilothouse visible, the photo dates to sometime in 1915; where the pilothouse has been completed, however, the photograph was taken during her 1916 service. (Clyde George Collection)

reality was that *Aquitania* and *Britannic* were merely subject to the constantly shifting demands made by the war. What is more, their return to service as hospital ships was just in time, as Allied casualties from the Salonika Front began to increase with the launch of the Central Powers' offensive there late in the summer.

In November 1916, the *Aquitania* encountered heavy seas and a severe storm while returning from a voyage to the Mediterranean. While it managed to press through the bad weather, the vessel was damaged significantly enough that she would be unable to make her next outward voyage on time. On 6 November, the *Britannic* had just returned from its fifth voyage as a hospital ship, when her crewmembers were unpleasantly surprised to discover that due to the *Aquitania* needing repairs, they would

have less than a week before their ship headed back to the Mediterranean. Casualty numbers in the region dictated that the Admiralty could not wait for the *Aquitania* to be repaired before sending another large hospital ship there. Because of this, *Britannic* set sail on 12 November. During this voyage, while the ship was sailing in the Kea Channel off the coast of Greece, she struck a mine which had been laid by a German U-boat, and sank in just fifty-five minutes. If the *Aquitania* had not been damaged in the storm, it is possible that she might have become a victim of the mine, rather than the *Britannic*.

Following the loss of the *Britannic*, and with the *Aquitania* still being repaired, the Admiralty was faced with a significant shortage of capacity to move wounded troops. This was compounded by the loss of the smaller hospital vessels *Galeka* in late October and *Braemar Castle* in late November. Both vessels struck German mines, suffered significant damage and had to be beached to avoid sinking. With the Admiralty's Transport Division stretched so thin, the remaining number of smaller hospital vessels had to be relied upon until *Aquitania* was fit to return to duty. Once repairs were made, the *Aquitania* immediately returned to service as a hospital ship, and not a moment too soon. However, likely owing to the loss of the *Britannic* and other vessels to mines, *Aquitania* would travel no further than Augusta, Sicily, to pick up wounded, rather than run the risk of sending her closer to the front.

By the end of 1916, the Mediterranean theatre began cooling down again. As a result, the Admiralty determined yet again that the services of the *Aquitania* were no longer necessary. This time, however, it was decided she should be laid up, rather than starting yet another conversion, in case she would be needed later. The *Aquitania* was thus laid up in the Solent for almost all of 1917. During her career as a hospital ship, she had transported no less than 25,000 wounded troops home from the Mediterranean.

Following Germany's declaration of unrestricted submarine warfare on 31 January 1917, U-boats began targeting Allied and neutral shipping alike, in an attempt to blockade and starve Britain out of the war. As the losses to submarines reached terrifying numbers in the spring, the increasing acts of aggression and lack of consideration for American lives finally forced the neutral United States into the conflict. However, even with the United States' declaration of war on Germany in April 1917, *Aquitania* remained laid up.

As it turned out, the *Aquitania* was not recalled for duty until November. Then she was enlisted to serve as a troopship once more, this time transporting Canadian troops and an increasing number of American troops across the Atlantic. As part of the conversion for this duty, *Aquitania*'s hull was emblazoned with 'dazzle' camouflage. The sharp angles, colours and bizarre patterns of dazzle camouflage were designed to confuse U-boats by disguising the silhouette, speed and heading of the vessels. *Aquitania*'s pattern of camouflage would be altered as the war proceeded.

With ever-growing numbers of American troops being deployed in Europe, *Aquitania* was paired up with the *Olympic*, *Mauretania* and *Leviathan* (ex-*Vaterland*), all of which were then serving as troopships. Improved anti-submarine countermeasures, the availability of escorts and travelling in convoys helped protect these large passenger liners from the constant threat of lurking U-boats. Ralph J. Anderson, a volunteer who served as a wagoner in Ambulance Company 113 of the US Army's 20th Division, wrote a vivid description of what it was like to traverse the Atlantic aboard the *Aquitania*. He began his crossing on 4 July 1918:

[On] the morning of July the 4th … a harbor tug ferried the men … up the Hudson to Pier 54, where lay the mammoth Cunard Liner, the *Aquitania* … [Gala] flags and decorations [were] everywhere displayed in honor of Independence Day … A marvelously simple yet efficient system … served equally well to get the soldier aboard, and to locate him once he was aboard … In single file the long lines arranged in alphabetical order marched toward the gangplank. Here an officer called aloud, from a roster … the soldiers' surnames … At the foot of the gangplank a card was thrust into the hand of each man. This printed slip showed the section of deck, compartment, number of berth, of dining saloon, and of seat at table. Guides, stationed everywhere, pointed the way.

… Next day, July 5th … the *Aquitania* began her perilous voyage across the submarine-infested Atlantic … The Statue of Liberty loomed large, then faded from view in the distance … Unlike the majority of troop ships the *Aquitania* sailed alone. She depended upon speed for safety. Directly ahead when the harbor was left behind, like some strange, unnatural bird of prey, a dirigible appeared, and from a commanding height searched the waters for U-boats.

… Every foot of available space aboard the giant liner was utilized as berths. These were uniformly swung upon a metal frame, two high, and were substantial, comfortable. The berths in first and second-class staterooms were occupied by officers and nurses, while many enlisted men fell heir to those of the third ... class. The men were crowded, but not uncomfortably so ... On board the *Aquitania* English cooks served English food, which was different from what the men were accustomed to; but of this there was an abundance.

... The dangers of ocean travel were strongly impressed upon those on board. ... Every man was issued a life belt and was required to wear it at all times. At night, this belt made an excellent pillow. ... Each day drills were held, in which everyone, at a given signal, fell into line and, firmly grasping the belt of the man ahead, lock-stepped up the many flights of stairs until the first deck was reached, and proper station taken beside life boats. Six 6-inch guns, at the loaded breeches of which, night and day gunners stood, frowned across the waters. ... Rumor was busy in reporting submarines sighted and destroyed; but no official confirmation was ever offered.

... [During] the six days of her passage across ... life aboard passed pleasantly. ... Moving pictures were shown nightly. ... A canteen aboard was well stocked with fruits and candies, but the line was too long for the patience of most.

About 8 o'clock in the morning of July 11th five American destroyers rose above the horizon. They advanced rapidly and took positions best calculated to protect the troop ship [*sic*] from U-boat attack. ... Belching smoke, Old Glory streaming to the breeze, it thrilled one deeply to meet these representatives of the United States in foreign waters. ... [Two] dirigibles appeared and maneuvered one before, one abaft. ... Without adventure of any kind, this most dangerous part of the voyage passed. About 2 p.m. ... July 12th, the engines of the *Aquitania* were shut down and she drifted slowly into the harbor of Liverpool, England.[57]

Ultimately, the *Aquitania* would safely make nine troop-carrying voyages across the Atlantic between January and November 1918, carrying around 6,000 American and Canadian troops on each trip. *Aquitania*'s personal record was 6,090 troops carried during a single crossing.

While the British Admiralty had decided not to seek a profit for transporting American troops overseas, they did expect the American Government to offset some of the costs of operating the liners, feeding and housing their troops, and so forth. Initially, the director-general of the British Ministry of Shipping suggested a flat rate of $125 per soldier be charged the United States for this transportation service. He then suggested that a larger fee of $150 per soldier should be charged for anyone being transported aboard the *Aquitania*, *Mauretania* or *Olympic*. The United States rejected those prices as too high.

Eventually, in what was termed the 'Reading-Hines Agreement', the two allies agreed on a per-soldier cost dependent upon which class the soldier was berthed

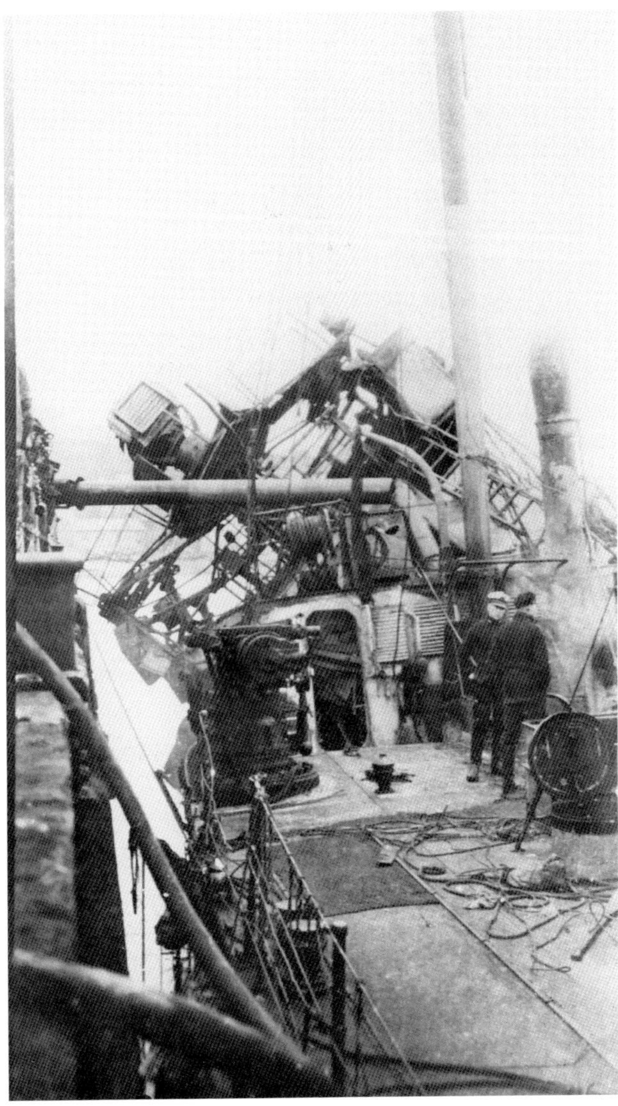

Taken aboard the USS *Shaw* in October 1918, this image shows the significant amount of damage that the destroyer sustained in the collision with the *Aquitania*. (Authors' collection)

in: $176.30 for First-Class; $128.65 for Second-Class; and $42.88 for Third Class. In general, commissioned officers were berthed in First-Class cabins, non-commissioned officers in Second Class and enlisted men in Third Class. The United States would pay the accrued amount for the transportation of its soldiers in full with cash in 1921, in lieu of being credited against overdue interest on loans.[58]

In October 1918, the *Aquitania* experienced its second serious accident of the war. Early in the morning of 9 October, the fully loaded troopship was in a convoy, approaching the English coast. To make it more difficult to be targeted by U-boats, the ships in the convoy were zig-zagging as they sailed. The American destroyer USS *Shaw*, which had a length of 315ft, was sailing off *Aquitania*'s side at an estimated speed of over 27 knots. The helm of the *Shaw* somehow became stuck, and suddenly the escort crossed over into the *Aquitania*'s path. Despite the efforts of both crews, there was not enough time to avoid a collision.

The sharp prow of the *Aquitania* blasted into the destroyer about 90ft from its bow, ripping through it like a tin can and puncturing one of its oil tanks. A fire broke out aboard the *Shaw*, threatening to detonate the ship's munitions. However, despite the catastrophic damage, the destroyer managed to stay afloat. After damage control efforts were implemented, the *Shaw* limped into Portsmouth for repairs. Twelve Americans died in the incident. However, as had been the case when the *Aquitania* collided with the *Canadian* in 1914, the former escaped with only minor damage and was able to continue its passage safely.

Commander William A. Glasford, who was commanding the helm of the *Shaw* at the time of the collision, described the horrifying accident as follows:

> I was called and went to the bridge and took the con of the *Shaw* … As it was early dawn, I followed the usual procedure, making wide zig-zag at full speed, which speed I purposed maintaining until about 6:30 A.M. I was on a right leg of my zig-zag, heading about 35 degrees converging to the right of the *Aquitania* and at a distance of more than one thousand yards from her, when I gave the

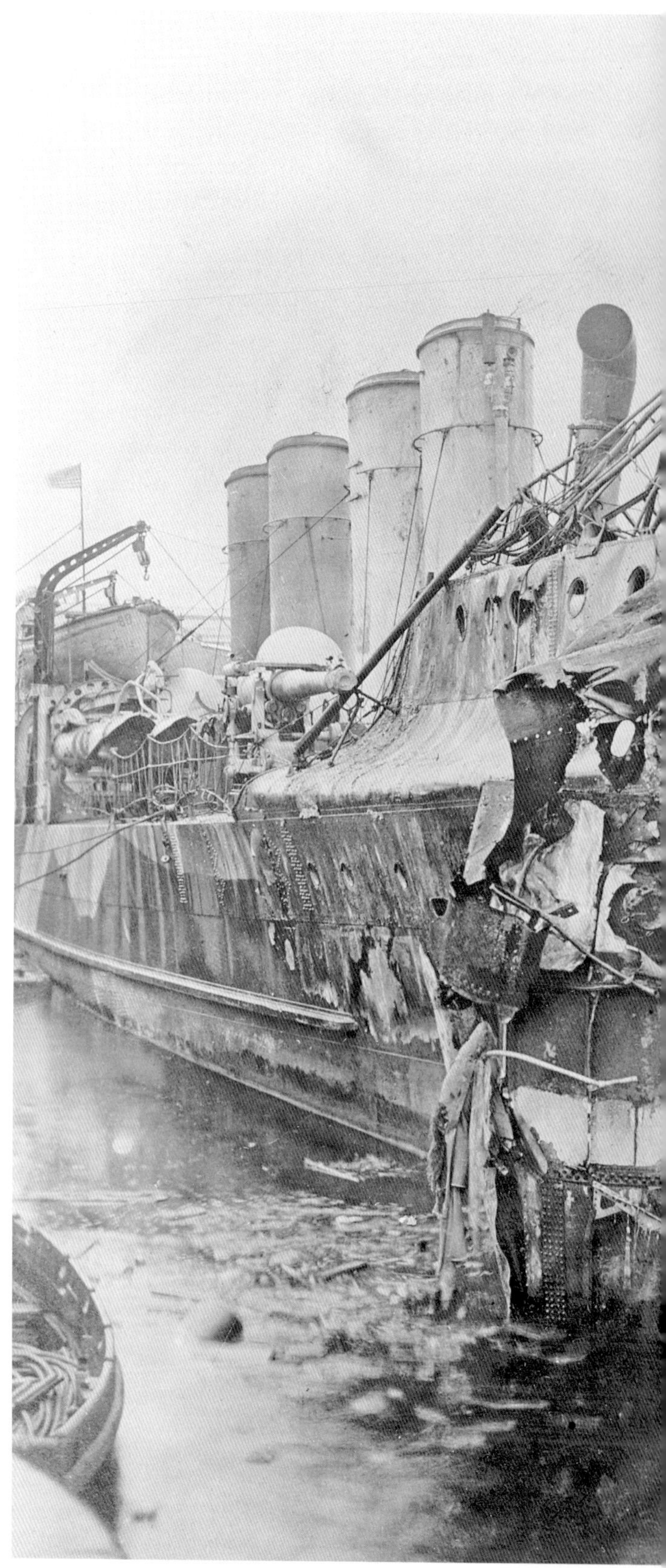

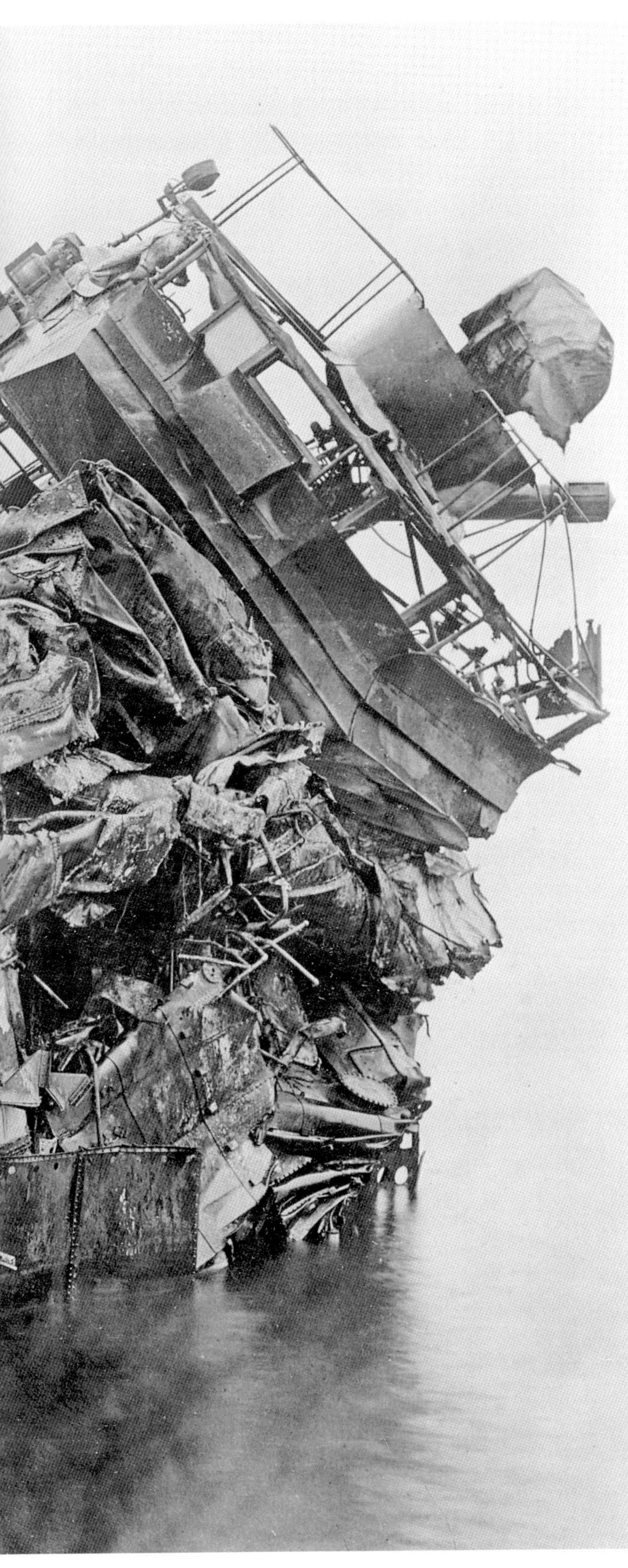

order 'left rudder' to begin my zig-zag. The rudder, at this time, I believe, was full right. The helmsman in obedience to my order swung the wheel and rudder left from a full right and reported to me that the wheel had stuck. I immediately jumped at the wheel which I tried to move myself and could not, and noticed that the helm indicator was at about 17 or 18 degrees right rudder.

... The ship started to swing rapidly toward the *Aquitania* and I saw immediately that one or the other of us were going to get it. My judgment at the time, in sizing up the situation rapidly, was that I was going to hit her. My immediate concern was the safety of the convoy, and such decisions as I made and acted upon were governed by my desire to avoid injuring the *Aquitania*. I immediately rung emergency full astern in order to take speed off the ship, so that if I did hit her, less damage or none at all would be done to her.

I ordered the general alarm sounded ... I passed the word to put the depth charges on 'safety' ... Just before the *Aquitania* struck me, I remember leaning over the side of the ship ... and I can remember a distinct impression of still rushing through the water at great speed.

The *Aquitania* struck me immediately forward of the bridge at an angle of about 45 degrees, covering to my right. She scraped the port edge of the bridge and took the helmsman overboard. I was then swung under her and the *Shaw*'s port side scraped along her port side, smashing my lifeboats after my main-mast fouled one of her life-boats and snapped off at the base and hung over my starboard counter. A hole about 20 feet by 4 feet was torn in the *Shaw*'s port side abreast the forward boiler room.

There was a tremendous flash of sparks as the *Aquitania* hit the *Shaw*, and I believe that this ignited the oil fuel in my forward tanks. The forepart of the ship was immediately in flames ... No orders were subsequently given by me concerning the engines at that time. The order was given to get the life rafts in the water, and as I got down from the bridge somehow, I noticed ... that my bow was floating about 200 yards from me.[59]

Viewed bow-on, the extent of the damage to the bow of the USS *Shaw* is stunning. Fortunately, the vessel remained afloat, and *Aquitania* sustained only minor damage. (US Naval History and Heritage Command)

By the time the Armistice was signed on 11 November 1918, finally ending the devastating global conflict, the *Aquitania* had managed to deliver around 120,000 troops to the European battlefields. Even though the Great War had ended, the world had changed dramatically, and just because the conflict was over did not mean that things were immediately going to return to some semblance of normal. Thus, 1919 was a strange time of limbo for the *Aquitania*. Rather than returning the ship to Cunard Line, the British Admiralty employed her in the repatriation of troops back to Canada, and then she began returning troops to her original American port of New York. A press article appeared wherein the author said:

> When I picture the *Aquitania* ploughing her way to New York, as she is at this moment, it is with one joyous thought in my mind – that the Atlantic is itself again ... [The] war is over.[60]

As reassuring as this might have sounded, things were not *entirely* back to normal. While the *Aquitania* was formally released from Government service on 10 January 1919,[61] she remained under charter to the Government, and for at least a portion of these post-war months she was also chartered to the United States Government. Although she was no longer in danger from enemy submarines or warships, she retained her camouflage 'dazzle' paint scheme as she carried many troops back to Halifax and New York. However, the situation had changed in that passengers could – and did – book passage in the liner, in addition to the troops.

One event of note took place on 28 February 1919, when the *Aquitania* was approaching New York with 5,893 troops aboard, in addition to a number of distinguished passengers. Among this latter group was J. Pierpont Morgan and his wife, as well as his son, Lieutenant Junius S. Morgan. The *Aquitania* had made

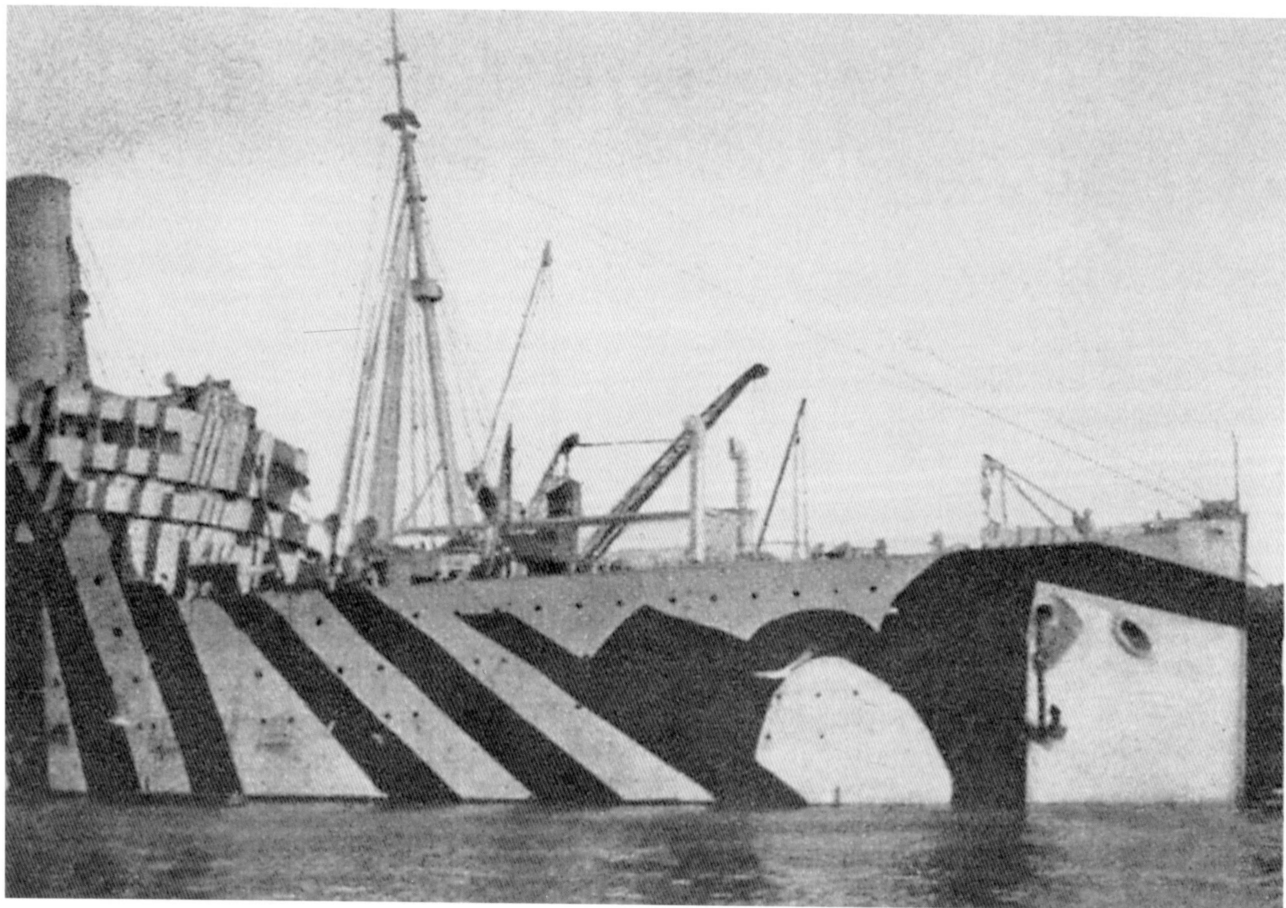

This image of the HMT *Aquitania*, taken in 1918, provides a nice view of the ship's dazzle camouflage, which was designed to protect the vessel from marauding U-boats in the North Atlantic. This pattern would be altered slightly as the war proceeded. (Ioannis Georgiou Collection)

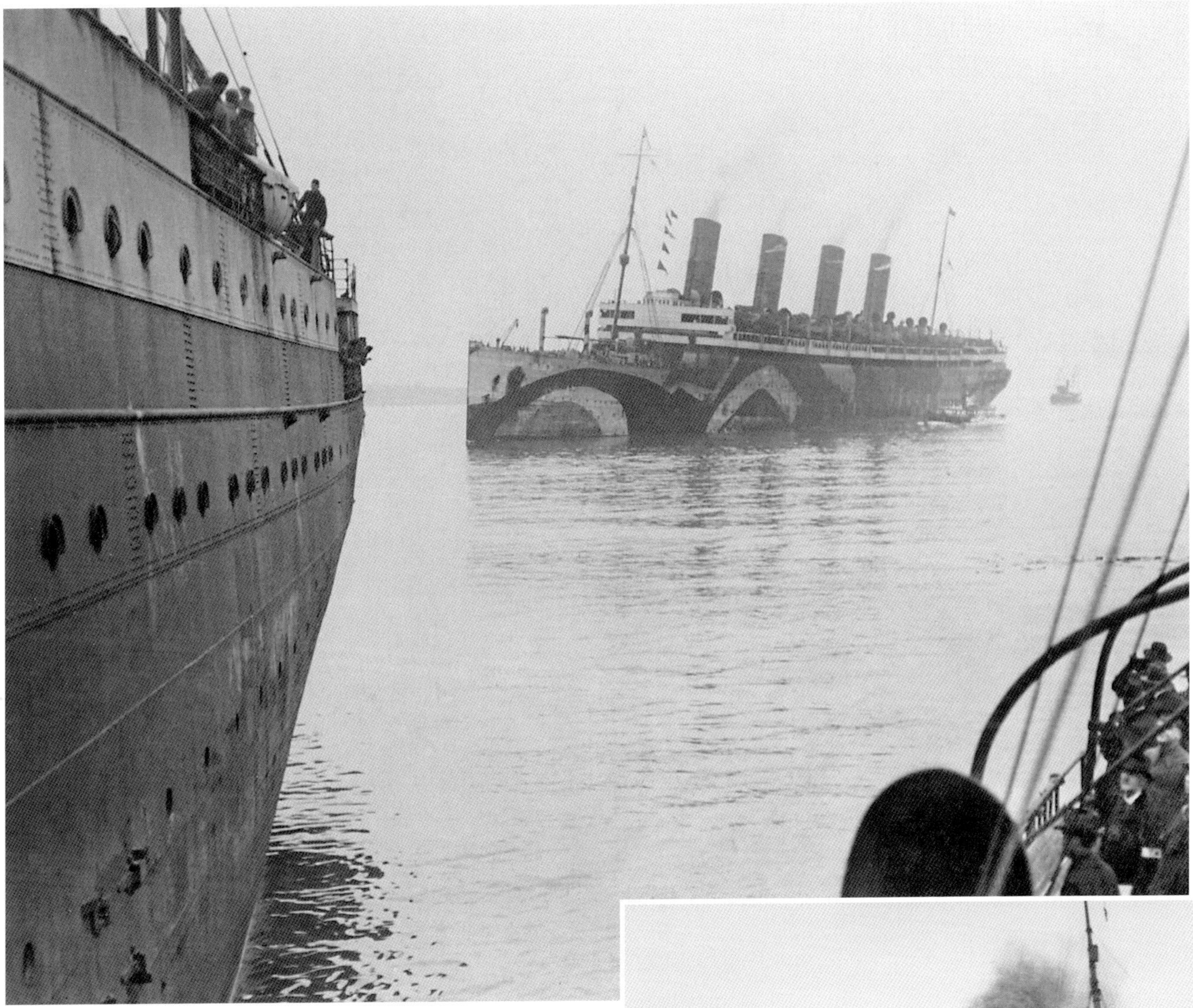

The *Aquitania* leaving quarantine in New York Harbor on 28 February 1919, just moments before colliding with the freighter *Lord Dufferin*. The ship on the left of the image is the French transport *Rochambeau*. (US Naval History Heritage and Command)

The *Aquitania* steams up New York water, after ramming the *Lord Dufferin* on 28 February 1919. Damage from the collision is visible on the bow. (Authors' collection)

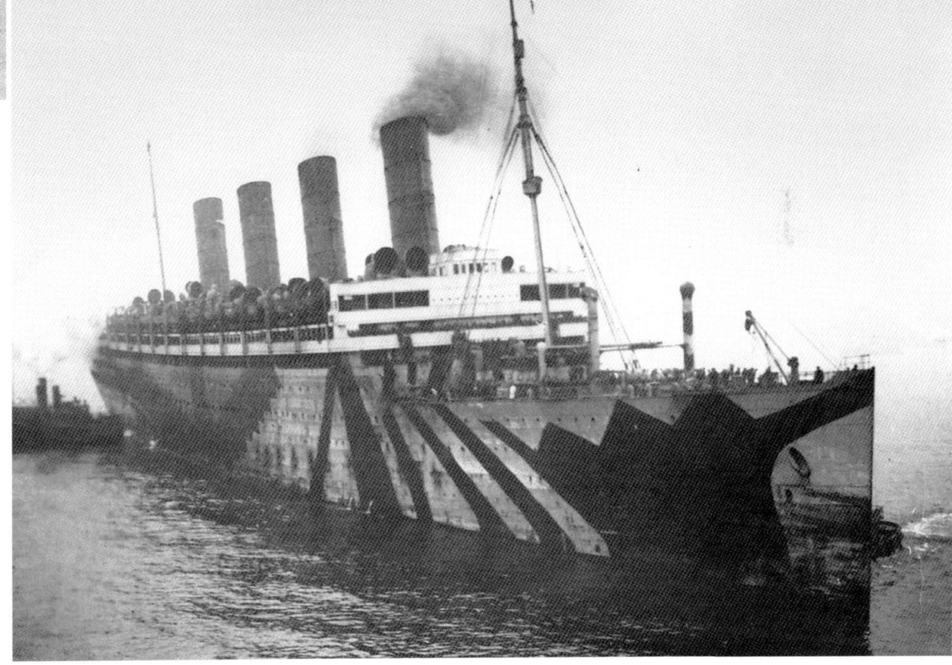

excellent time in the run from Brest, France, to New York, under the command of James T.W. Charles. At about 3 p.m., the *Aquitania* was steaming up the Upper New York Bay, about to pass between Governor's Island on her starboard side and Liberty Island on her port side. The Canadian freighter *Lord Dufferin* was then at anchor in the fairway, just south-west of Governor's Island. The crew of the *Lord Dufferin*, some forty-four men, were then being mustered in the officers' mess by two men from the US Customs Intelligence Service, Richard Ruger and Joseph Cassidy; the small vessel was preparing to depart port that night.

As the *Aquitania* approached the *Lord Dufferin*, the French transport *Rochambeau* was not far to the other

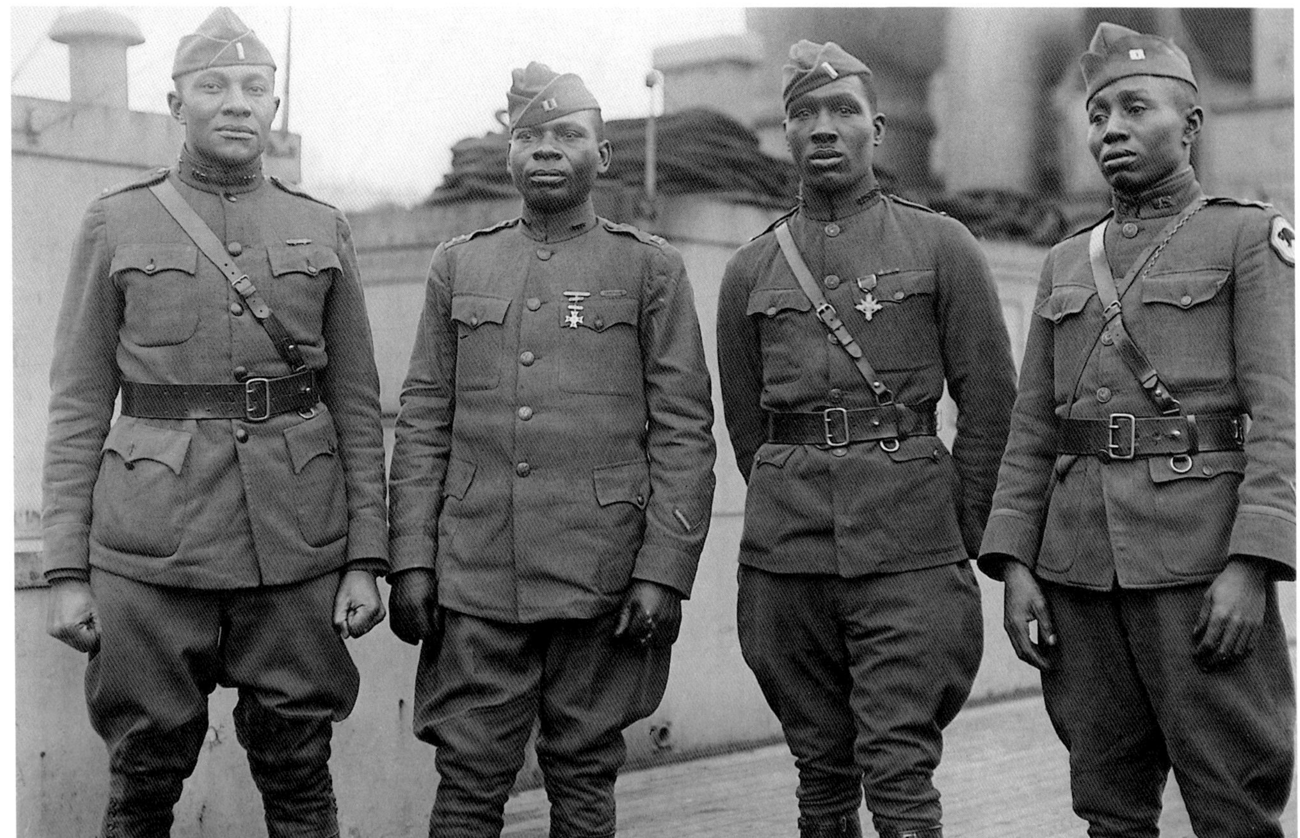

African-American officers of the 366th Infantry Regiment aboard the *Aquitania*, on the way home from Europe. They are, from left to right: Lieutenant C.L. Abbot; Captain Joseph L. Lowe; Lieutenant A.R. Fisher, winner of the Distinguished Service Cross; and Captain E. White. The photograph was taken on the same day as the collision with the *Lord Dufferin*. (National Archives & Records Administration)

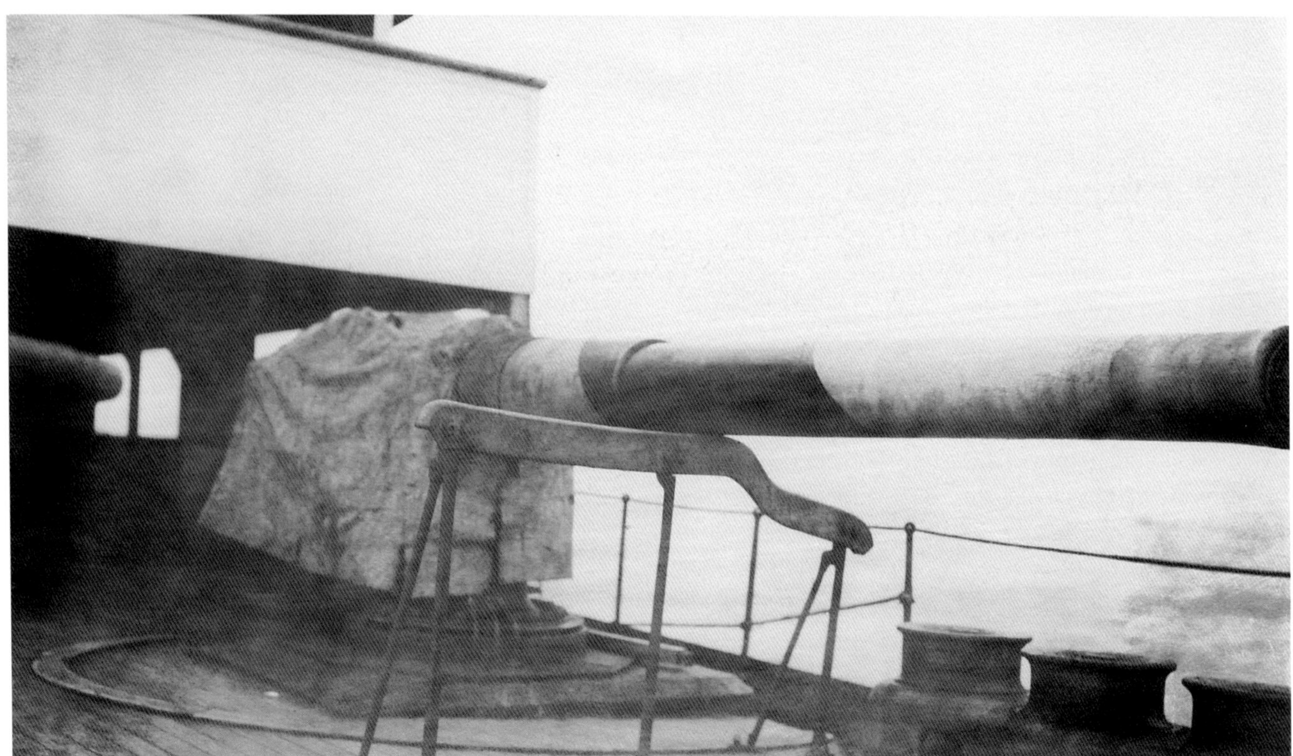

A 6in deck gun, as viewed aboard the *Aquitania* in February 1919. (Authors' collection)

side of the Cunarder. Captain Charles manoeuvred to give the French ship a wider berth, but the *Lord Dufferin*'s stern was then swinging in the strong ebb current toward the *Aquitania*. The tide caught the Cunarder's bulk and turned her prow suddenly toward the Canadian vessel's stern. The turbines were reversed, but the collision was by that time unavoidable. It was said that the 'accident happened so quickly and quietly that few on the *Aquitania* knew it until told'.

The damage was much more apparent to those on the *Lord Dufferin*, for the Cunarder sheared 30ft of her stern clean off. As it dropped away, it 'tore loose and bent downward, fourteen feet below the keel of the freighter, four of the lower strakes of plating on the port side'.[62] The freighter immediately began to sink, settling in six fathoms of water. Those aboard her went overboard, where they were subsequently rescued. One man, 40-year-old Fireman George Eperus, perished, but fortunately none of the other men ended up needing medical attention. The *Aquitania* stood by to render assistance, but there were enough other vessels in the area who made a tidy job of the rescue for her to proceed to her dock.[63] Fortunately she was able to continue in her chartered service, mixing troop repatriation with carrying passengers.

The liner's last arrival in New York during this phase of her career was on 20 July 1919.[64] When she returned to Southampton on 2 August, the liner, now slightly shabby looking from all of the wartime crossings, was placed in dry dock for a partial overhaul. During this stay, her hull was repainted in civilian livery. Interestingly, the very tip of her prow, painted black before the war, was painted white during this 'freshening up'. She departed Southampton on Saturday 6 September, arriving in New York on 13 September.[65] She was not to remain in service for long, however, for the summer stay in dry dock had been nothing more than a temporary patch. It was reported in September:

The *Aquitania* has not been entirely refitted as a passenger-carrier and still has accommodations for troops.[66]

In November, with all of her wartime and post-war responsibilities concluded, the liner was sent to Newcastle for a thorough, and much-needed, refit.

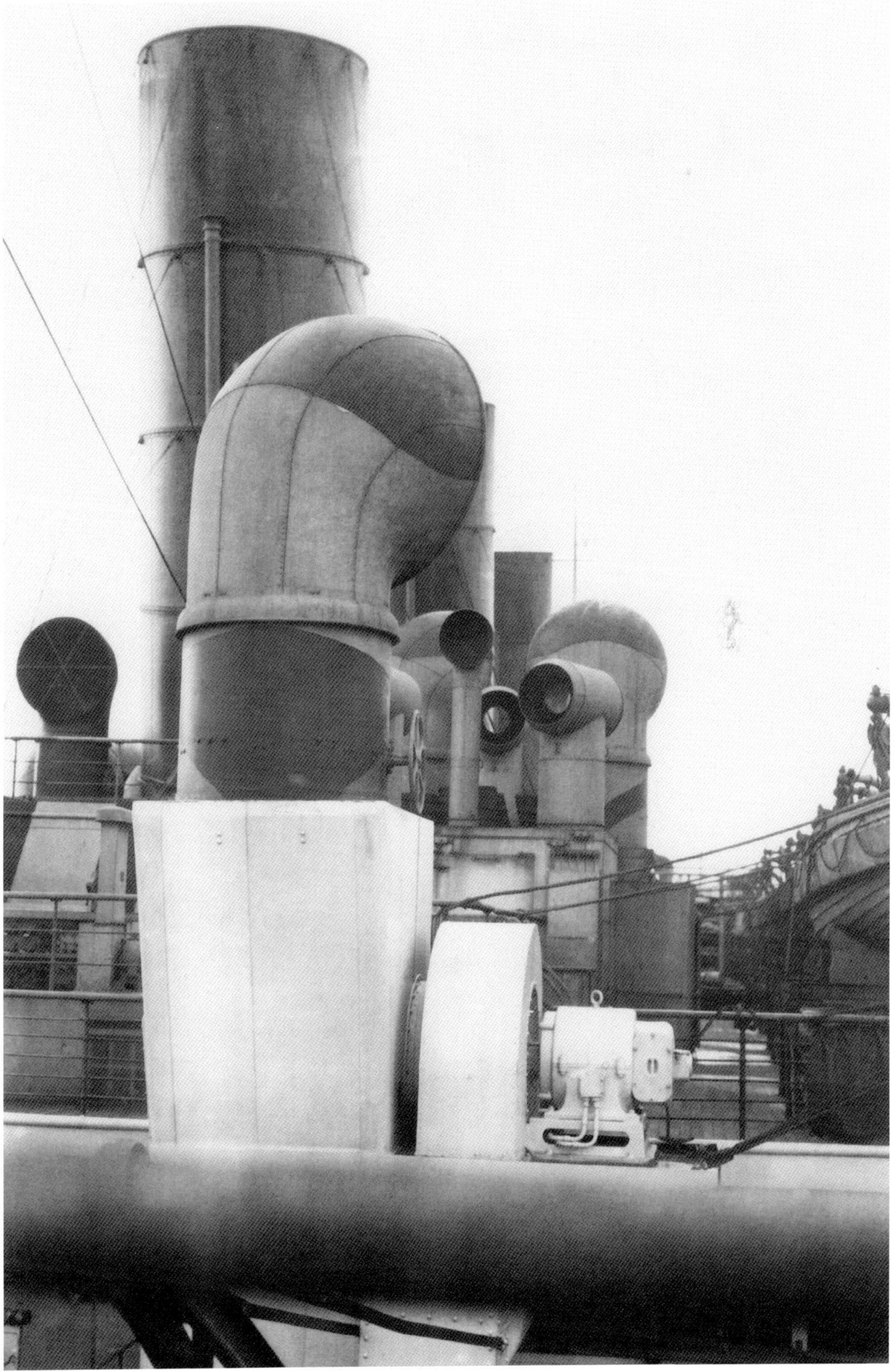

The deck and funnels of the *Aquitania* were still emblazoned with dazzle camouflage in February 1919, when this view was taken. (Authors' collection)

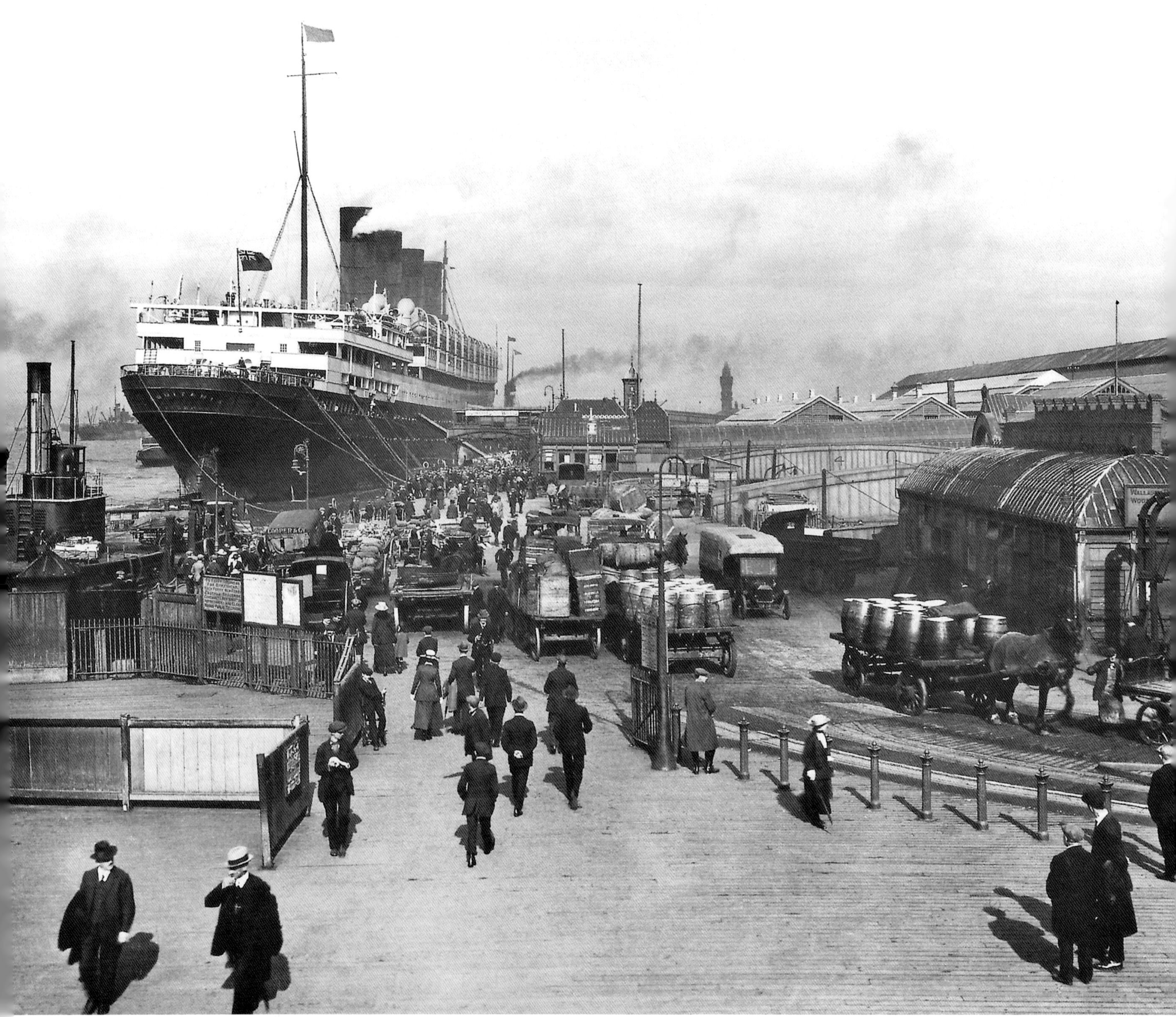

The *Aquitania*, repainted in its civilian livery, at the Liverpool Landing Stage in May 1919. (Clyde George Collection)

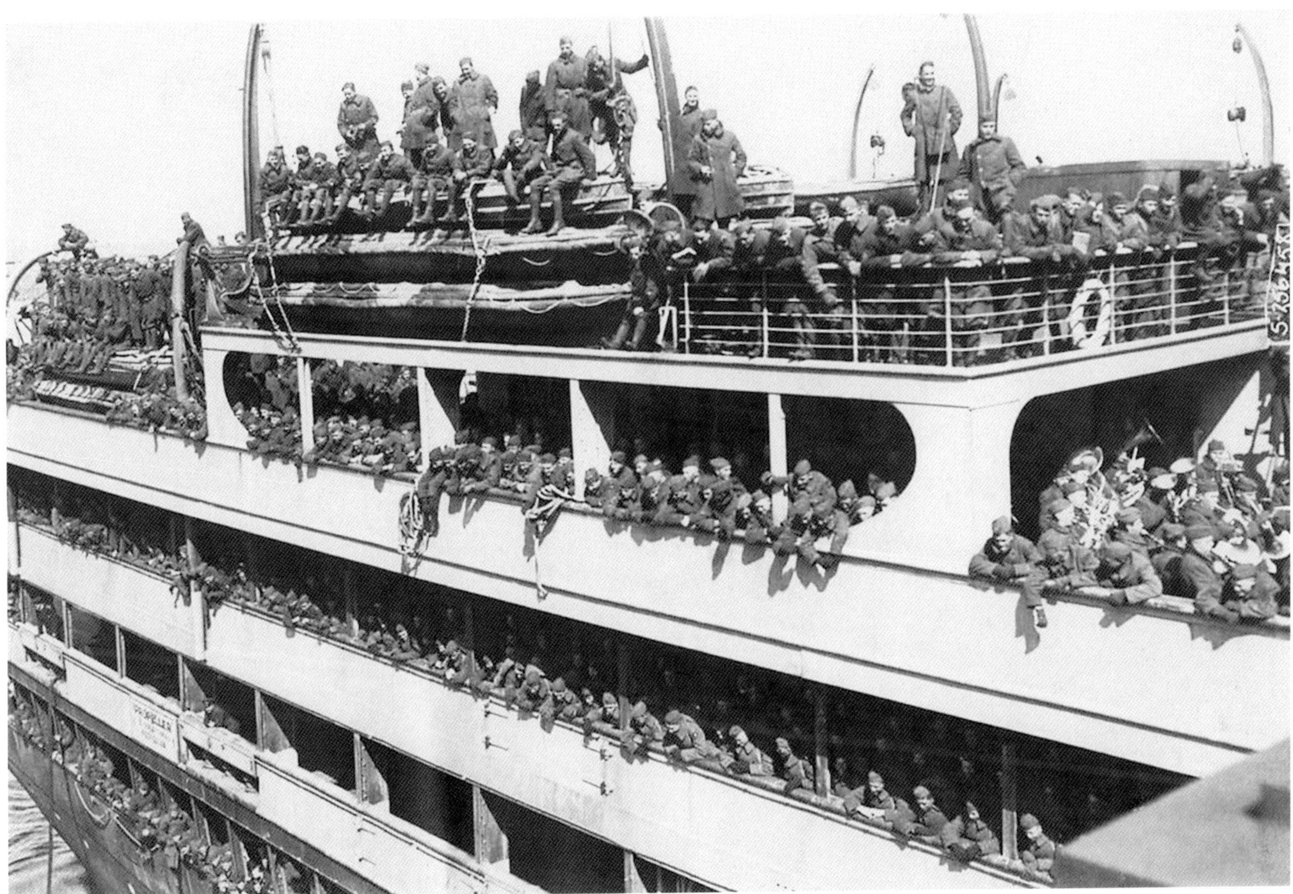

Troops and rafts crowd the deck of the *Aquitania* in this image from late 1919. (Clyde George Collection)

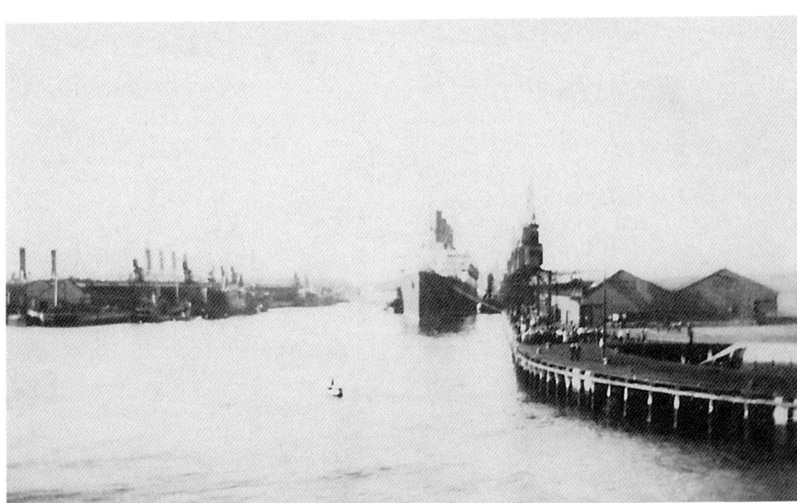

Aquitania docked at Southampton on 15 August 1919. This image shows the curved section of the pier, which was rarely photographed. (Steven B. Anderson Collection)

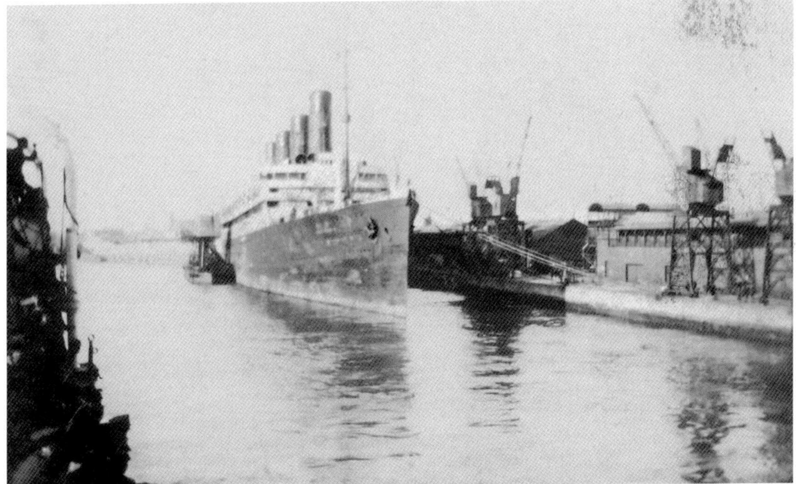

Another image of *Aquitania* at Southampton on 15 August 1919. The ship, looking shabby from its wartime service, was dry-docked and partially overhauled while in Southampton. (Steven B. Anderson Collection)

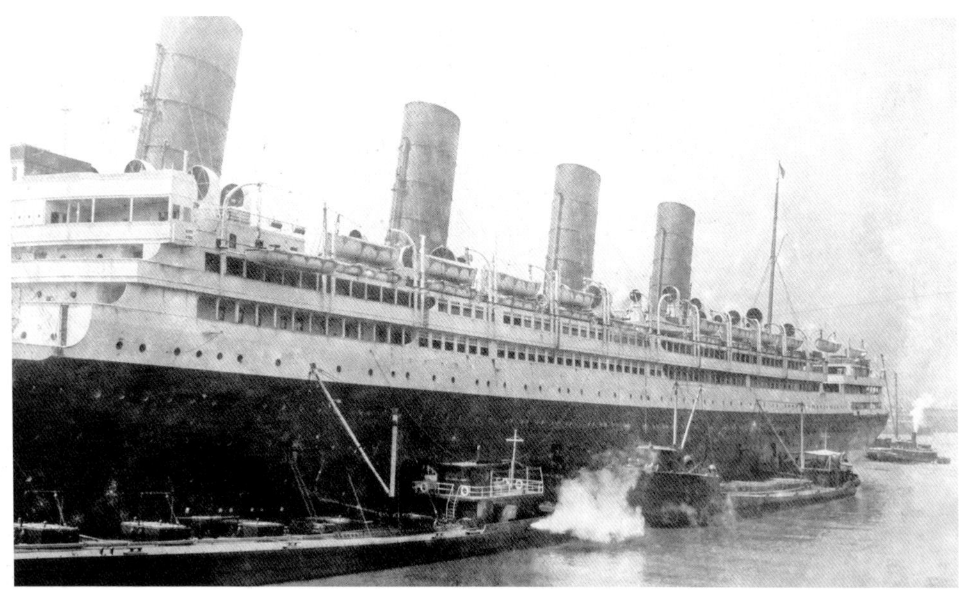

The *Aquitania* taking on coal, probably at her New York pier, sometime in 1919. (Steven B. Anderson Collection)

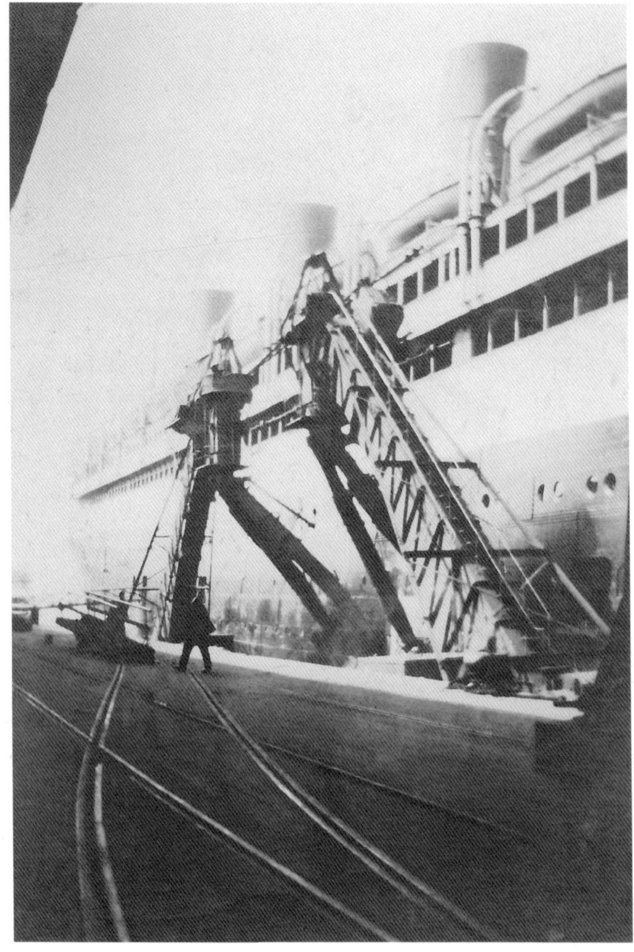

Aquitania being coaled at Southampton in 1919. These mechanised coaling derricks were a tremendous improvement over the days when coal was hand-hoisted in small buckets from the barge to the coal doors on the ship's side. Yet even this progress was not enough to prevent the *Aquitania* and all other major liners from converting to burn oil. (Steven B. Anderson Collection)

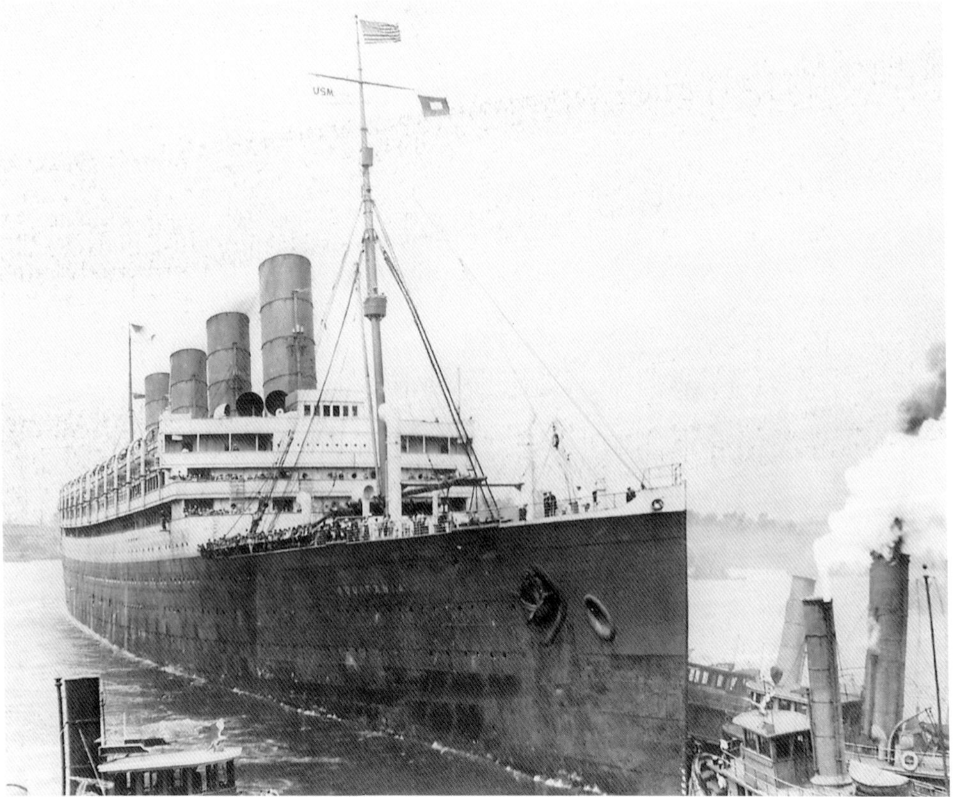

Departing New York in September 1919. The rushed and haphazard quality of her repainting into civilian livery is apparent. Of note, the tip of the prow, which had been painted black during *Aquitania*'s military service, has been painted white. (Clyde George Collection)

THE ROARING TWENTIES

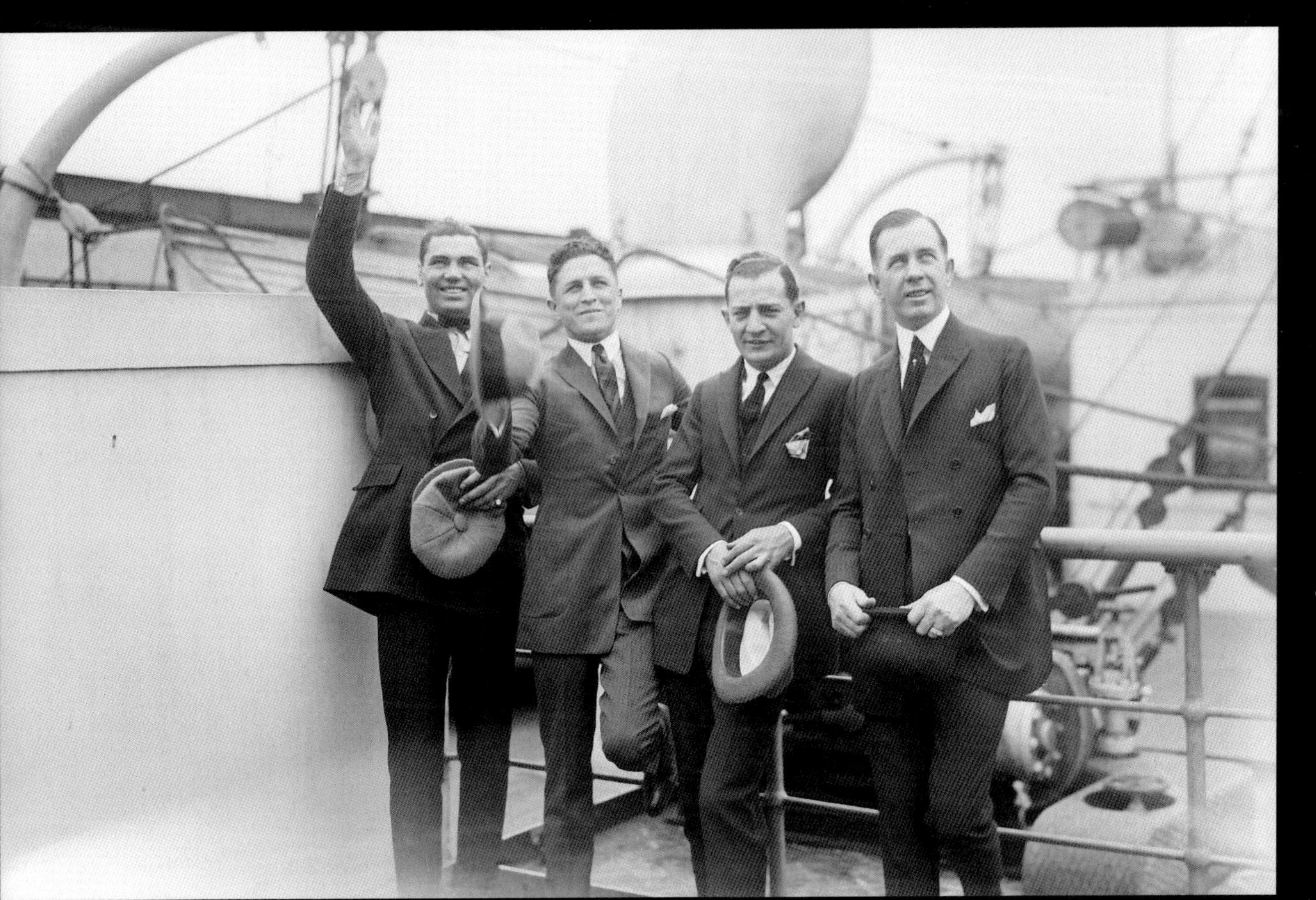

Famous boxer Jack Dempsey (left), along with (from left to right) sparring partner Joe Benjamin, trainer Teddy Hayes and manager 'Doc' Kearns board the *Aquitania* in April 1922. (Library of Congress, Prints & Photographs Division)

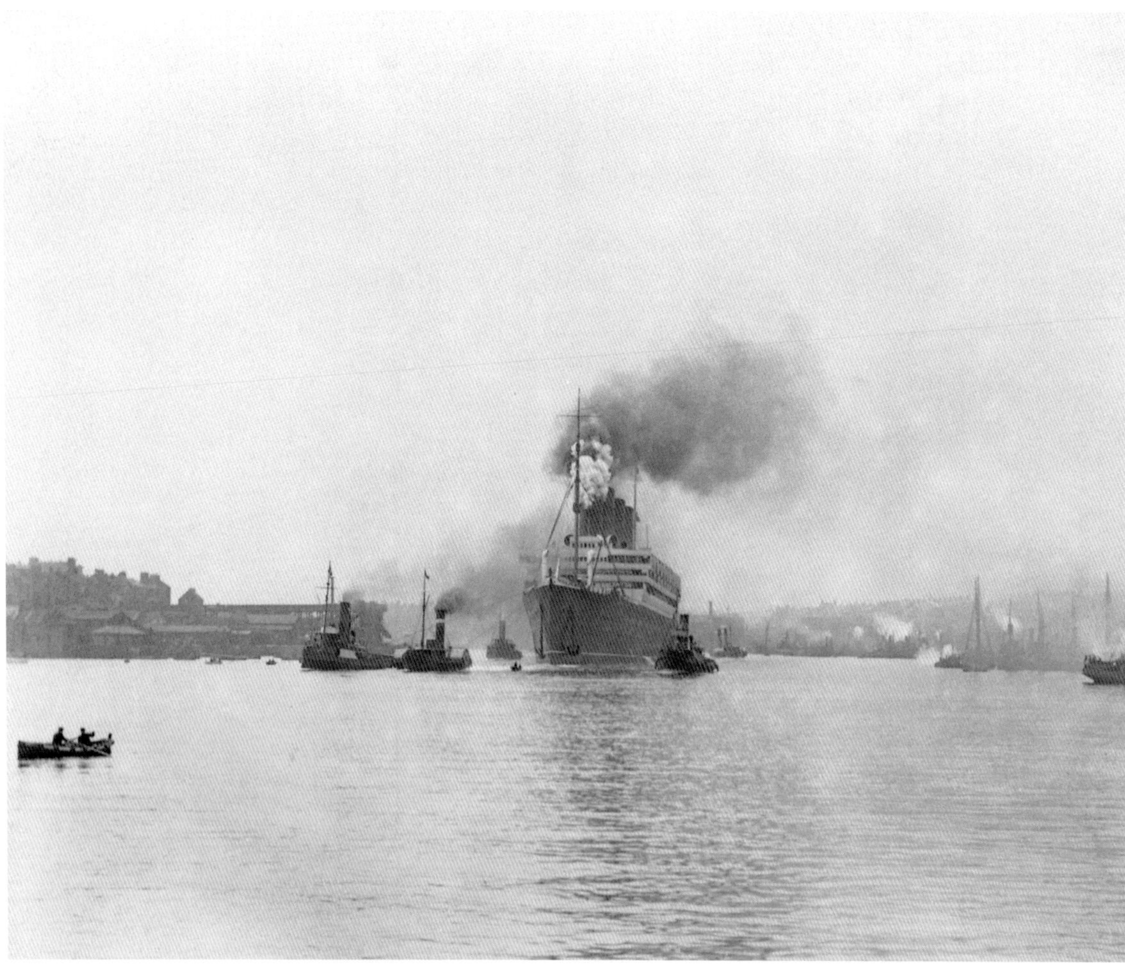

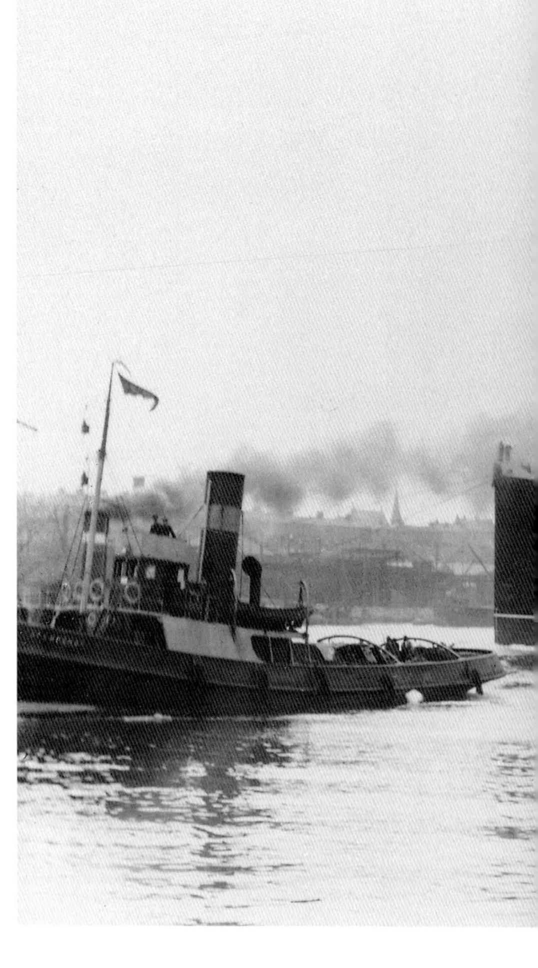

Above: *Aquitania* makes her way up the tight confines of the River Tyne River on 30 November 1919, headed for a thorough refit and overhaul. She is shepherded by a bevy of tugs. (Beamish Museum)

Above right: Two men in a rowboat are dwarfed by the enormous *Aquitania* as she continues her journey up the Tyne to the Armstrong, Whitworth & Company yard where she will undergo her post-war refit. Note the white tip of her prow, a painting feature that was removed by the time the ship had returned to service. (Beamish Museum)

The refit at Newcastle was arguably the most involved that the ship would ever undergo. Her interior spaces, which had seen just three peacetime voyages, but which underwent conversions and hard use during the war, would be completely restored. In some cases, they would emerge even better than new; meanwhile, her mechanical equipment received thorough attention. Finally, her coal bunkers would be converted to carry oil, and her furnaces converted to burn the liquid fuel. Changing over to oil would be expensive, but it would dramatically reduce the manpower needed in the engineering department; it would also reduce the time it would take to refuel the liner while she was in port. Finally, it would turn refuelling from a job that left the ship's interior spaces a mess, no matter how well they had been sealed for protection, to a quick and efficient job with little clean-up.

Just before the overhaul, the man who was to become one of the *Aquitania*'s most famous pursers, Charles T. Spedding, joined her. He was a survivor of the sinking of the *Laconia* in 1917. His friend, Chief Steward Frederick Jones, a survivor of the *Lusitania* disaster, joined him. Spedding recalled:

Newcastle was chosen as the place for her repairs and alterations, principally because there was so much labour trouble on the Clyde. This was a heavy loss to the Clyde, where the ship was originally built, for the job cost over six hundred thousand pounds, [*sic*: £400,000] most of which was expended in labour.

One morning the general manager sent for me and appointed me to the *Aquitania* just before she left Liverpool, on her way round to Newcastle … I felt very proud of being sent as purser to the

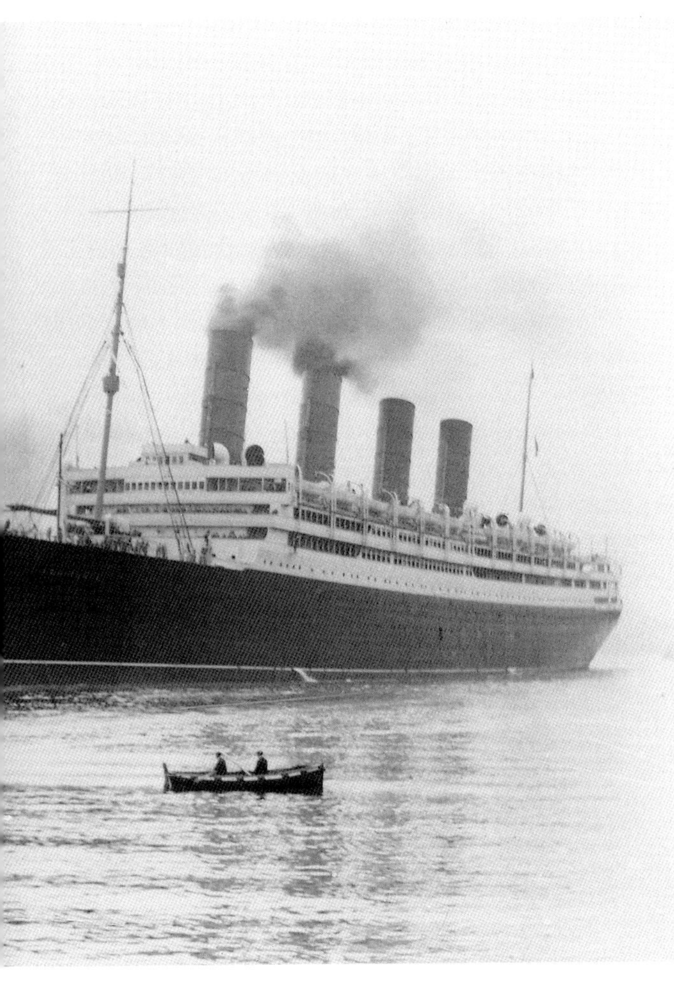

The magnitude of work that needed to be done is nearly incomprehensible:

> [I]n the comparatively short space of 30 weeks it may be mentioned that 300 tons of old material were removed, 650 tons of new material, including 95½ tons of rivets, were put in, and 11,300 yards of steel pipes were used, in addition to the steam and exhaust piping to pumps and heaters, while the labour force employed rose from over 800 at the beginning to a total of about 3,000, including several hundred women.[68]

Spedding recalled that towards the end of the ship's stay in Newcastle, there were as many as 'five thousand men working on the ship':

> To see the men leaving work at five o'clock in the evening, streaming down the gangways, was like watching the hands coming out of a huge factory.

One journal described just a segment of the technical work involved in this conversion to carry oil fuel:

The *Aquitania* is seen here in early 1920 on the River Tyne. Her post-war overhaul, which simultaneously saw a restoration to her full glory as a civilian passenger liner and her conversion to burn oil fuel, appears to be coming down the home stretch. The small upper pilothouse, added in late 1915, has been retained; work on the bulwark running from port to starboard beneath it is evident. (Clyde George Collection)

Commodore Cunard Liner, at the age of thirty-seven, the youngest man to hold that position in the history of the Company.

We sailed from the Gladstone Dock the following day at noon, passing round the north of Scotland, through the Pentland Firth, coming to anchor off Leith in the Firth of Forth. At our anchorage we could see the masts and funnels of the old *Campania* sticking up above the water. During the war she had been converted into an aeroplane carrier and was sunk where we saw her, after a collision with a battleship.

We remained at anchor for about a couple of days, then proceeded up the Tyne to a wharf which fronted the shipbuilding yard of Armstrong, Whitworth & Co. at Walker-on-Tyne, a suburb of Newcastle.[67]

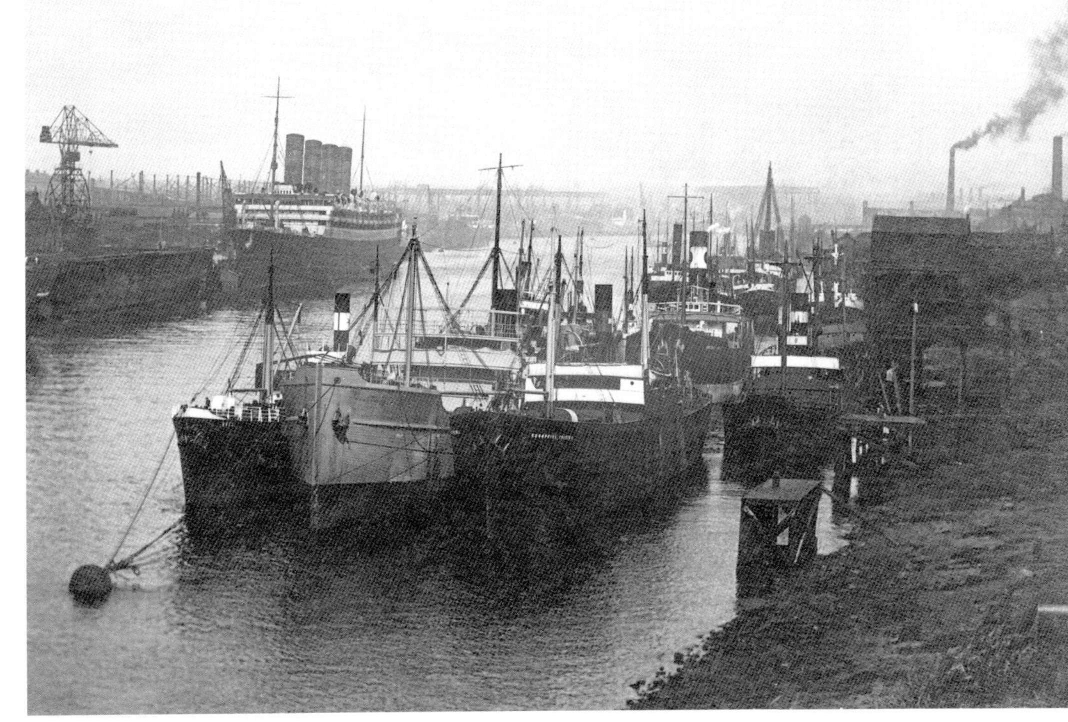

In this ship, with a beam of 97 feet, with three large double-ended boilers arranged abreast, there remains a space of about 18 feet amidships, gradually reducing to 6 feet at the forward end of the boiler space, which formed a coal bunker on each side, and in this space, extending the full length of the boiler space, a matter of 369 feet, the oil is now carried. These side tanks do not form a continuous group on each side, but are broken into two groups, port and starboard, by Frahm's anti-rolling tanks, which take up 32 feet of the length. In addition to these side bunkers, arranged forward of No. 1 boiler room, between Nos. 1 and 2, and between Nos. 2 and 3, have also been made suitable for oil carrying.

The large storage capacity in these side tanks and cross bunker tanks, however, did not prove sufficient to enable the ship to perform the double trip without requiring replenishing, so that six of the double bottom tanks have been made suitable for oil, to enable the 7,600 tons of oil required to be shipped at New York. Of the total quantity, 5,200 tons will be carried in the fore and aft side bunkers, 1,900 tons in the athwartship cross bunkers, and 700 tons in the double-bottom tanks. All the storage tanks, including the six double-bottom tanks, will be used solely for oil, no water ballast connections being provided. These connections are usual where double-bottom tanks are intended for oil, and involve many complications.[69]

Another reported:

The change from coal to oil involved the reconstruction of the coal bunkers to enable them to serve as oil tanks. To this end the bulkheads were stiffened and all joints were made oil tight. 'Wash' bulkheads were fitted to prevent surging of the oil.

... The system of burners ... is known as the White low pressure, after its inventor, Mr. W. A. White. This is the system used in the new installation on the White Star liner 'Olympic.' ... In this system the oil is sprayed by pressure, no compressed air being used. To prevent the possibility of a shut down, a duplicate set of pumps and heaters has been provided for each group of boilers.

Along both sides of the ship, for the length of the boiler rooms, there extends an 8-inch suction and filling main, and four connections to the filling pipes are provided on each side of the vessel. The capacity is such that the whole supply of 7,800 tons can be taken aboard in 6 hours from a tanker lying alongside. The value of this in shortening the turn-around of the ship in port will be evident.

... [T]he paramount advantage of oil fuel is the reduction of the boiler room force and the prevention of the loss of steam pressure, which occurs whenever fires have to be cleaned in coal firing. The 'Aquitania' has 21 double-ended Scotch boilers with a total of 168 furnaces. It was the practice to clean 28 of these fires in each watch, and the loss from this cause amounted to 8,000 horse-power every 4 hours. With oil firing there is a constant flow of oil; a constant furnace temperature; and contraction stresses, due to the inrush of cold air through the open fire doors, are avoided. The steam pressure is constant; the average speed is higher; and the life of the boilers is prolonged.[70]

It was not only machinery that saw attention. It was reported:

Many improvements have been made in the passenger accommodations ... among which the following are noticeable: On *D* deck, in the reception room opposite the restaurant, a bank and inquiry office, or information bureau have been installed. To the swimming pool and gymnasium have been added a large sun bath room. The staterooms on the boat deck have been converted into one-berth rooms, each fitted with a bed and settee. The whole of *C* deck amidships has been rebuilt, the staterooms having been greatly enlarged, reducing the number to 32 rooms. Many of these staterooms now have private dressing rooms.[71]

It was also said:

New carpets have been provided throughout the cabins and saloons and new rubber tiling for all passageways. The furniture, however, is in most cases the same as that carried by the ship on her first voyage in 1914. The furnishings have been stored in the company's warehouses since that time and have in no way deteriorated; in fact, it would be impossible to replace them at the present time.[72]

There were also changes to the scheme of decoration:

[The] *Aquitania* has been reconditioned throughout. The color scheme has been changed and the red and gold decorations in the dining saloon and some of the other public rooms have disappeared. The present scheme is of the Pompeian period.[73]

Despite the great amount of work going on in the ship, there really was not a lot for a purser to do, with no passengers aboard. Thus, Spedding recalled that the seven-month layup was a 'rather monotonous sojourn'. One highlight, however, came when Princess Mary and the Duke of York paid the ship a visit. But the greatest excitement came when the ship's conversion was completed. The Cunard company invited a large group of guests – who included members of the press, agents and veteran company officials and officers – as passengers for the trip around the north of Scotland, and then down to Liverpool. The departure was on Saturday morning, 26 June 1920, and was quite an event:

At various points down the river crowds assembled on the banks to watch her pass, and at South Shields she was greeted with a veritable babel, ships of all degrees and sizes combining to use whatever means of making a noise they possessed. At the Walker shipyard of Messrs. Armstrong, Whitworth, and Co., who undertook the work both of reconditioning and of converting her coal furnaces to burn oil, she was lying with her bow up stream, and at that point the Tyne is not wide enough to permit a ship of her length over 900 ft. to be turned. She had, therefore, to be taken down stern first for four or five miles by six tugs,

three at each end, aided occasionally by her own engines, to a point below Jarrow Slake. Even there the … space available is none too great, and the operation of turning her huge mass round was one of considerable delicacy. It was, however, performed with consummate skill on the part of her navigators, and soon after half-past 1 she passed between the pierheads at the mouth of the river, having taken about 2½ hours to travel the distance of 7½ miles down from Walker.[74]

Purser Spedding recalled of the departure:

As we steamed backwards down the Tyne, which is too narrow for the ship to turn, the high sloping banks were black with people. It was the biggest ship that had ever been in the river, so the Lord Mayor had proclaimed a public holiday. I have never seen so many people gathered together at one time. Two million people live on the banks of the Tyne, and it seemed as though they were all there to see us off.[75]

On Sunday morning, as the ship 'ran under easy steam', there came an unusual event. Spedding continued:

We had a very merry trip round to Liverpool, passing through the Hebrides and sighting the magnificent scenery of the west coast of Scotland. The 'Press Gang' gave an exhibition of Father Neptune coming on board. This ceremony should of course take place crossing the line (equator) but any old line suited the gentlemen who direct public opinion and are the guardians of our liberty. It was done in top-hole style, and in accordance with all the traditions; they were just as rough as they were in my old sailing ship days; the lather that Neptune's barber used to shave the candidates with was also just as vile, except that we had no pigs on board. I do not know what it was made of, but some of it blew on to my brand new uniform, which cost twenty pounds, and burned so many holes in it that it was ruined. It was certainly *some* lather.[76]

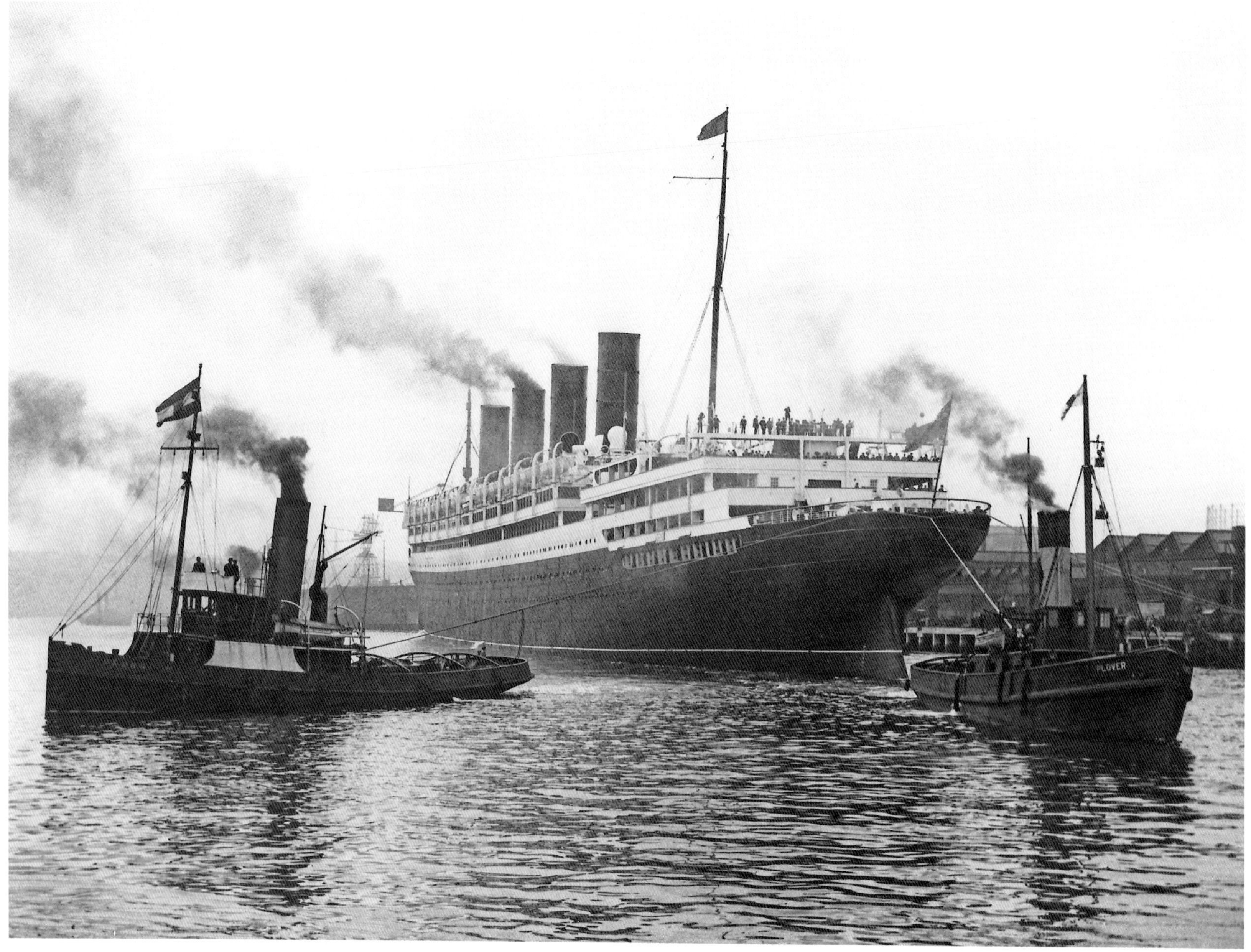

Saturday, 26 June 1920: *Aquitania* departs the Armstrong yard after spending seven months there. She is towed out stern-first, as there was no room in that part of the river to turn her around. (Clyde George Collection)

A press account of the trip stated:

> On Sunday afternoon … her engines were opened out to full power. At no time were more than 18 out of her 21 boilers in commission, and the results indicated that … will be sufficient to drive her across the Atlantic at full speed, so that there will always be three in reserve. With the old coal-firing all the boilers had to be continuously employed. Oil, again, has transformed the stokehold. Instead of an army of stokers, grimy with coal dust, perspiring with physical exertion, and scorched by the blaze of the furnaces as the doors are opened for firing, one sees a mere handful of attendants, cool and clad in clean overalls, who seem to have nothing to do beyond adjusting a valve or very occasionally scraping a burner. From the point of view of the passenger oil has the advantage that it produces no cinders or grit to dirty the decks and get in his eyes. There is also no smoke to trouble him. It is preferred that there shall be a slight trail of smoke issuing from the funnels, because that shows that the correct amount of air for the combustion of oil is being supplied to the burners, but there are none of the volumes of black smoke that commonly mark the course of a coal-fired steamer.[77]

During this speed run, some claimed *Aquitania* had grazed 24.5 knots, while Captain Charles later said that she had 'made nearly 24 knots'.[78] Regardless, it was a tremendous performance for a liner that was only designed to run at 23 knots. The ship arrived in Liverpool on Monday 28 June 1920. She entered dry dock, where final preparations were made for her return to service.

The reduction in stokehold staff caused a strike that delayed sailing: union officials demanded a stoker for every nine oil fires, but Cunard wanted to stick to the previously agreed upon ratio of one stoker for every twelve fires. Things worsened when members of other unions threatened to join the strike if Cunard tried to move the ship without an amicable resolution to the dispute.[79]

Fortunately things were resolved, and *Aquitania* was able to sail for New York on Saturday 17 July. The ship gleamed as if brand new. Her captain was still James Charles, but he was by then Captain Sir James Charles, as he had been knighted at the end of March. She carried 2,433 passengers, an incredible showing. Among these were American golf champion Walter Hagen and Scottish golf champion Tommie Armour. Hagen had just suffered defeat in the British Open at Kent, and he admitted that he was 'not at his best' for that event. However, he had come off victorious at the French Open at La Boulie. Another passenger for the trip was William White, the man who had designed the oil-burning system which had been installed in the ship; Cunard personnel aboard for the crossing included Director M.H. Maxwell and Supervising Engineer John Austin.[80]

On Sunday 18 July, off the coast of Ireland, there was a tragic accident. A high-pressure steam valve broke, killing Seventh Assistant Engineer Scott Barkway,[81] and badly scalding Fireman James Curran. The ship was forced to slow while repairs were made; the lost engineer was buried at sea. The remainder of the crossing was uneventful. The ship performed well, and great economies had been achieved in fuel consumption. Cunard's supervising engineer, John Austin, later announced that she had

This famous photo of the *Aquitania* has been shown before, but the story behind it has never been told. It was taken on Saturday 24 July 1920, as the liner approached New York on the first crossing after her refit. She carried over 2,400 passengers, among them famous golfer Walter Hagen. Despite bad weather on the last night of the crossing, which kept Captain Charles on the bridge, the liner averaged 22.04 knots. The liner seems completely unfazed by the rough seas she was steaming through. (Authors' collection)

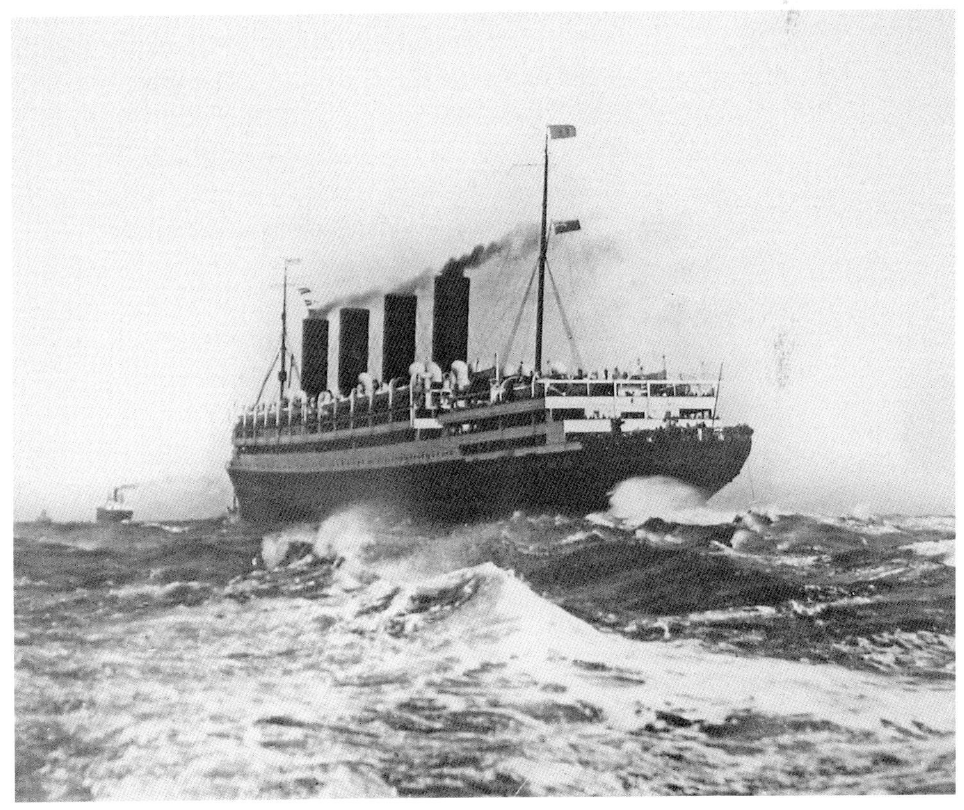

'burned 610 tons of oil daily instead of 960 tons of coal. The installation of oil fuel has allowed No. 3 hold, formerly used for bunker coal, to be used for freight purposes and the number of firemen and coal passers has been reduced from 290 to 100.'[82] It was clear that the oil conversion had taken well.

Some work was performed on the ship while she stayed in New York, finishing up her conversion. It seems that it was during this stay that her Frahm's anti-rolling tanks were turned into oil-carrying tanks; in their absence, the ship was fitted with bilge keels.[83] Also during that stay, over 100 newspapermen were invited aboard, and treated to a luncheon in the Louis-XVI Restaurant. Captain Charles was their host, showing 'overwhelming hospitality'. One reporter recalled that they also joined heartily in singing popular songs of the day. Then the guests were given a tour of the liner's updated machinery spaces:

And now to the engine-room which the Company threw wide open to the newspaper men! First of all cigars and cigarettes had to be left behind, as smoking is rigorously forbidden in oil-fired power-houses.

... Nothing seemed to be simpler than the tending of the oil fires beneath the great boilers. The oil is fed to a single burner through a pipe about half an inch in diameter and the burner itself can be inserted into the furnace or pulled out of it by hand. The flow of oil can be stopped instantaneously and started the same way. We saw a burner extinguished and relighted by means of a torch, a kind of oil-soaked brush at the end of an iron rod. Part of the furnace door was removed for us, and looking in, we saw lumps of incandescent carbon, which, we were told, dissolved away of their own accord.

... In the signal-room was a bewildering array of indicators, gauges, levers and other controls. We were in the nerve centre of the ship, the place where its awful and wonderful structure was made to articulate, its glands to function, and the vital fluids to flow through its veins and arteries.

... It was one thing and a very inspiring thing to see this monster at rest. It would have been even more inspiring to see the vast mechanism in motion with the levers all alive.[84]

The *Aquitania* quickly settled into the routine she should have enjoyed from the start, had the war not intervened. Times had changed, but the liner proved very successful and popular with passengers. By mid-July of 1921, she had steamed 100,000 miles and carried a record number of over 90,000 passengers.[85] During 1921, she carried over 60,587 passengers, averaging 2,019 per trip. However, between 1921 and 1925, that peak began to sequentially shrink until, in 1926, she carried only 26,220, at an average of 1,008 per crossing – still more than respectable.[86] Several factors were at play in this ongoing reduction. First, more large ocean liners were entering service after completing their post-war conversions: White Star's *Olympic* just before the *Aquitania*, their *Majestic* and *Homeric* in 1922, the *Aquitania*'s old running mate *Mauretania* that same year, and the United States Line's *Leviathan* in 1923.

What was more, each of these ships was taking a larger share of a shrinking total figure. This was because in 1921, the United States passed the Dillingham Immigration Restriction Act, which was also known as the 'Three Percent Act' or the 'Quota Act'. Its basic effect was to restrict the numbers of immigrants annually allowed into the country to 3 per cent of the immigration population from that country living in the United States in 1903. This meant no more than 360,000 immigrants per annum; in 1924, the number was cut again, down to only 160,000.

To clarify a common misconception, while Third-Class tickets were an important portion of revenue for steamship lines on the Atlantic, at least insofar as the greatest of Atlantic liners was concerned, they were not the primary revenue that kept their books 'in the black'. Actual numbers tell a slightly different tale.

As an example: on the *Aquitania*'s eastbound crossing from New York to Southampton, which started on 31 July 1920, ticket revenue from First Class was nearly three times that from Third Class: £59,555 from 686 First-Class passengers, as opposed to the £21,139 made in Third Class from 862 passengers. In other words, roughly 80 per cent of the number of passengers carried in Third Class had brought in revenue roughly three times greater. If spread evenly across the price of every ticket per class, the average First-Class ticket was nearly £87, while Third-Class tickets averaged less than £25. When Second-Class tickets, which were also largely unaffected by the immigrant restrictions,

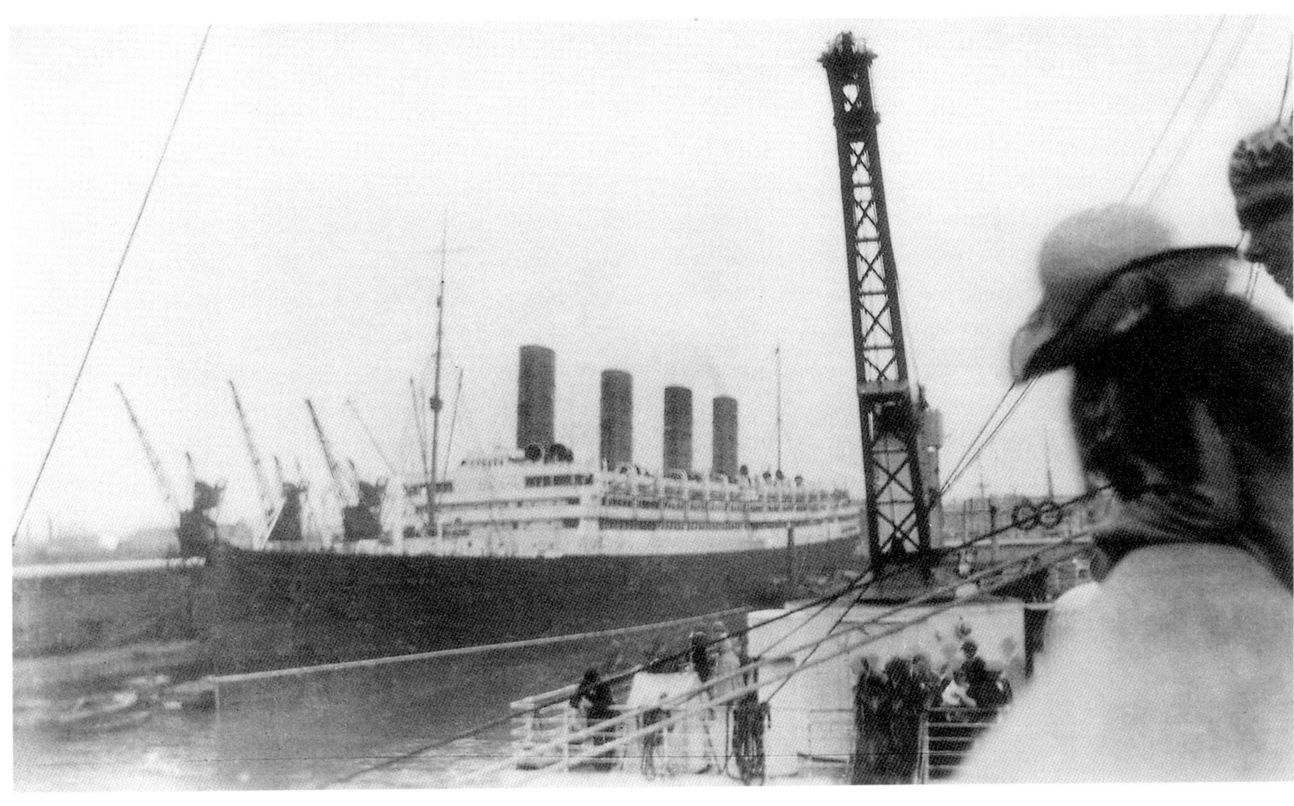

A splendid candid view of the *Aquitania* at Southampton on 8 September 1920. It was taken by a group of travellers starting a crossing to New York on the *Olympic* as that ship departed. (Ioannis Georgiou Collection)

were taken into effect, this added another £22,524 to the revenue from 637 passengers, at an average ticket price just over £35. This meant that of a total revenue on this crossing of £103,218, 60 per cent of the passengers (the total of First and Second Classes) had generated roughly 80 per cent of the revenue. And this single crossing is but a case in point.[87]

Certainly, it was more expensive for the steamship lines to provide First-Class passengers the top-quality meals; it took more stewards, waiters, cooks and pursers to cater to them; and it cost much more to initially decorate and maintain their interior spaces. However, to say that Third-Class passengers contributed the main profits for the companies does not adequately portray the picture, particularly as it applied to the large steamships. On the other hand, the biggest of liners, like those of the *Aquitania*'s calibre, had been designed to carry large numbers of immigrants in Third Class; this restriction on immigration would not help improve their profits.

The steamship companies would work, over the coming decade or so, to find ways of coping with the changing clientele. A new class, named Tourist Third

Cabin, was introduced in 1925 to help fill unused spaces of the ships. For ticket fares slightly higher than those of an ordinary Third-Class ticket, passengers could travel abroad and enjoy some adventure. Beginning in April 1931, the designations were shuffled again for all Cunard ships, and Tourist Third simply became Tourist; they would occupy former Second Class sections of the liners.[88] In 1936, the North Atlantic conference redesignated First Class as Cabin Class. Along with these changes, many alterations in the ship's interior layout were carried out.[89]

Meanwhile, the *Aquitania* continued her regular transatlantic service. People from all walks of life chose to take passage on her, but she proved particularly popular among some of the most famous people of the day. One of these was actor Rudolph Valentino, who boarded in New York on 24 July 1923. He was accompanied by his new wife, Natacha Rambova, previously known as Winifred Hudnut. The couple had been married on 14 March, and were taking a belated honeymoon to Europe, including his home country of Italy. It was Valentino's first trip back since leaving there in 1913. It was later said that:

[F]or three days before [*Aquitania*] was due to sail Rodolpho was almost too excited to eat or sleep. He could hardly believe it true that this particular dream was being fulfilled. To be recrossing the ocean to the world of his boyhood with a beautiful wife by his side was something before which all his problems faded into insignificance. Many times he had dared to hope for it, even during the months of patient wooing of Natacha. Now it was becoming a reality.

The day of sailing arrived and they went down to the quay ... [He] and Natacha boarded the liner amid a welter of fuss and excitement. Hundreds of reporters and cameramen were there to cover the departure and many friends had come to say farewell. Crowding the quayside were the legions of Valentino fans. [90]

Valentino borrowed a handy motion picture camera, apparently brought by a newsreel team, and posed for pictures while he cranked out footage of his new wife. After the ship had sailed, they found that their suite was 'crowded with flowers, baskets of fruit and other gifts and littered with letters and telegrams, the majority of them from people he did not know – "unseen friends" as he had come to call his fans'.

Tired, the couple spent much of their time during the upcoming voyage avoiding fellow passengers and taking their meals in their suite. They spent the remainder of that day reading all of the cards from their *bon voyage* gifts. Valentino even took the time to dash off a 'thank you' letter – written on *Aquitania* stationary – to a Mr Rosen. Rosen had apparently come down to see them off, and had a gift basket sent to their suite. However, his gift had not been discovered until later because the couple had stayed on deck until after departure. [91]

The couple did not maintain complete seclusion during the trip. One morning, while taking a walk on deck, Valentino happened across British actor George Arliss. Arliss and his wife dined with the Valentinos in their suite that evening.

However, the seclusion that the Valentinos indulged in during the crossing did not go over well with passengers who had wanted to see them. Some accused the couple of being 'aloof and upstage'. For the ship's concert, both Arliss and Valentino were asked to perform. Both men declined and instead donated outright to the seamen's charities that were being benefited. This caused the woman co-ordinating the affair to claim he was being 'outrageous' because he 'owed so much to the public'. [92] Yet Purser Spedding recalled him with greater fondness: 'Although spoiled by the ladies, [Valentino] is a very likeable and interesting man, and can make himself very agreeable in any company.' However, he added that he wished that Valentino 'would not tell everybody he only lives for "My art! my art!" He has been successful enough without that affectation.' [93]

Despite this moderate level of persecution, it seems that the Valentinos enjoyed their crossing. They were greeted by a crowd at Southampton, and by that evening had arrived at their destination in London, from which point they would continue their trip. When they returned to the United States later that year, it was aboard the *Leviathan*.

In late 1920, the famous Antarctic explorer Sir Ernest Shackleton selected the *Aquitania* when he needed to cross to America. He had been a friend of *Aquitania*'s chief purser, Charles Spedding, for years. The reason for his trip was actually quite exciting, for the Canadian Government was interested in helping to fund Shackleton on an expedition in the Arctic. During this crossing, Shackleton had to deal with the ordinary 'hero worshippers', but there was also one lady who, Spedding recalled, was 'old enough to know better' yet who fell 'violently in love with him'. Shackleton was greatly embarrassed by her attentions, even asking Spedding to remind her that he was a married man. This was done, but without any variation in her affections. She:

chased him everywhere, and when he resorted to a plan of dodging her, she wandered all over the ship, asking everybody she knew if they had seen 'my dear Sir Ernest Shackleton'. At last he grew so frightened that he planted scouts to warn him when she was approaching.

Somebody suggested he should pretend to be ill, to which he replied, 'My God, no! She'll insist upon coming to nurse me.' Everybody derived a lot of amusement out of the affair except poor

Shackleton, and I think he was very pleased when the voyage ended.[94]

The trip to Canada seemed to bode very well for the Arctic expedition – so well, in fact, that Shackleton returned to England via the *Aquitania*. He purchased a ship for the voyage in Norway, the *Foca*, which he renamed *Quest*.[95] His next trip to Canada was also made on the *Aquitania*, arriving in New York on 30 January.[96] The reason for his passage was not made public at the time, and his name did not appear on the passenger list, either. Although the winter was mild – a fact that Shackleton took time to comment on to reporters after the ship had docked[97] – there was a problem with ice that day. Spedding recalled:

Arriving off the Cunard Pier, the dock as so jammed with ice that it was very difficult to get the ship in. Sir James Charles, on the bridge, got exasperated, and said to the pilot: 'You had better send for Shackleton; he knows all about ice. Perhaps *he* can get her in, for I'm damned if I can!'[98]

Unfortunately, during this trip to Canada, it became clear that plans for this expedition were beginning to unravel; the Canadian Government was showing signs of hesitancy. Shackleton sailed on 27 February for England aboard the *Aquitania*.[99] Among the other travellers on that trip was Lady Flora Eaton, wife of Toronto businessman Sir John Eaton, who had been involved in the planning for this expedition. At departure, Sir John

A merry party of famous people about to set sail on the *Aquitania* on 3 May 1921. The ship was booked to absolute capacity, and this party posed for photographers. Present (left to right) are actor Arnold Daly, songwriter Irving Berlin, aviator Claude Grahame-White and his wife, Ethel Levey, Mrs Georgette Southern (*née* Cohen) and her husband J.W. Southern. (Library of Congress, Prints & Photographs Division)

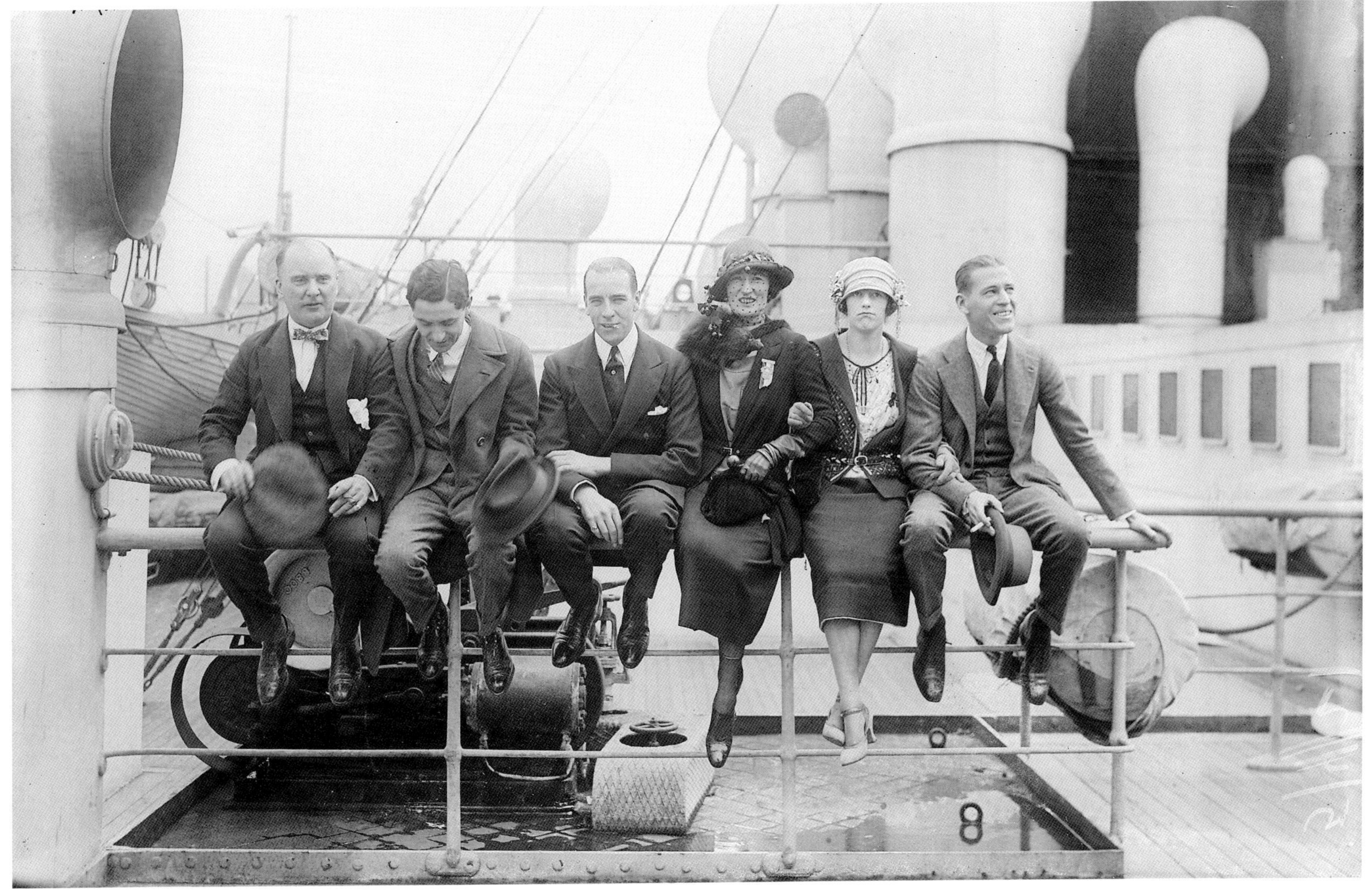

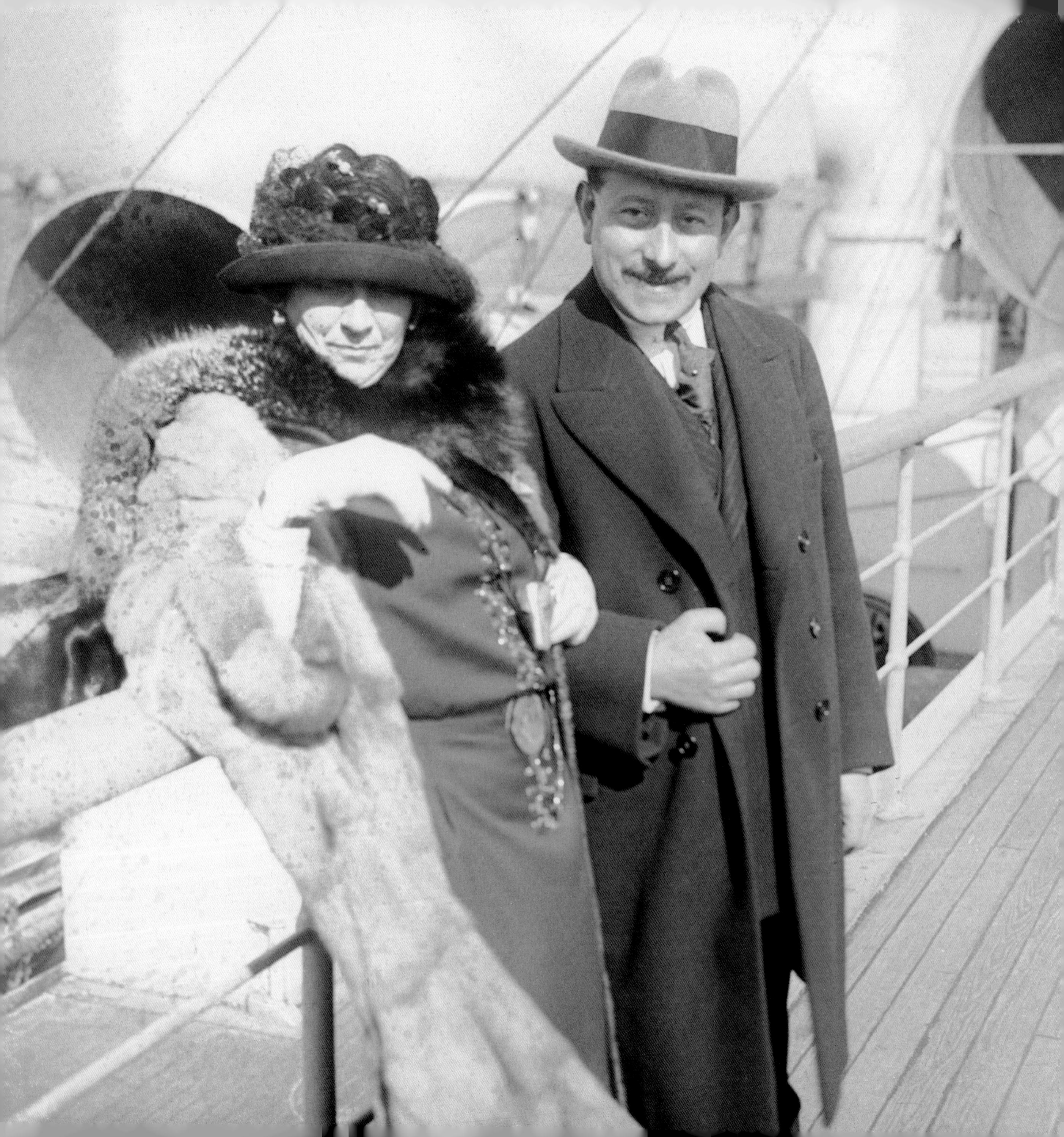

and noted tenor Edward Johnson boarded the ship to see Lady Eaton and Ernest Shackleton off. Lady Eaton joked that Johnson's voice would be sufficient to melt even the polar ice to which Sir Ernest was accustomed.[100]

Shackleton travelled to America again by the *Mauretania* in late March. During this crossing, he wrote a letter home to his wife in which he said that while his cabin on that ship was comfortable, the *Aquitania* had spoiled him for any other ship.[101] However, his plans for the upcoming Arctic expedition continued to fall apart. When the Canadian Government backed out completely, a friend of Shackleton's named John Rowett offered to fund an expedition, and Shackleton changed his plans to make a fourth expedition to the Antarctic, rather than the Arctic. Spedding recalled that when the *Quest* was finally ready to depart, a farewell luncheon was held aboard the *Aquitania* for the expedition members, and that the *Aquitania* later passed the *Quest* at Spithead.[102] Sir Ernest recorded in his diary for Saturday 24 September, that they were seen off by many 'cheering crowded boats'. At about 1.00 a.m. on Sunday morning, he mentioned that:

> [T]he glare of the *Aquitania*'s lights became visible as she sped past a little to the southward of us, going west, and I received farewell messages from Sir James Charles and Spedding.[103]

Sadly, the expedition continued to be a troubled one. The ship encountered rough weather, and in the early morning hours of 5 January 1922, Shackleton himself passed away, apparently of a heart attack. The news reached his friend, Purser Spedding, in mid-Atlantic, while he was in the Louis XVI Restaurant socialising with another group of passengers. One of those at Spedding's table was Sir Philip Gibbs, another friend of Shackleton; the three men had been together at the table during Shackleton's last crossing of the Atlantic on the *Aquitania*. Both men later wrote touching tributes to the Antarctic explorer.[104]

Opposite: Miss Mary Garden, director of the Chicago Opera Company, and Giorgio Polacco, the former conductor of the Metropolitan Opera House, aboard the *Aquitania* in New York in October 1921. (Library of Congress, Prints & Photographs Division)

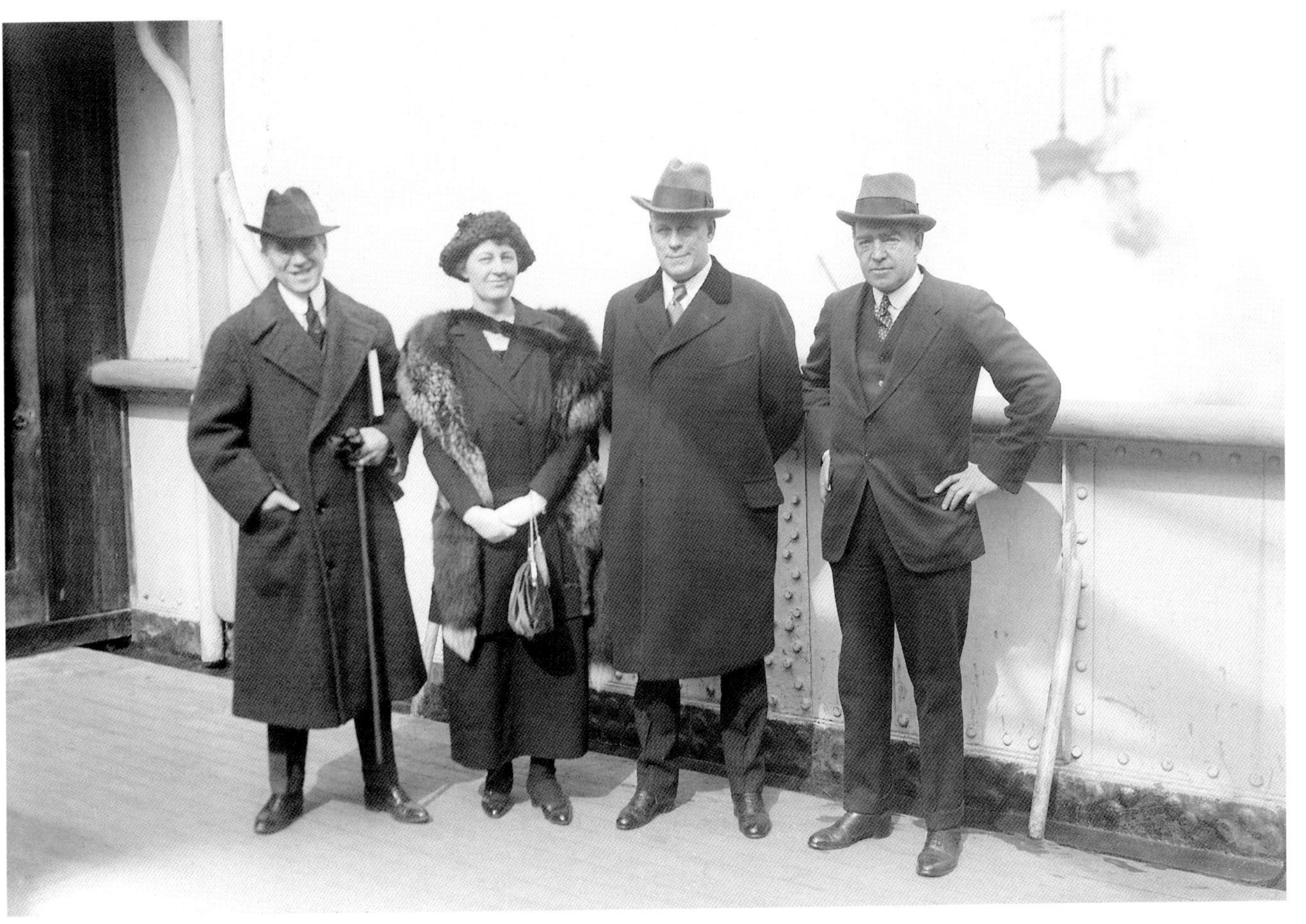

On 27 February 1921, Sir Ernest Shackleton – then in the midst of trying to plan his next expedition – sailed for Europe on the *Aquitania*. He is seen here (on right) just before the liner left New York. In the middle are Sir John Eaton and his wife, Lady Eaton, prominent Toronto citizens and friends of Shackleton. On the left is Edward Johnson, leading tenor of the Chicago Opera Association. Lady Eaton reportedly joked that Johnson's voice would be 'sufficient to melt even the polar ice to which Sir Ernest was accustomed'. (Library of Congress, Prints & Photographs Division)

Lillian Russell, the noted actress and singer, was another famous passenger. She sailed to New York with her husband, Alexander Moore, arriving on 17 March 1922. Russell had just finished a study of European countries and the immigration conditions there, and how they were likely to affect the United States. It had been an official trip, for she carried 'credentials from [Labor] Secretary [James] Davis'. What she witnessed had convinced her that it would be wise for the United States to suspend immigration for five years, or until 'the aliens in this country [had the] chance to become Americanized'. During the trip, she made history by becoming the first woman to act as chair-person for the *Aquitania*'s concert.[105] She later met with Secretary Davis to report her findings, but sadly passed away on 6 June at the age of 61.

Boxer Jack Dempsey was another famous passenger on the *Aquitania*. On 11 April 1922, Dempsey and his entourage boarded the *Aquitania* at New York for his first transatlantic trip. The press mobbed the sailing, but both Dempsey and his manager, 'Doc' Kearns, were said to be 'affable and sensitive to the desires of the whirling throng' about them. They, along with his sparring partner Joe Benjamin and trainer Teddy Hayes, posed for pictures wherever asked. Dempsey had booked stateroom B-46, and told reporters there: 'I'm going to be seasick as hell in this place.'

Reporters asked Dempsey about rumours then circulating that he was engaged to one of the Dolly Sisters; twins named 'Rosie' and Jenny, who were famous dancers and actresses. Dempsey denied the engagement with a laugh, claiming that he would never dare marry one of them, 'because he could not tell one from the other'.[106] Dempsey later said in one of his autobiographies that they had held a 'big party on the *Aquitania* before we left and with three girls who were there – Mary Lewis of the 'Follies', Florence Walton, the dancer, and Mary Sherman, who was in the movies – discussing how it was to kiss me'.[107]

His crossing was not without incident. A number of card sharps had come along, and Purser Spedding suspected they intended to fleece Dempsey; the purser advised Dempsey about them, however. Yet one still

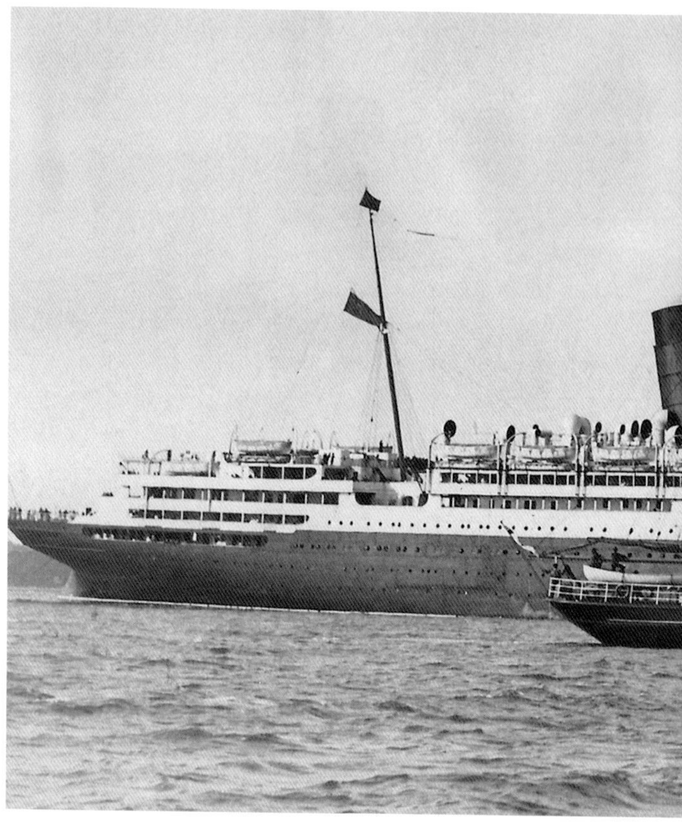

caused trouble when he went down to Dempsey's stateroom in the middle of the night in a drunken stupor, waking the boxer up and demanding a bottle of whisky. The nightwatchman was summoned and the card sharp was removed, but the incident distressed Dempsey. The boxer went down to Spedding the next morning, explaining that he had nothing to do with instigating it all, and that he didn't want passengers in neighbouring staterooms thinking that he was 'a ruffian'.[108]

The Dolly Sisters, who were rumoured to be attached to Dempsey, were also frequent travellers on the *Aquitania*. Charles Spedding recalled that they were 'vivacious' and 'just as much alike in private life as they are on the stage'. They presented two white Persian cats to the crew of the *Aquitania* to serve as mascots, and while they were aboard the ship, the little felines were called the Dolly Sisters because no one could tell them apart.[109]

One interesting detail that Spedding recalled was that the *Aquitania* not only carried stars of the cinema, but sometimes also starred in them:

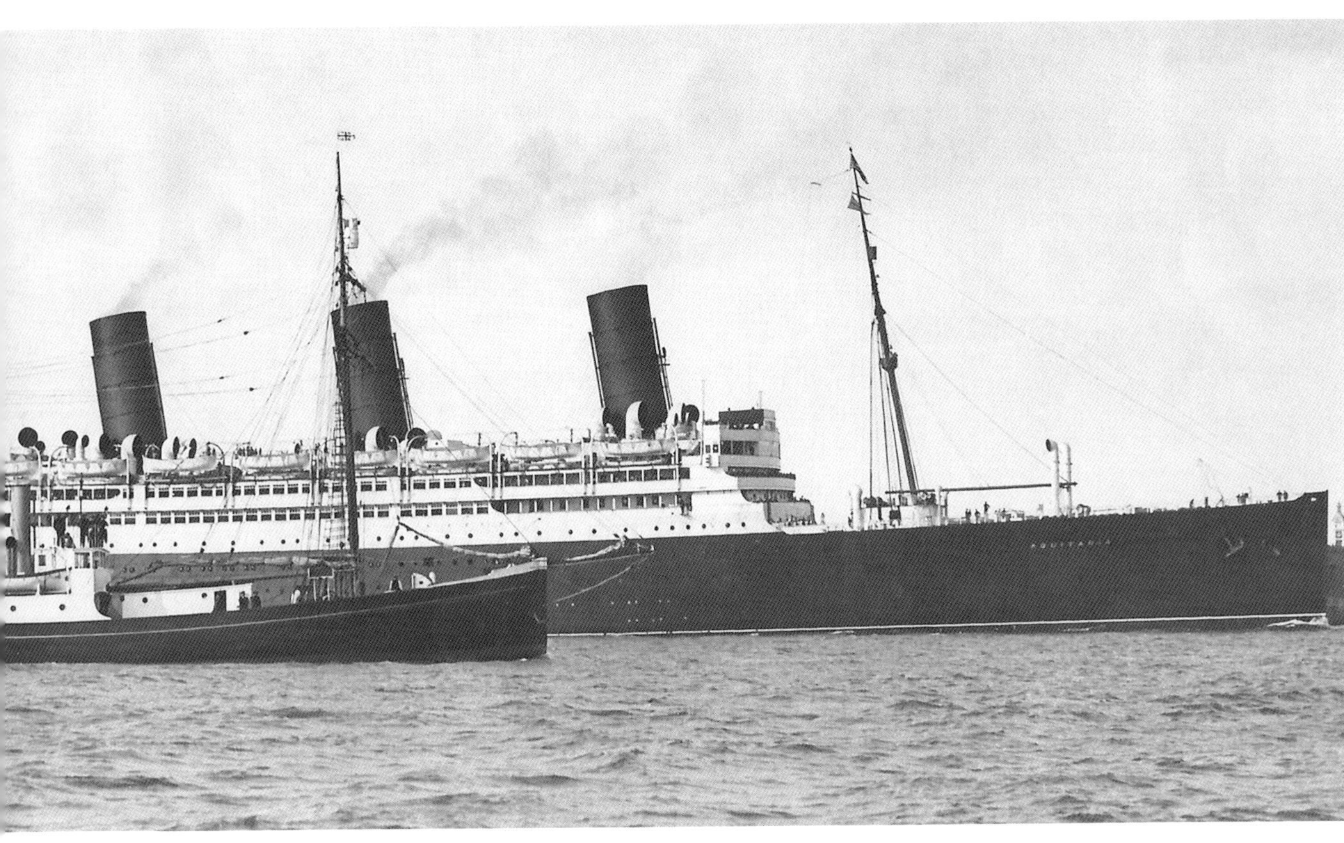

On Saturday 3 September 1921, the *Aquitania* passed the RV *Quest* as the liner was departing Southampton for a westbound crossing to America. Sir Ernest Shackleton, who had travelled on the *Aquitania* more than once in preceding months, was about ready to embark on his last expedition. The *Quest* sailed on to London to provision, then departed for Lisbon, the first stop of the doomed trip. Shackleton and the *Quest* encountered the *Aquitania* once more at about 1.00 a.m. on 25 September, as the liner was beginning her next westbound crossing to New York. (Clyde George Collection)

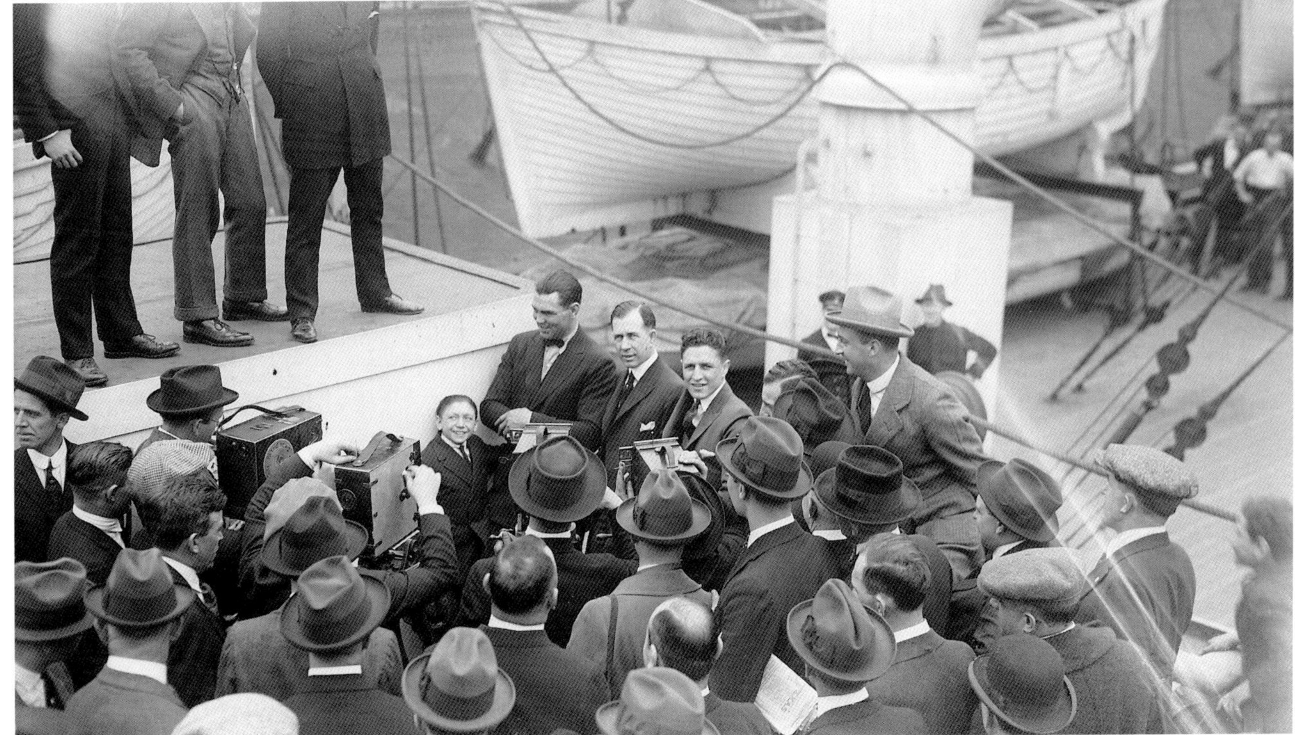

On 11 April 1922, boxer Jack Dempsey and his entourage boarded the *Aquitania* for a crossing to England. He was mobbed by reporters, and posed happily for many photographs. (Library of Congress, Prints & Photographs Division)

The Cunard company had a film made of a round voyage showing ordinary ship life, all kinds of scenes were taken, in the dining saloons, lounges, or deck, the ship's concert, the swimming pool, the crew at various duties, etc. A copy of this film has been put away so that it can be shown a hundred years from now … [A] hundred years from now the *Aquitania* will seem, to the generation of the day, even more quaint than Nelson's ship would seem to us of the present day.[110]

The *Aquitania* has also been closely connected with the early Alfred Hitchcock film *Champagne*, and exterior sequences were purportedly filmed against either her or the *Berengaria*. However, existing footage does not appear a match to either vessel. Scenes set in the interior spaces of the ship, while not actually filmed aboard the *Aquitania*, clearly approximate the decor of her famous Palladian Lounge. A photograph seen in the film, showing couples dancing, was even captioned 'New Year's Eve, *Aquitania*'.

Whether her passengers were stars, everyday businessman or a poor immigrant, stormy weather levelled the social barriers by treating everyone aboard equally. In late February 1922, while westbound, the *Aquitania* 'took a beating' from weather that some of her crew and officers said was 'the worst she has had' since entering service. 'Fifteen port lights made of 1¾-inch glass gave way before the crash of waves that broke over her main deck forward and hurled tons of water over the bridge,' it was said. Although some passengers were uneasy, lively jazz and dancing '[whiled] away the six stormy nights of the passage'. Chief Surgeon Sydney Jones was forced to perform an emergency appendectomy at the height of the storm, but said that the ship need not slow down because she was 'steady enough despite the weather'.[111]

On 24 September 1922, the ship hit a hurricane two days out of Cherbourg. Shortly after church services that Sunday, it became clear that she was heading into a powerful storm, but no one had any idea how violent it would become. By six o'clock that evening, they were in a south-east gale; by midnight, the wind had become westerly and was increasing. At its worst, Captain Charles saw the barometer register 28.23", the lowest he had

seen since March 1898. For hours, speed was reduced to 4 knots. Despite this, a number of the ports on both sides of B Deck were smashed, flooding some of her finest accommodations, including the Raeburn Suite. The occupant of that suite, American oil magnate Charles Peters, said that water 'poured through the ports' and that he and his son were showered with glass. They had to flee the suite in ankle-deep water. He said that the ocean looked like a 'Rocky Mountain sea with hundreds of Pike's Peaks'.

In total, some thirty-five to forty soggy passengers were forced to trudge up to sleep in the Lounge; meanwhile, stewards tried to calm them down, working for up to thirty-six hours straight to bail out and mop up the drenched staterooms. A door to one of the B Deck passageways was torn out, replaced and then smashed out again; a strip covering one of the expansion joints was also torn off. By 8 a.m. on Monday, the worst of the storm had passed and the ship was able to pick up speed again, but the sea remained rough until Thursday. By the time she docked safely in New York, the ship's average speed for the crossing was only 19.55 knots.[112]

The North Atlantic continued to pummel the *Aquitania*: in January 1924, heavy seas reportedly cracked a steel plate 20ft above the waterline; a First-Class passenger had also sprained her wrist when the ship lurched while she was on the stairs.[113] On Thursday, 25 December of that year, *Aquitania* hit another particularly foul storm, and Captain Charles reported that from then through to Sunday, the ship encountered one gale after another, 'with hurricanes in between'. He added: 'The seas were terrific.' At about 8 p.m. on 26 December, two waves had come over the bridge, smashing two windows on the starboard side and two of the small upper windows forward on B Deck. The ship's speed was reduced to only 9 knots, and the liner was thirty hours late in reaching New York.

In January 1926, while steaming west for New York, the ship hit absolutely ferocious storms, with snow reducing visibility to only 100 yards. Both *Aquitania* and the liner *France* were two days late in reaching port. Captain Charles and the master of the French liner agreed that it was the roughest crossing that they had experienced in the last forty years. Her next crossing to New York in February brought another surprise: while the ship was

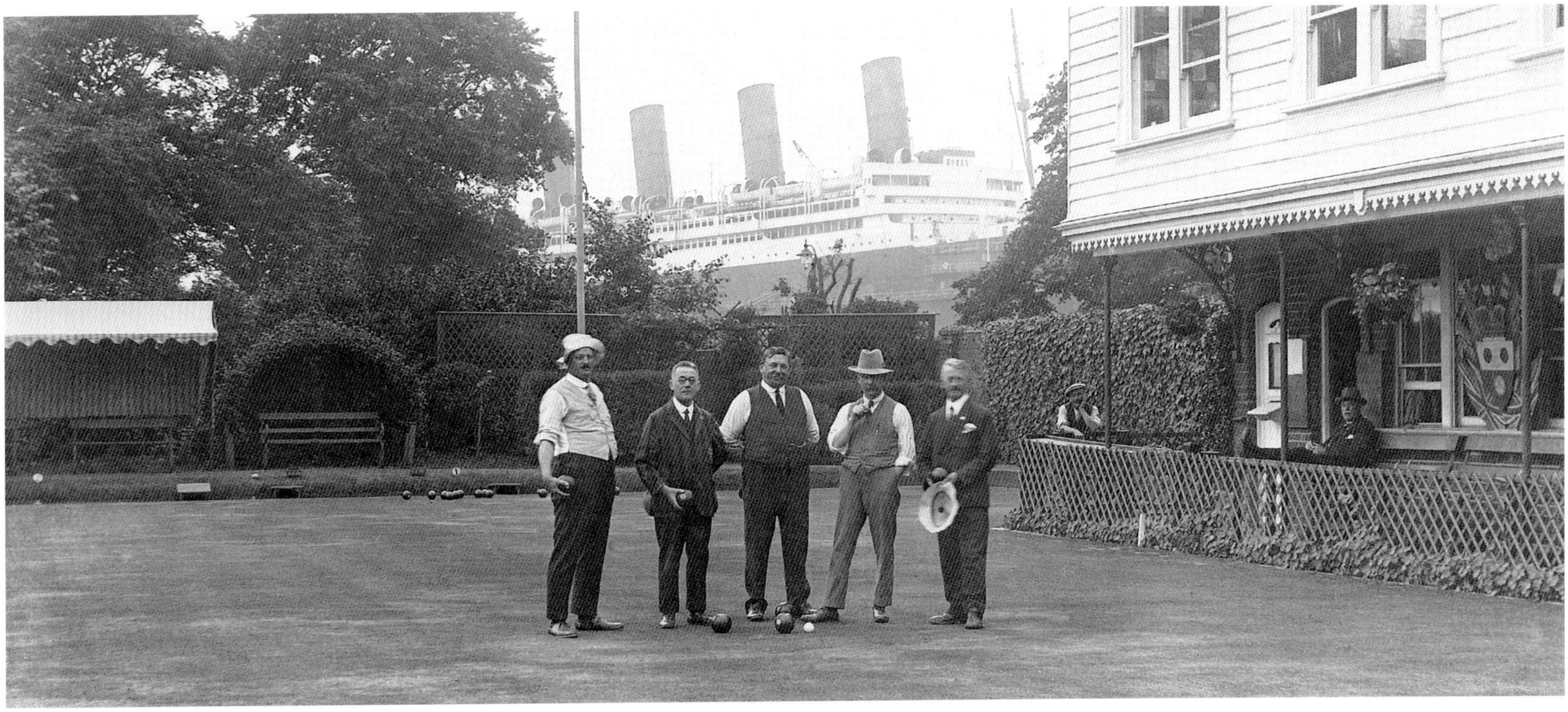

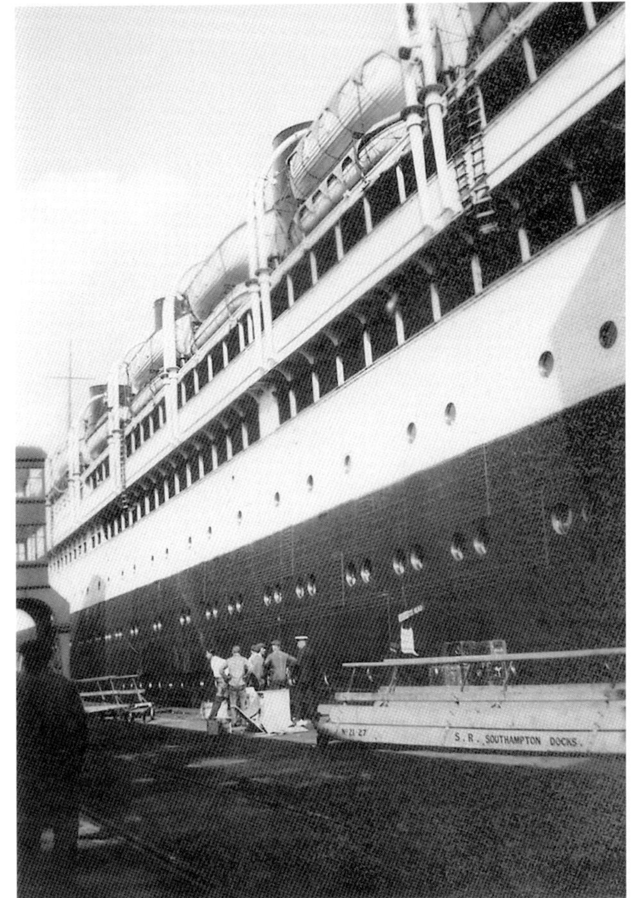

This mid-1920s photograph shows the *Aquitania* featuring prominently in the background of a group portrait taken at Southampton. (Brian Hawley Collection)

steaming through 'a moderately calm sea', a 'great wave' suddenly rose up 'and dashed against the ship sixty feet above the waterline'. It smashed her cargo boom into three pieces, and literally tore the brass 'Q' of her name from her port hull plates.[114]

In 1922, while steaming westbound, Captain Charles had promised his passengers that he would dock the liner in New York '[at] or before 6.30 p.m.' on Friday 28 April. The ship had been held up for over an hour at Quarantine, however, only casting off at 5.30 p.m. Her skipper had been seen impatiently waiting on the bridge for the last of the mail boats to cast off. When she got moving again, he showed 'what she could do'; the circumstances were ideal for making fast time up to her pier, with clear weather and 'little or no traffic' in the fairway. The ship was 'hooked up to high speed', and proceeded up the river at what appeared to be a fast clip — Captain Charles said he never exceeded 10 knots. She was abeam of the pier by 6.10 p.m., and precisely at 6.30 p.m., the gangplank was landed.[115]

In September 1920, *Aquitania* was the setting for the start of an interesting romance for Alicia du Pont, adopted daughter of Alfred du Pont, a millionaire manufacturer from Wilmington, Delaware. While aboard,

The mountainous sides of the *Aquitania* towering over Ocean Dock in Southampton. Interestingly, several workmen and an officer stand on the dock beside what appears to be a broken shipping crate. (Mike Poirier Collection)

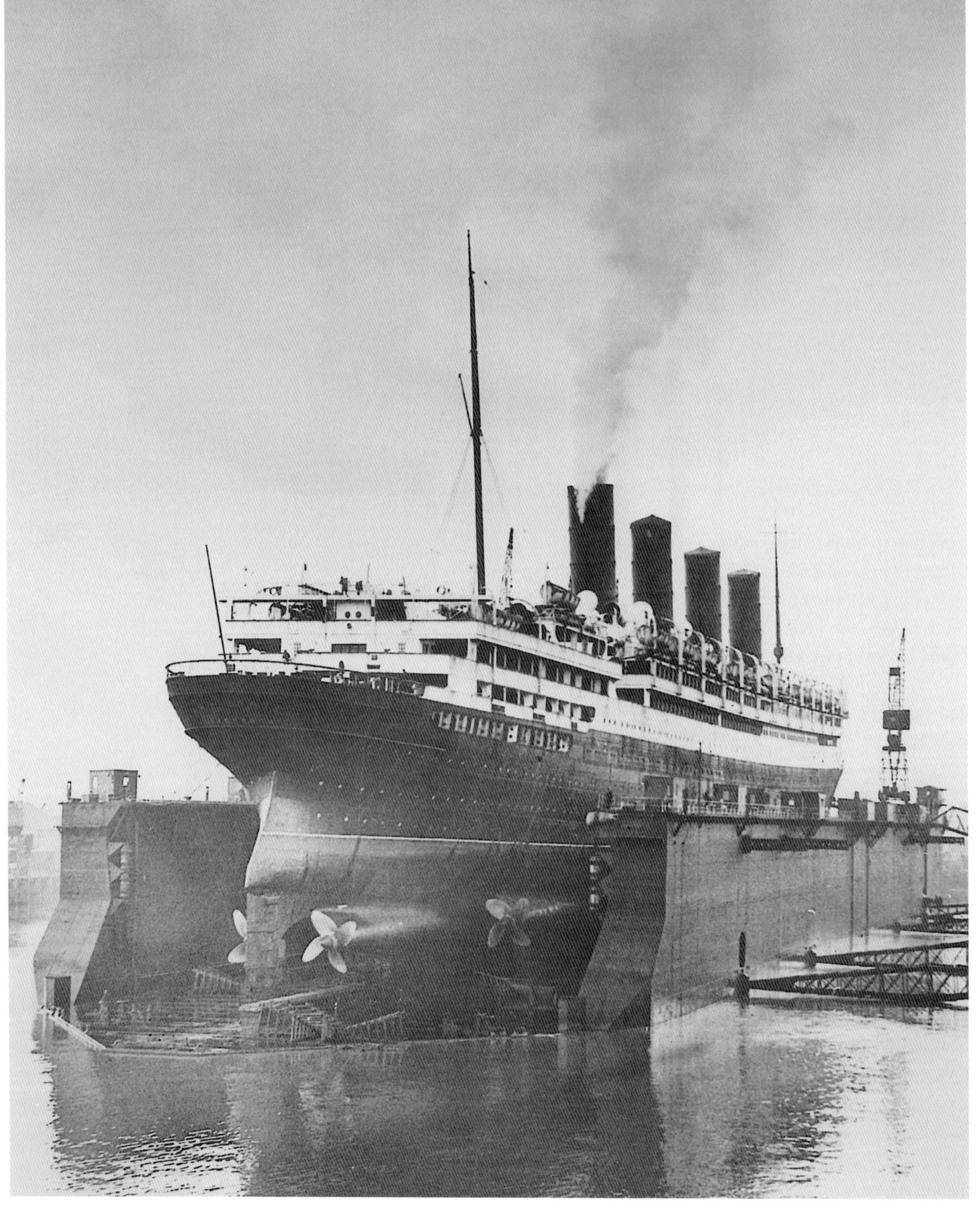

Aquitania resting in the floating dry dock in Southampton. Ordered in 1922 by the Southern Railway, and built by Armstrong, Whitworth & Co. on the Tyne, this dry dock was 960ft in length, with a 130ft 6in inside breadth. It was officially opened by the Prince of Wales on 27 June 1924; the first trials of the dock were made on 7, 8 and 12 July; finally, the White Star liner *Olympic* was raised on 12 July for her summer overhaul. It was used extensively thereafter by all of the crack liners, including the *Aquitania*. (Steven B. Anderson Collection)

she met young Harold S. Glendenning, who during the war had worked as a chemist for the du Pont company's Carney's Point smokeless powder plant. A romance blossomed. Their engagement was announced in early June 1922, and her father and stepmother gave their approval despite the fact that young Glendenning's family was only of 'comfortable circumstances', for his father had been a mail carrier. They were married on 28 June, less than a month after their engagement was announced. The couple had a son, Alan, shortly thereafter. Unfortunately, their marriage did not last. They were divorced in Reno, Nevada, in 1925, and there was a protracted custody battle over Alan; by 1935, Alicia had been married three times.[116]

In December 1923, forty-two beautiful young English girls boarded the *Aquitania*. They were headed for New York to appear in French impresario André Charlot's *London Revue of 1924*. Among its stars were Jessie Matthews, who was to make her acting debut, and Jack Buchanan. The show would prove to be a smash hit, running for nearly nine months and having 298 performances. However, one of the English beauties, Marjorie Sexton, disappeared before the morning of the first rehearsal. After some investigation, it was found that she had travelled to South America to wed wealthy horse-owner John Contenti. Contenti had been a fellow passenger on the *Aquitania* when they came over, and a romance had blossomed quickly. In fact, on the first day of the voyage, Mr Contenti's attentions to Miss Sexton had been noticeable.[117]

On 23 May 1922, the *Aquitania* sailed from New York with Lord and Lady Astor; Nancy Astor had been the first woman elected to Parliament, three years earlier. Also aboard was William Randolph Hearst, the American newspaper magnate with clear anti-British feelings, his wife and their two sons; Hearst was then campaigning for political office. Chief Purser Charles Spedding suspected that there could be some 'hair flying' if they met, for while he liked Lady Astor very much, 'when she gets really angry with me I am going to disappear out of her reach'. They narrowly missed each other on the gangplank when boarding. She boarded only twenty minutes before departure time, hopping 'nimbly from the gangplank to the deck', and then remarking with a twinkle: 'When I

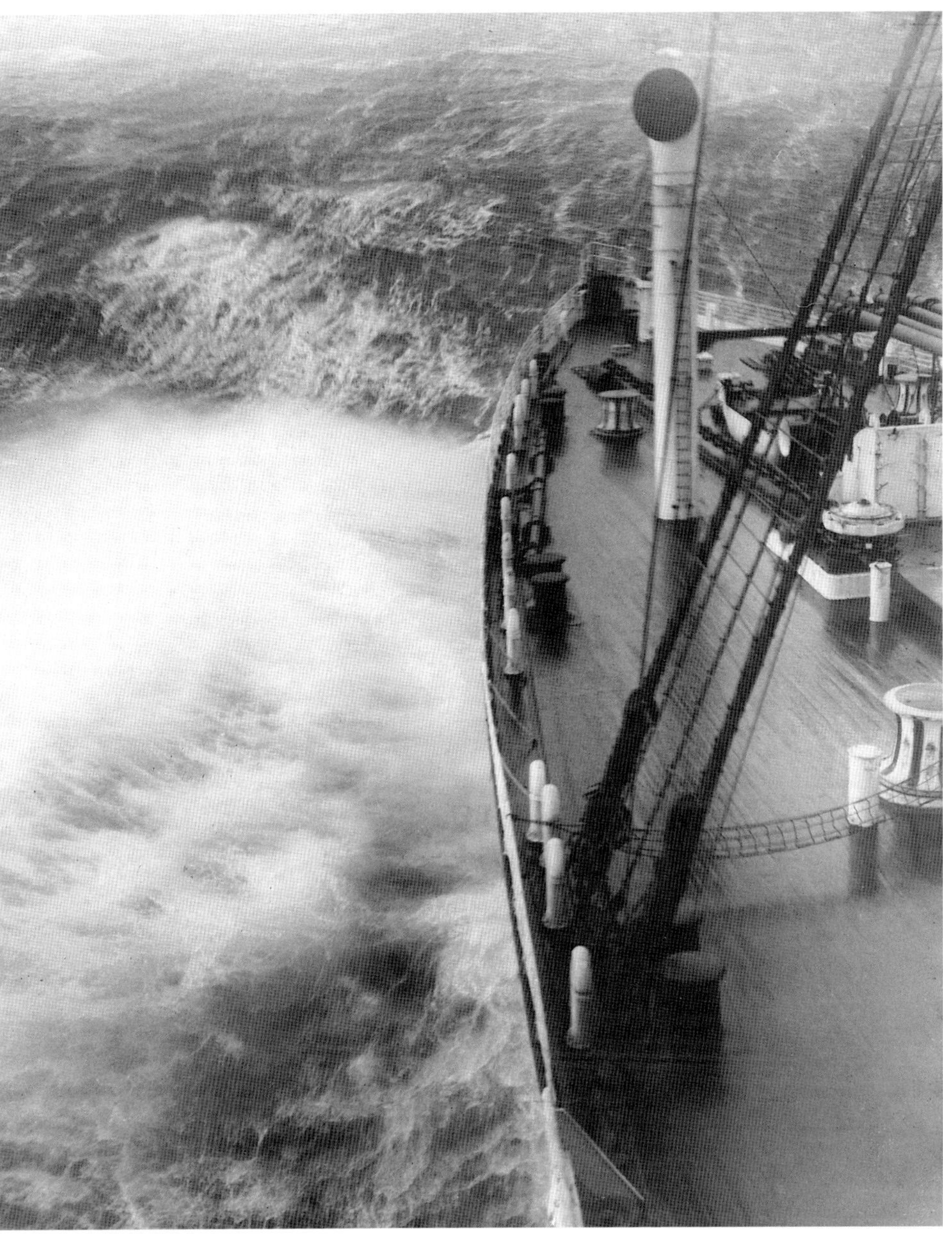

The *Aquitania*'s bow carves its way through mountainous seas on 1 January 1925. Seas reached heights of 50ft, and the ship was buffeted by 80mph winds. She was forced to decelerate to a mere 9 knots as the seas 'piled up like watery hills' about her. (Authors' collection)

am running for office I don't run away.' It was a clear dig at Hearst's campaign. Reporters surrounded her, asking many questions. One asked: 'Well, are you going to put Mr Hearst "right" on the way over?' She replied, 'Certainly not'. However, Spedding recalled:

To the great delight of both English and American passengers, also the stewards, who repeated to me as many of Lady Astor's remarks as they could remember, Nancy gave William Randolph the

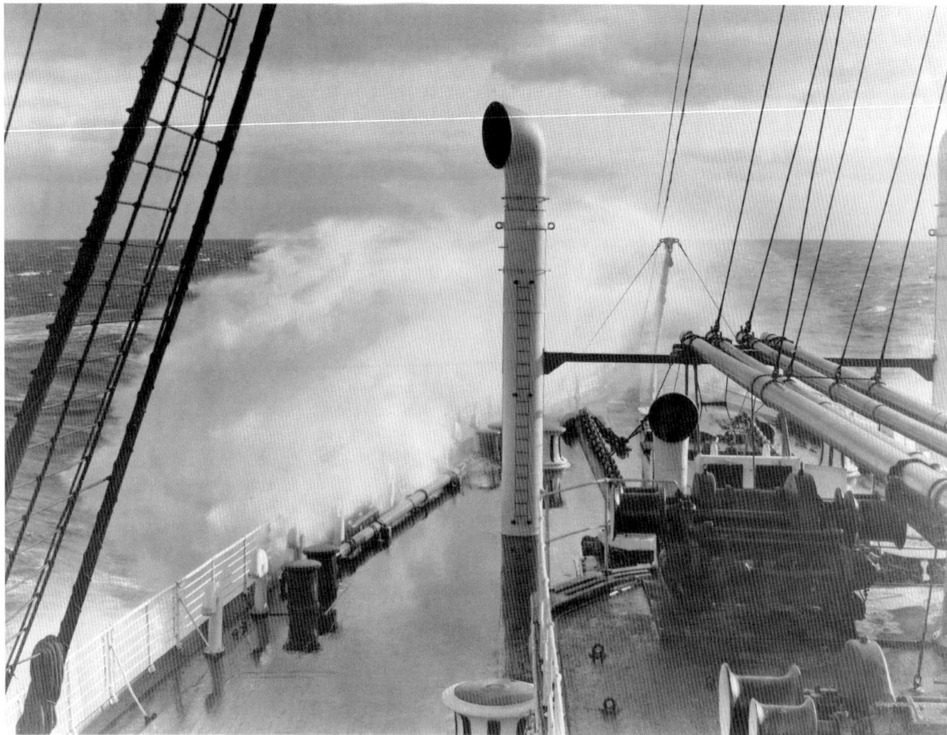

Although the seas are not as high in this photograph, it shows the ship's bow crashing through an unsettled sea. (Ioannis Georgiou Collection)

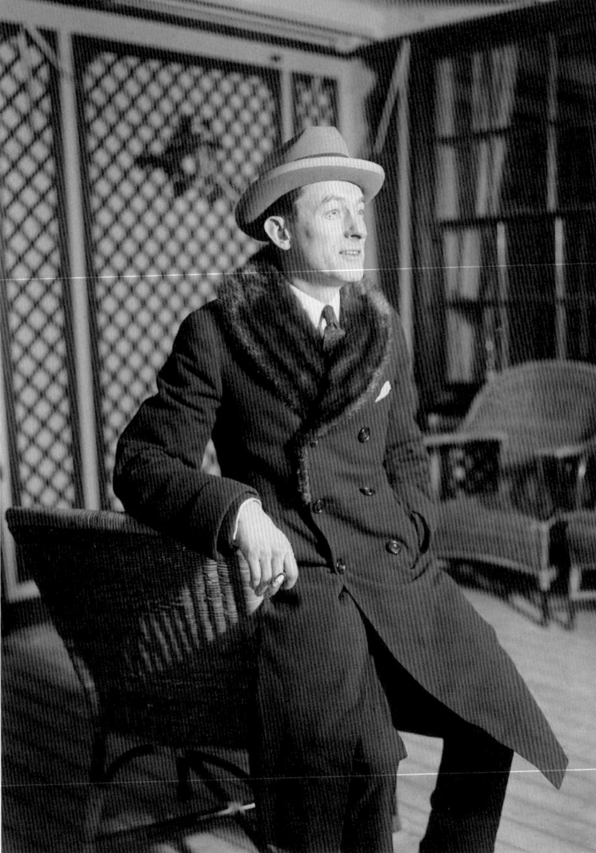

New York State Senator James J. Walker, photographed in one of the *Aquitania*'s Garden Lounges at the end of her tempestuous crossing to New York in January 1925. (Library of Congress, Prints & Photographs Division)

rounds of the kitchen, telling him *exactly* what she thought about him; the meeting occurred on deck one bright sunny morning in the presence of Mrs Hearst and her two sons.[118]

Sailing on the *Aquitania* was a life-changing event for one 19-year-old debutante named Rosamond Pinchot. While sailing to New York in early November 1923, theatrical producer and director Max Reinhardt spied the young girl at Cherbourg, while she and her mother were boarding from a tender. For the previous nine months, Reinhardt had been scouring Europe in the hunt for a girl to play the part of the nun in his upcoming play, *The Miracle*, a role that required an energetic performance. During the crossing, he watched the young girl take her daily constitutional, watched her dance and 'regarded with interest her broad shoulders, like those of her uncle, her lithe movements, betokening the sportswoman, and her healthy five feet ten and a half inches of strength and beauty'. After three days of this, he requested an interview with her mother, and asked if she would play the role. The offer was a tremendous surprise; Miss Pinchot had been thinking of her upcoming debut party, and had thought nothing of becoming an actress, but the event launched her career as a star, and she earned the nickname, 'The Loveliest Woman in America'. Reinhardt later said that finding her had been 'a miracle'.[119]

There was a sad event on 16 August 1926; 51-year-old able-bodied Seaman Ernest Bandfield killed himself by drinking a stain-removing chemical used on the ship's decks. He was discovered wandering around on deck early that Monday morning, reportedly acting 'dazed', and died the following morning. On the previous eastward crossing, he and two other crewmen had been questioned about envelopes found on the deck which had purportedly been taken from one of the locked mail bags. Bandfield had been acquitted of any wrongdoing; he was a trusted member of the crew who had been aboard for six years, according to Captain W.A. Hawkes. His shipmates, however, said that he had been 'brooding' over the incident ever since. Bandfield's body was buried at sea.[120]

Times were clearly changing. Charles Spedding retired as the ship's purser in May 1925; his first replacement was

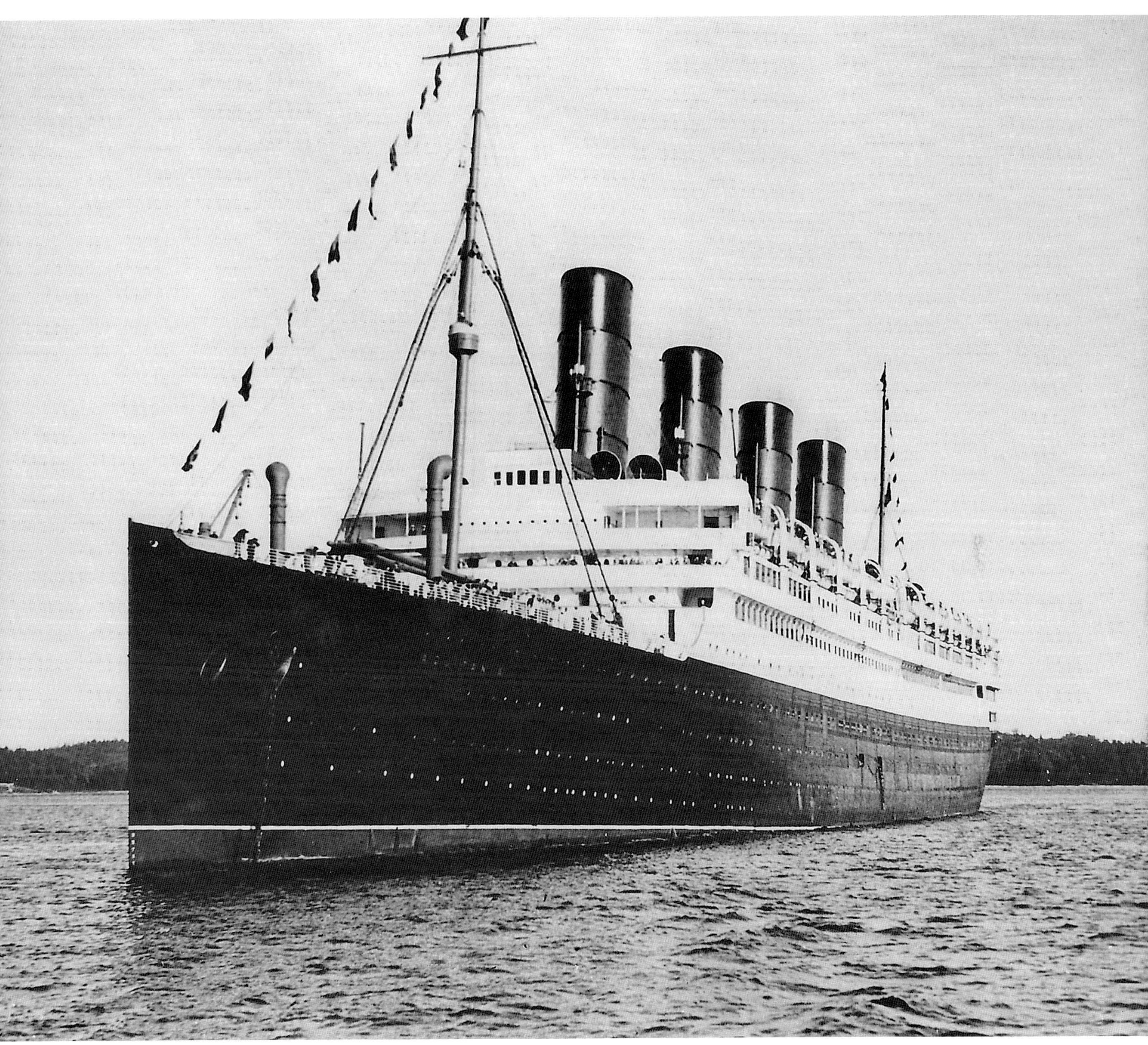

Aquitania in Halifax in 1925, commemorating the 85th anniversary of the arrival of the *Britannia* in that port. (Clyde George Collection)

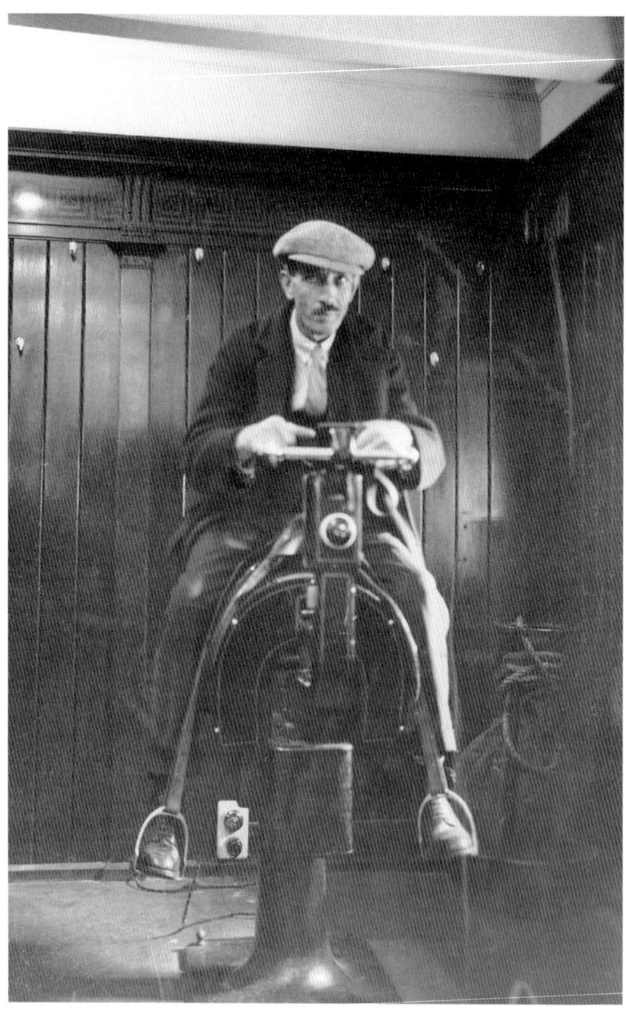

A passenger rides an electric horse in the *Aquitania*'s First-Class Gymnasium. (Ioannis Georgiou Collection)

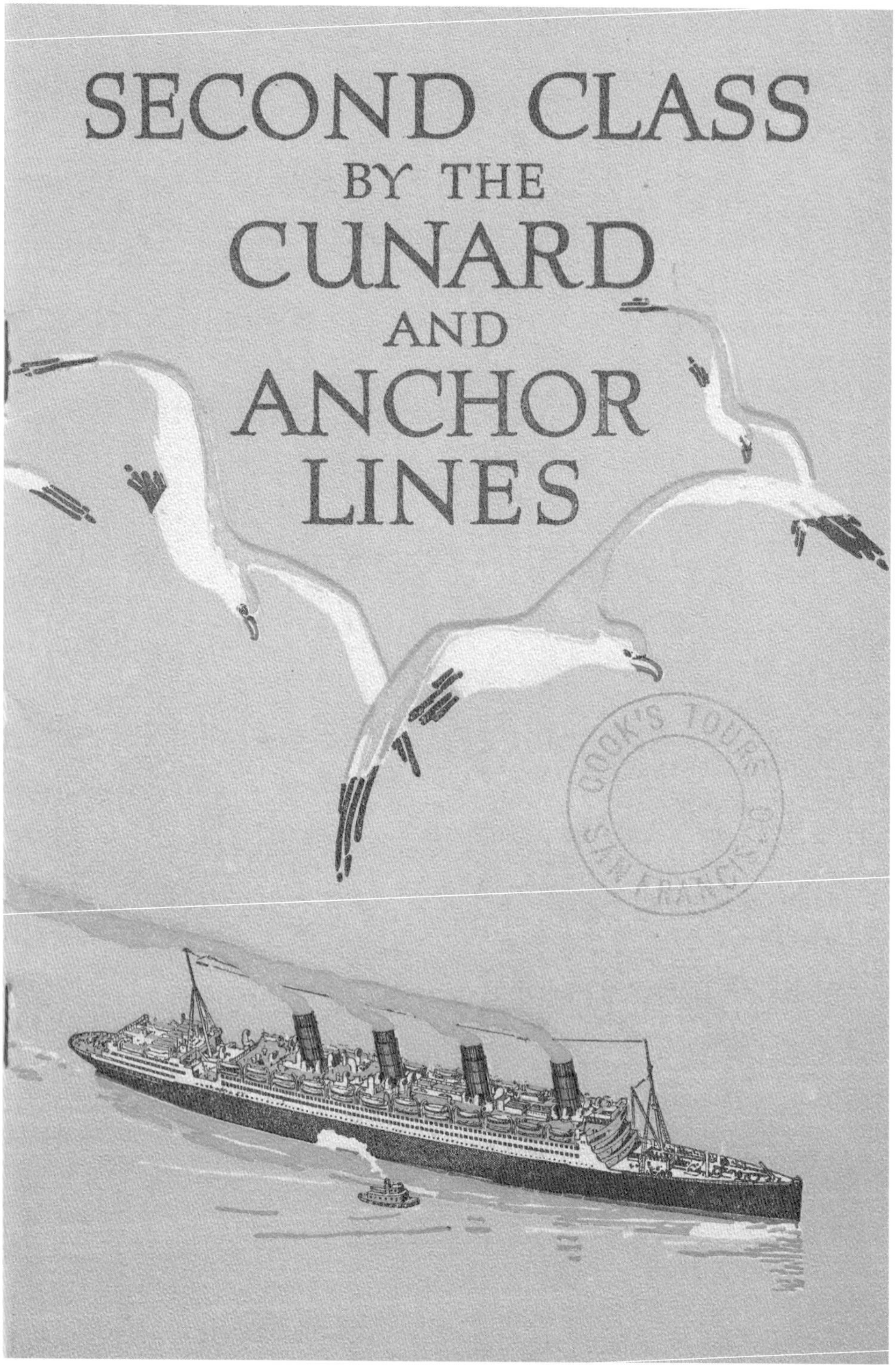

The cover for a brochure advertising Second-Class passages by the Cunard and Anchor Lines. The *Aquitania* is shown with a tiny tug alongside, being overflown by a group of seagulls. (Russ Willoughby Collection)

James 'Jimmy' Lawler of the *Mauretania*.[121] Lawler would remain chief purser of the *Aquitania* until 5 January 1938, when he retired from the sea to take a job as a travel agents' representative. He would earn the nickname 'Ambassador of the Atlantic' during his tenure.[122]

The great ship herself also underwent many changes. Not all of them were positive. With Prohibition taking hold in the United States, there seemed to be some serious questions as to whether a foreign-flag liner, such as the *Aquitania*, could actually carry liquor into an American port and then return without being boarded and having the liquor seized. As late as June 1923, American authorities were not making the point clear, and during July, *Aquitania*'s chief surgeon, Dr Sydney Jones, found very long lines of passengers outside his office, complaining of a 'remarkable variety of illnesses' that could apparently only be cured by the stock of medicinal alcohol he was allowed to carry aboard ship.

What came next was even more bizarre. In late August, a delicately carved mahogany bar with a marble top and brass rails was installed in the Garden Lounge while the ship was in New York. It was a soda fountain, and two fresh-faced young lads stood behind the bar waiting to serve thirsty passengers with such exotic-sounding concoctions as the 'Typhoon Snifter', 'Burmese Bracer', 'Wallaby Boomeroo', 'Bachelor's Blush' or 'Widow's Smile'.

Soda fountains were not unheard of on either side of the Atlantic at the time, so the reaction from passengers and crew alike seems odd from a remove of nearly a century. Chief Purser Fred Jones was very uncertain as to whether it would be a success; Purser Charles Spedding was dubious; Captain Charles was reportedly 'blushing for shame at the idea'; Dr Jones said it certainly was not going to affect the number of malady-stricken passengers outside of his office looking for refreshment of a stronger type. Boarding passengers were not impressed. The English thought it an aberrant addition to the space, declared they couldn't drink soda water and shook their heads in dismay; American passengers were less restrained in venting their dislike for the new addition. Some of them said it gave them 'the willies', and others were heard openly cursing over it. One exasperated passenger exclaimed: 'Have we come to this?'

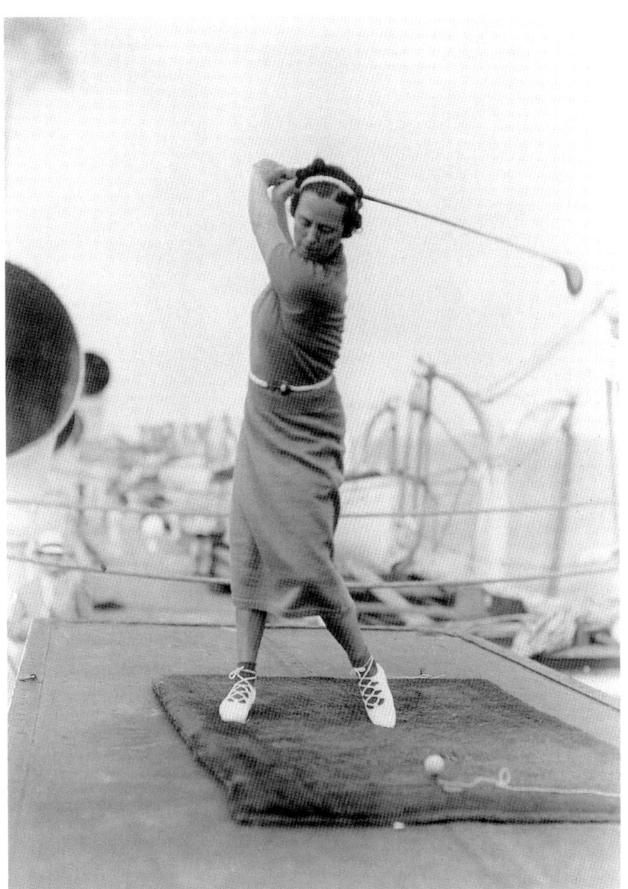

Left: Famous golfers were frequently on board the *Aquitania*. Here one lady practises her swing from atop the liner's Boat Deck. (Ioannis Georgiou Collection)

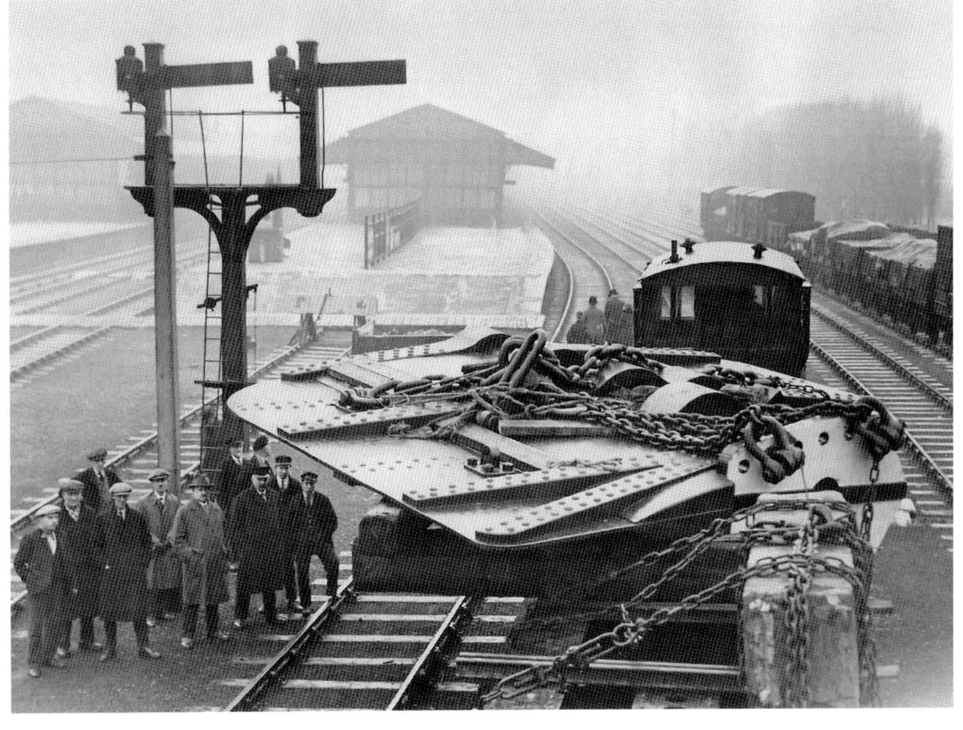

Below: On 30 January 1928, the *Aquitania*'s new 54-ton rudder was in transit from the Darlington Forge Company, where it had been created, for installation on the ship in Southampton. The photograph was taken at the Eaglescliffe Railway Station, at Stockton-on-Tees, and shows how awkward the move by rail was, since it overhung the car so much. (Clyde George Collection)

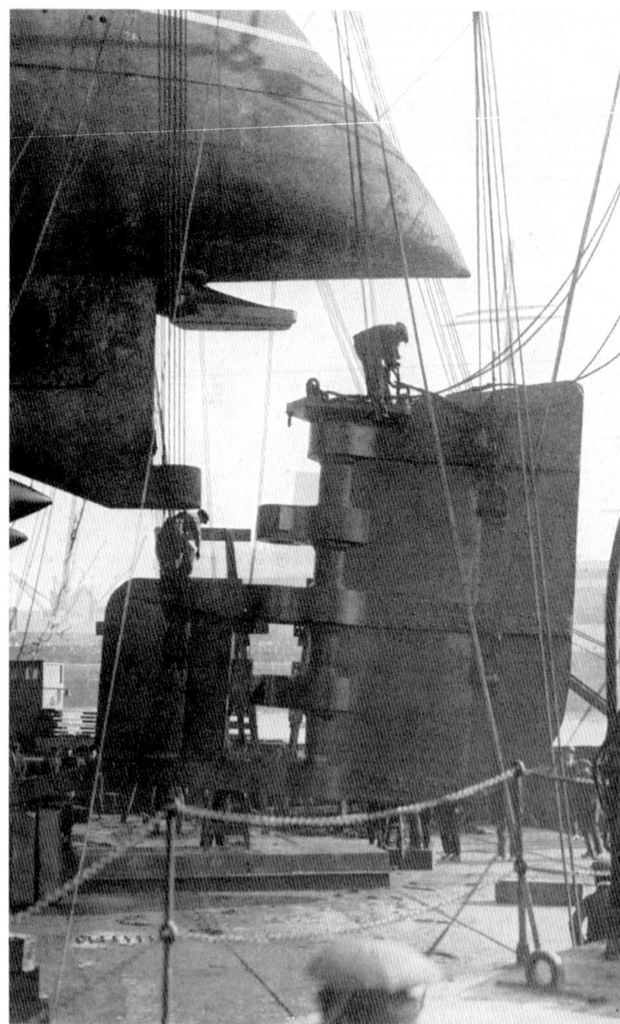

The new rudder arrived safely, and was installed on the *Aquitania* in the floating dry dock at Southampton during her annual overhaul. (Ioannis Georgiou Collection)

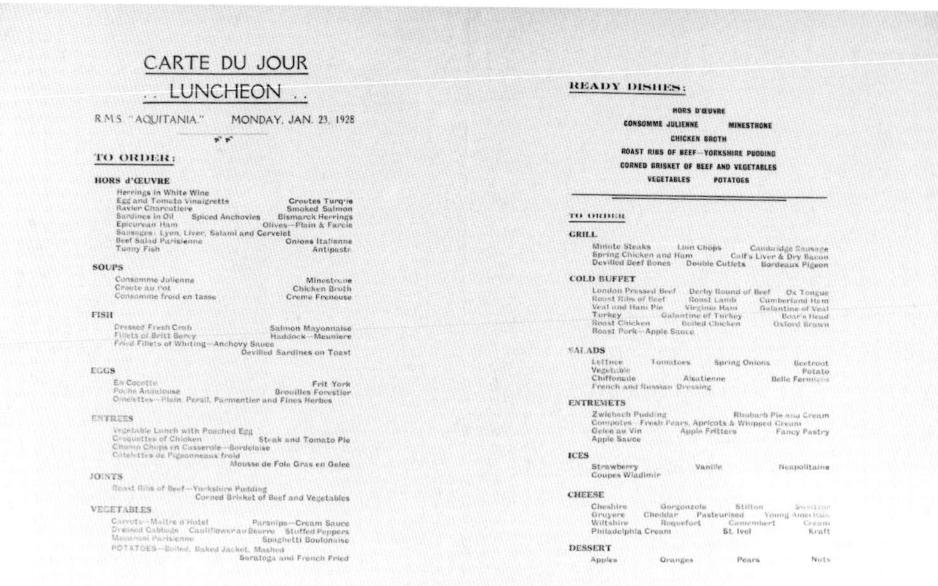

CARTE DU JOUR
.. LUNCHEON ..

R.M.S. "AQUITANIA."   MONDAY, JAN. 23, 1928

TO ORDER:

HORS d'ŒUVRE
Herrings in White Wine
Egg and Tomato Vinaigrette   Croutes Turque
Ravier Charcutiere   Smoked Salmon
Sardines in Oil   Spiced Anchovies   Bismarck Herrings
Epicurean Ham   Olives—Plain & Farcie
Sausages: Lyon, Liver, Salami and Cervelet
Beef Salad Parisienne   Onions Italienne
Tunny Fish   Antipasta

SOUPS
Consomme Julienne   Minestrone
Croute au Pot   Chicken Broth
Consomme froid en tasse   Creme Freneuse

FISH
Dressed Fresh Crab   Salmon Mayonnaise
Fillets of Britt Bercy   Haddock—Meuniere
Fried Fillets of Whiting—Anchovy Sauce
                    Devilled Sardines on Toast

EGGS
En Cocotte   Frit York
Poche Andalouse   Brouilles Forestier
Omelettes—Plain, Persil, Parmentier and Fines Herbes

ENTREES
Vegetable Lunch with Poached Egg
Croquettes of Chicken   Steak and Tomato Pie
Chump Chops en Casserole—Bordelaise
Cotelettes de Pigeonneaux froid
                    Mousse de Foie Gras en Gelee

JOINTS
Roast Ribs of Beef—Yorkshire Pudding
          Corned Brisket of Beef and Vegetables

VEGETABLES
Carrots—Maitre d'Hotel   Parsnips—Cream Sauce
Dressed Cabbage Cauliflower au Beurre Stuffed Peppers
Macaroni Parisienne   Spaghetti Boulonaise
POTATOES—Boiled, Baked Jacket, Mashed
                    Saratoga and French Fried

READY DISHES:

HORS D'ŒUVRE
CONSOMME JULIENNE   MINESTRONE
CHICKEN BROTH
ROAST RIBS OF BEEF—YORKSHIRE PUDDING
CORNED BRISKET OF BEEF AND VEGETABLES
VEGETABLES   POTATOES

TO ORDER

GRILL
Minute Steaks   Loin Chops   Cambridge Sausage
Spring Chicken and Ham   Calf's Liver & Dry Bacon
Devilled Beef Bones   Double Cutlets   Bordeaux Pigeon

COLD BUFFET
London Pressed Beef   Derby Round of Beef   Ox Tongue
Roast Ribs of Beef   Roast Lamb   Cumberland Ham
Veal and Ham Pie   Virginia Ham   Galantine of Veal
Turkey   Galantine of Turkey   Boar's Head
Roast Chicken   Boiled Chicken   Oxford Brawn
Roast Pork—Apple Sauce

SALADS
Lettuce   Tomatoes   Spring Onions   Beetroot
Vegetable   Potato
Chiffonade   Alsatienne   Belle Fermiere
French and Russian Dressing

ENTREMETS
Zwiebach Pudding   Rhubarb Pie and Cream
Compotes—Fresh Pears, Apricots & Whipped Cream
Gelee au Vin   Apple Fritters   Fancy Pastry
Apple Sauce

ICES
Strawberry   Vanille   Neapolitaine
Coupes Wladimir

CHEESE
Cheshire   Gorgonzola   Stilton   Swettner
Gruyere   Cheddar   Pasteurised   Young American
Wiltshire   Roquefort   Camembert   Cream
Philadelphia Cream   St. Ivel   Kraft

DESSERT
Apples   Oranges   Pears   Nuts

By the end of January 1928, the *Aquitania* had returned to service. This menu is from her first trip after the refit. She departed on 18 January, and ran into heavy seas almost immediately. By Wednesday, 25 January, she was in the worst of it. Two bridge windows were smashed out, although the Chart Room aft of it emerged relatively unscathed. Captain Diggle later said it was the worst weather he had ever seen on the Atlantic. (Beamish Museum)

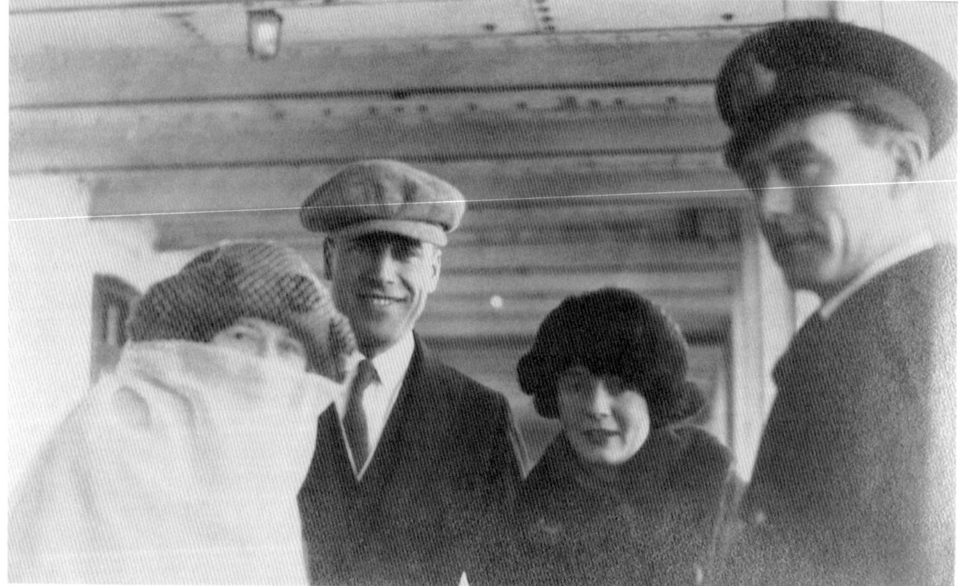

One of the ship's officers poses for a rare photograph with three First-Class passengers. (Mike Poirier Collection)

This view by maritime artist Charles Dixon was published in *The Graphic*
April 1913. It shows the *Aquitania* taking shape amidst a busy scene at
the John Brown shipyard. The many cranes along the slipway are hoisting
heavy hull plates into position, while the fires from riveters' furnaces light
the scene dramatically. (Authors' collection)

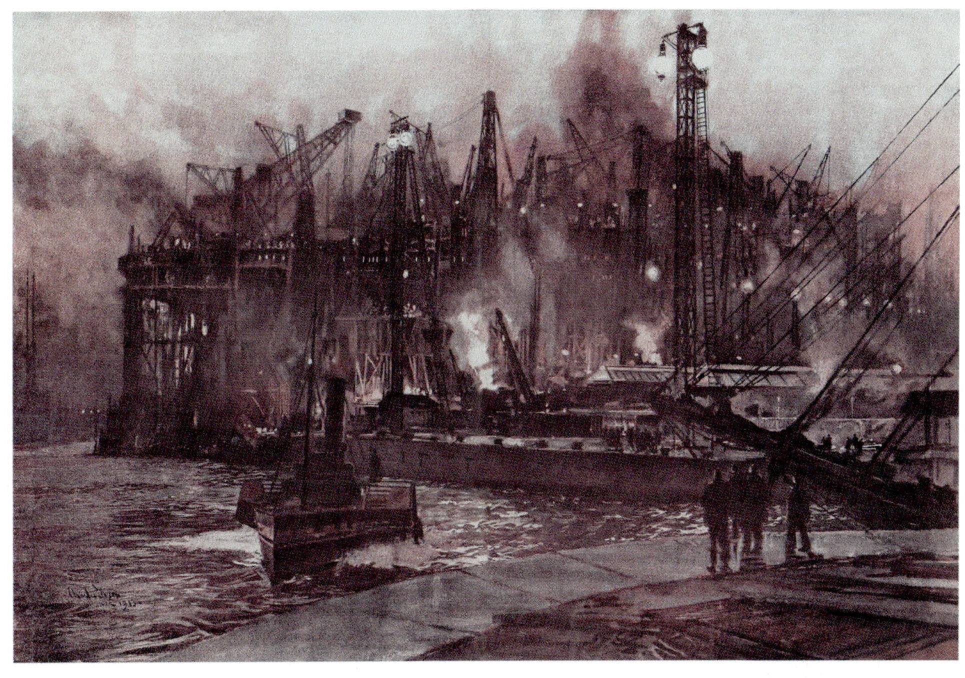

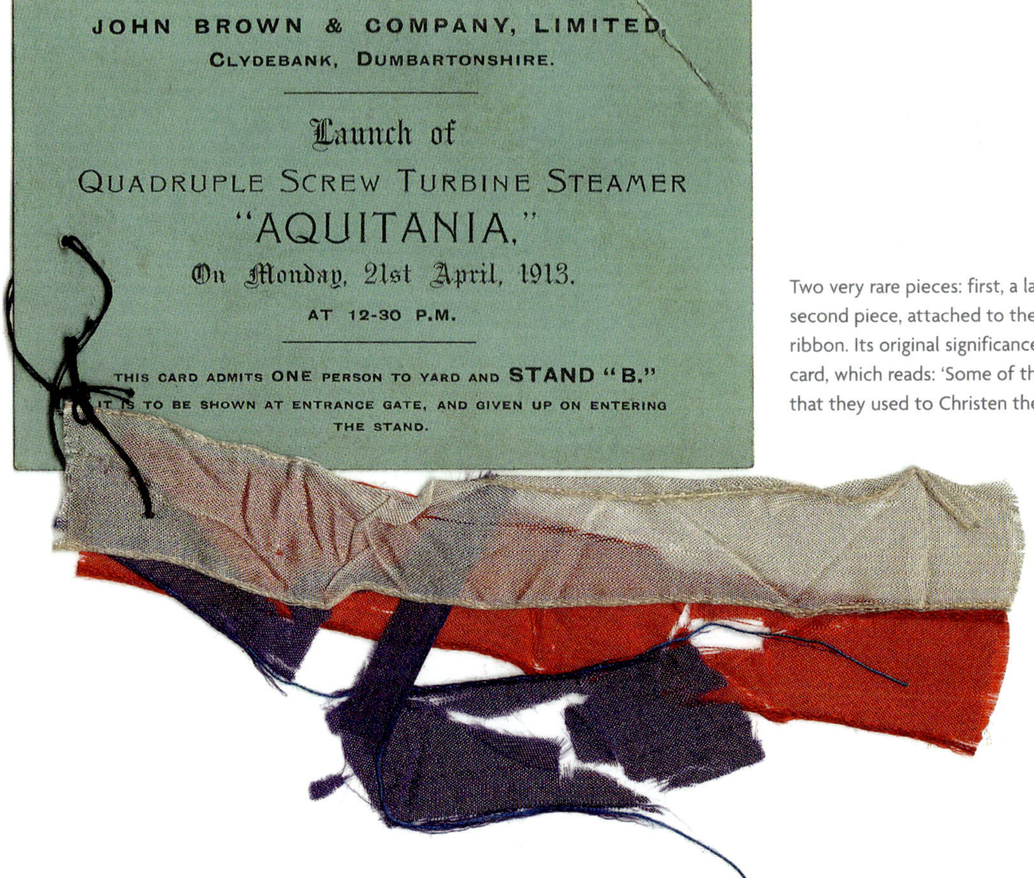

Two very rare pieces: first, a launch invitation card that would admit the holder to 'Stand "B."'. The
second piece, attached to the card by a bit of thread, is a frayed segment of red, white and blue
ribbon. Its original significance would be lost, were it not for the inscription on the back of this
card, which reads: 'Some of the ribbon that was on the crown that hung above the bottle of wine
that they used to Christen the ship.' (Kalman Tanito Collection)

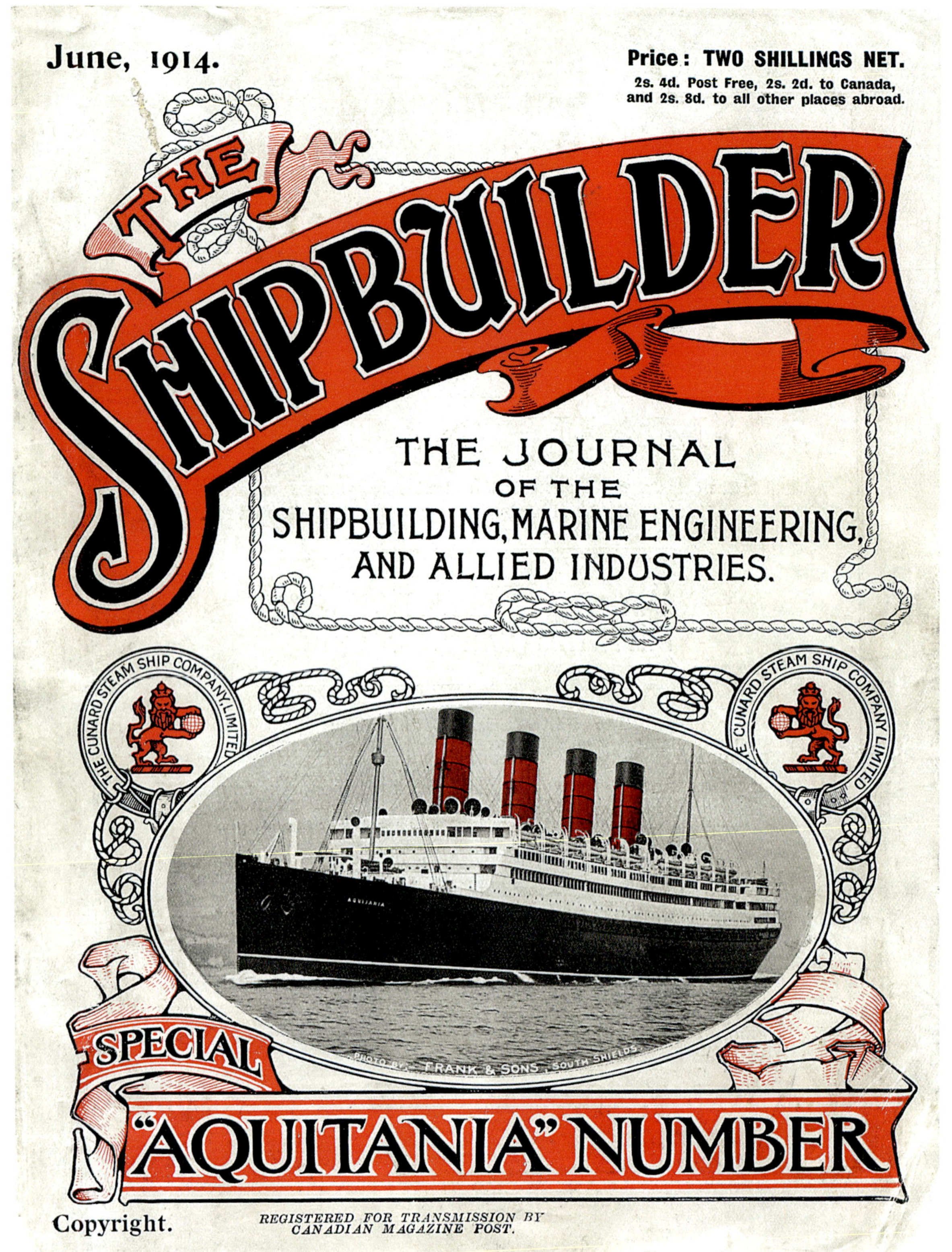

The two most famous technical journals regarding all things nautical were *The Shipbuilder* and *Engineering*. Each devoted extra space to the new *Aquitania*. This is the cover for the special *Aquitania* number of *The Shipbuilder*, June 1914. (Ioannis Georgiou Collection)

Opposite page: In the era of black-and-white photography, colour artist's perceptions were used in publicity materials to help people grasp the true beauty of the *Aquitania*'s public rooms.

Top left: This view shows the First-Class Restaurant in all its splendour – albeit not very thoroughly populated with tables and chairs. (Brian Hawley Collection)

Top right: This portrait shows an empty Drawing Room, with the view amidships looking forward beneath the dome. The alcove on the starboard side of the finished room was not as deep as that seen in this image. (Eric Sauder Collection)

Centre left: This view shows the port side salon which ran from the Entrance to the Palladian Lounge. The uptake for the No. 2 funnel was located amidships here, and on the opposite side was a mirror image of this salon. This view looks forward toward the Entrance, with the doors to the Drawing Room visible beyond. (Eric Sauder Collection)

Centre right: Another artist's rendering of the First-Class Palladian Lounge, again devoid of furniture. This view looks toward the forward end of the room, with the fireplace. (Authors' collection)

Bottom left: A fine rendition of the Second-Class Drawing Room. Unlike some of these views, this portrait shows the room furnished and with palms and flowers. (Brian Hawley Collection)

Bottom right: The Second-Class Dining Saloon, looking forward. The opening in the ceiling leading up to C Deck was in reality much smaller than that seen here. One remarkable thing about all of these illustrations is the very 'manor home' feeling they give. It would appear that these rooms were in a grand land-based home with the furniture cleared away for dancing. There is very little to indicate that one is at sea. (Ioannis Georgiou Collection)

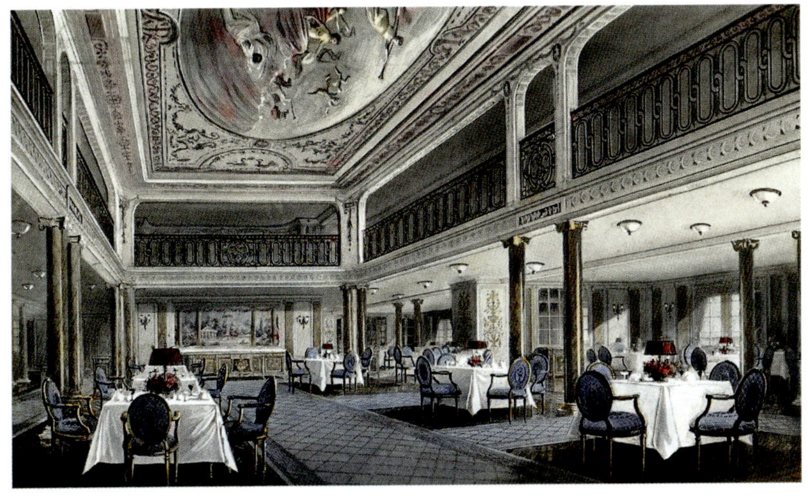
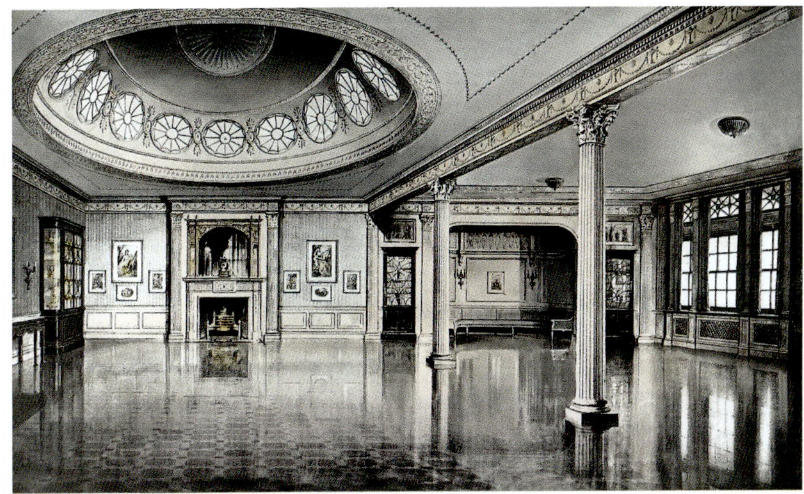
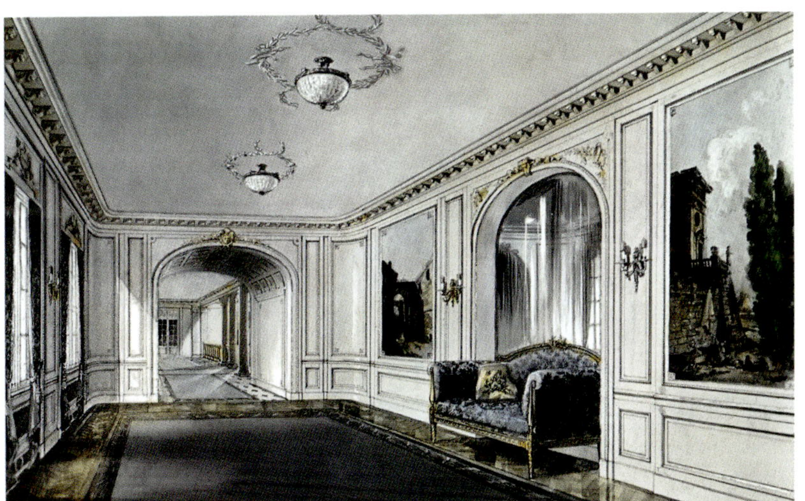
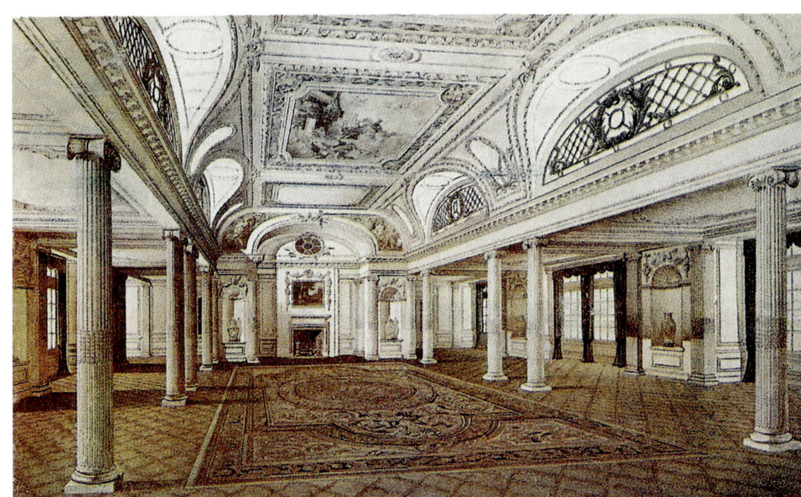
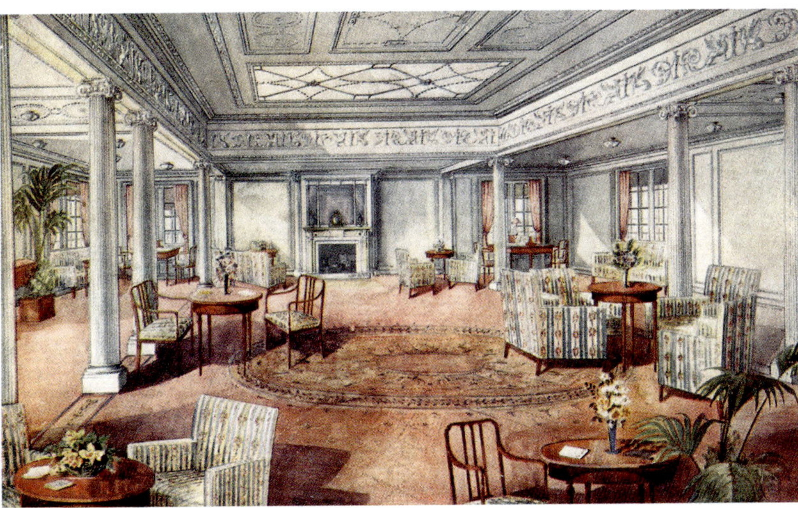
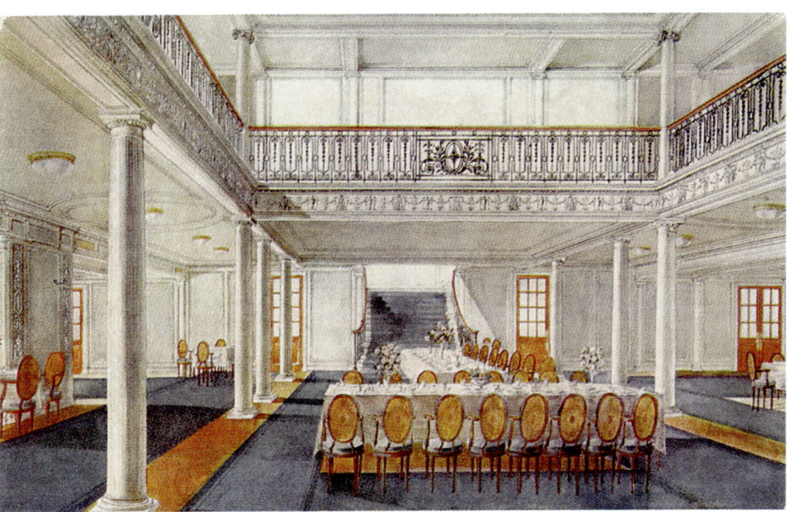

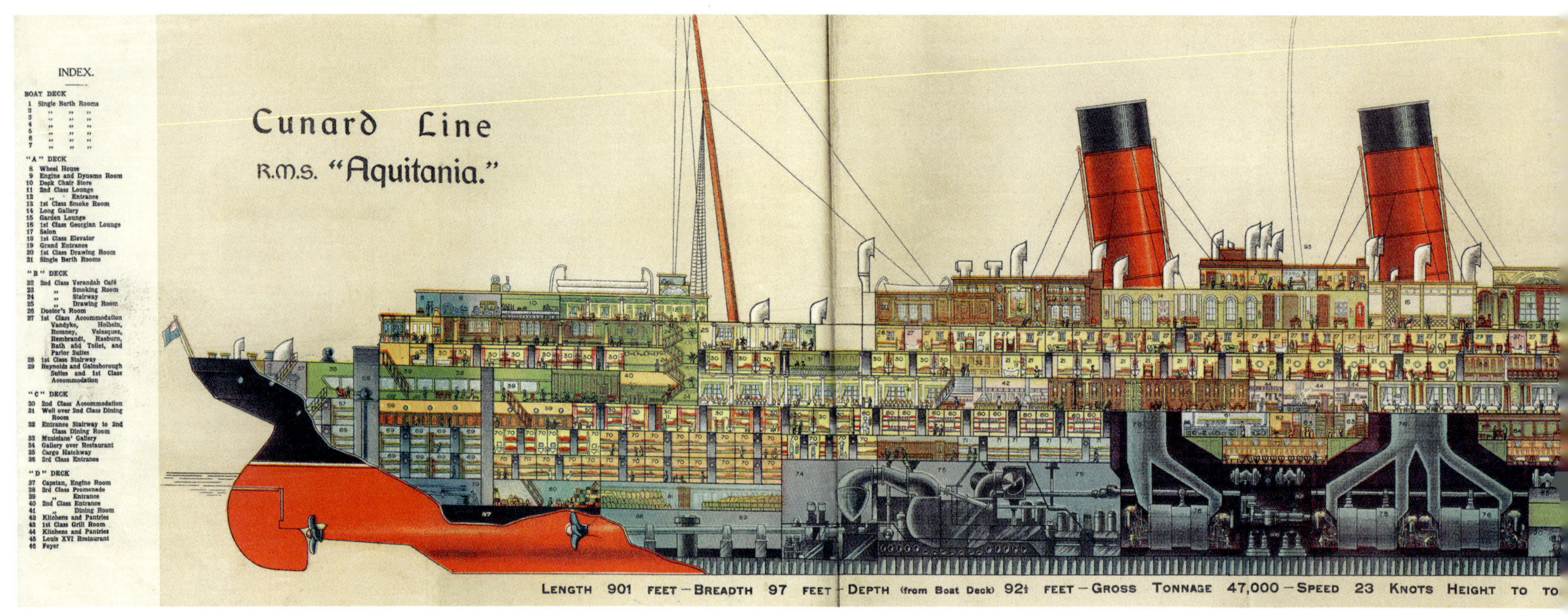

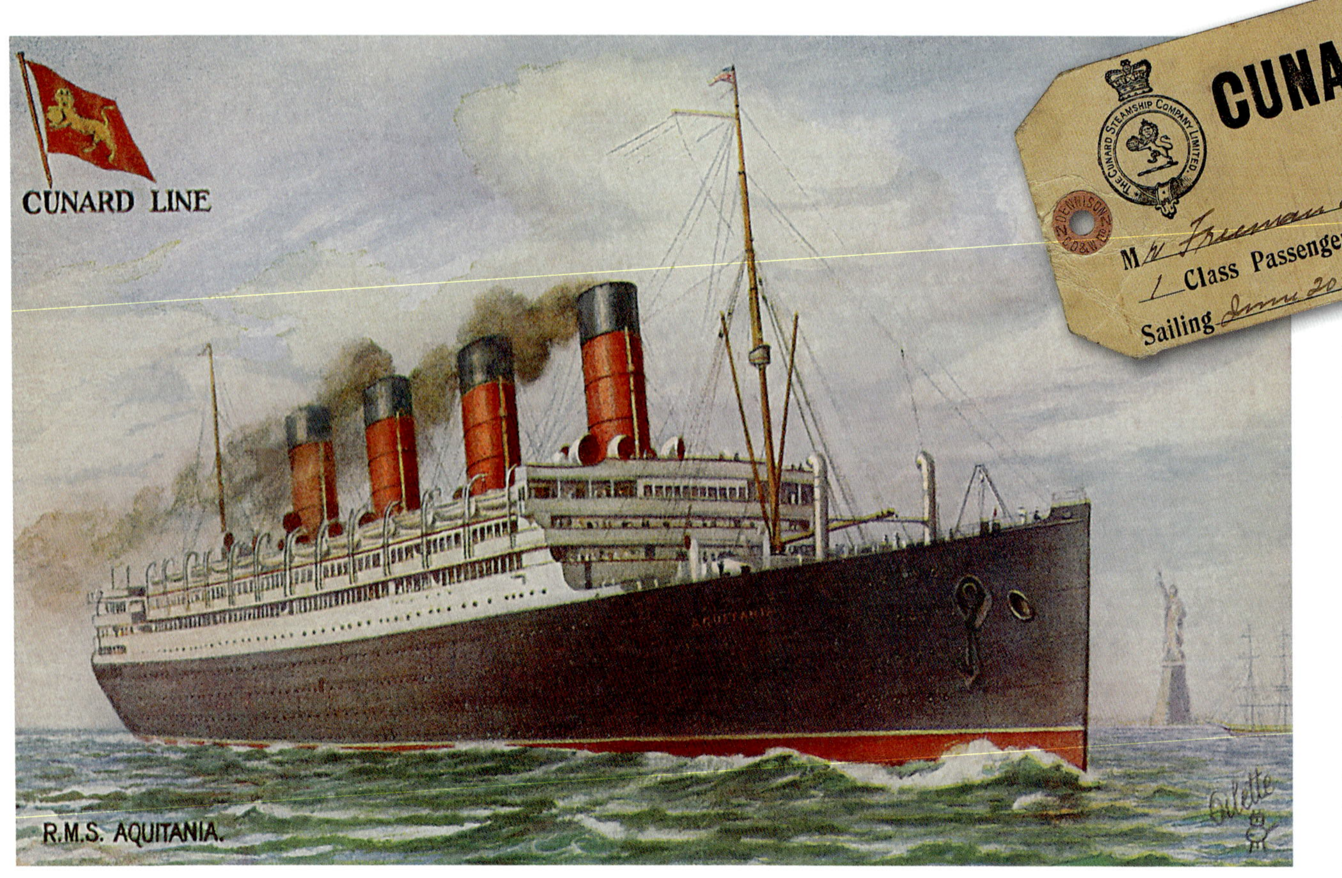

This very rare baggage tag was from the maiden voyage of the *Aquitania*, and belonged to First-Class passenger Freeman Davison. (Mike Poirier Collection)

One of the finest artistic portraits of the *Aquitania* ever done, this depicts the *Aquitania* entering New York as she appeared on her maiden voyage. It was released as a Tuck's Oilette card. (Authors' collection)

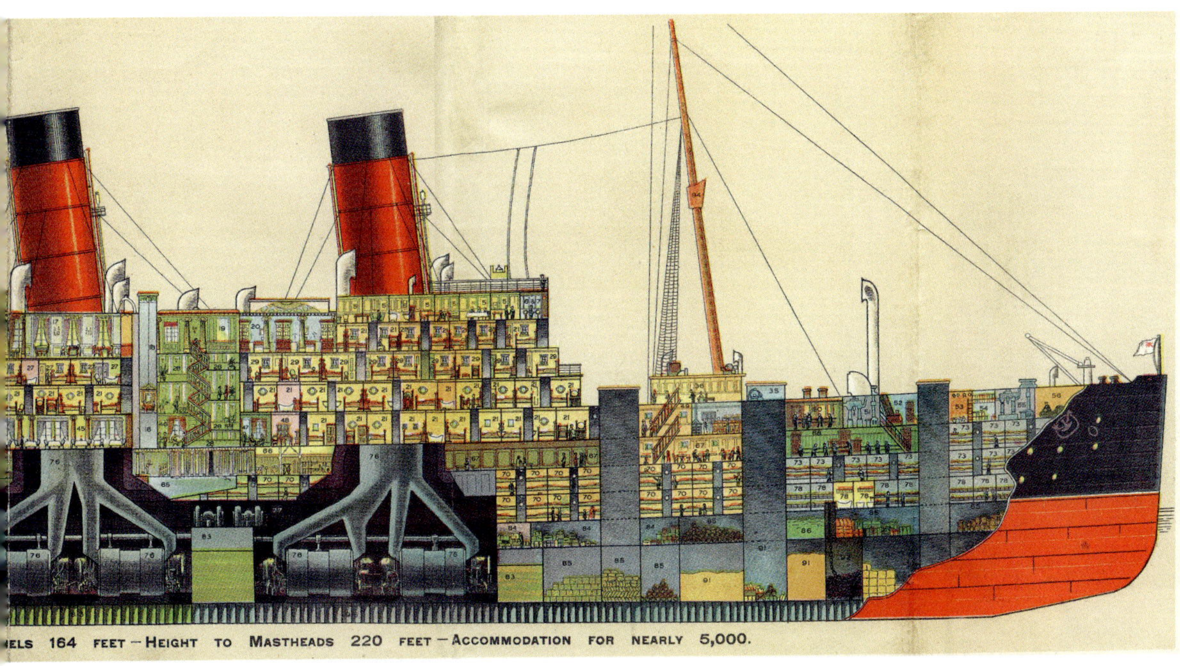

This starboard cutaway illustration of the *Aquitania* came in foldout form in a brochure from the 1920s. Every public room, suite, stateroom and major machinery space is both shown and labelled. It gives some idea of how impressive the liner was, both when she originally entered service and throughout her career. (Authors' collection)

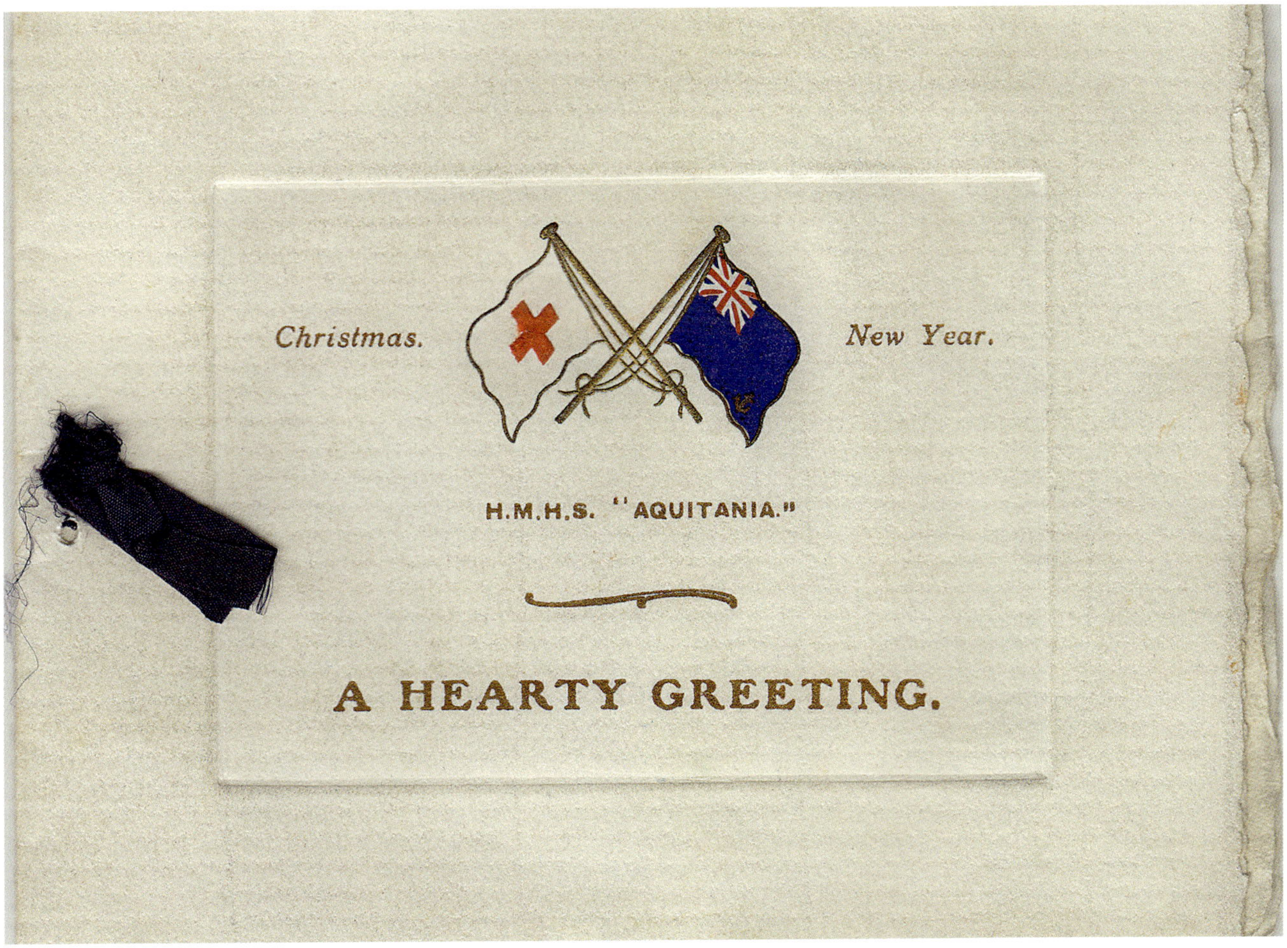

A Christmas greeting card from the *Aquitania*'s time as a hospital ship. (Authors' collection)

Before 1915 had ended, noted artist Norman Wilkinson had released a book filled with sketches and watercolours he made of the action at the Dardanelles. This piece showed the *Aquitania* with an unnamed collier of about 8,000grt alongside. (Authors' collection)

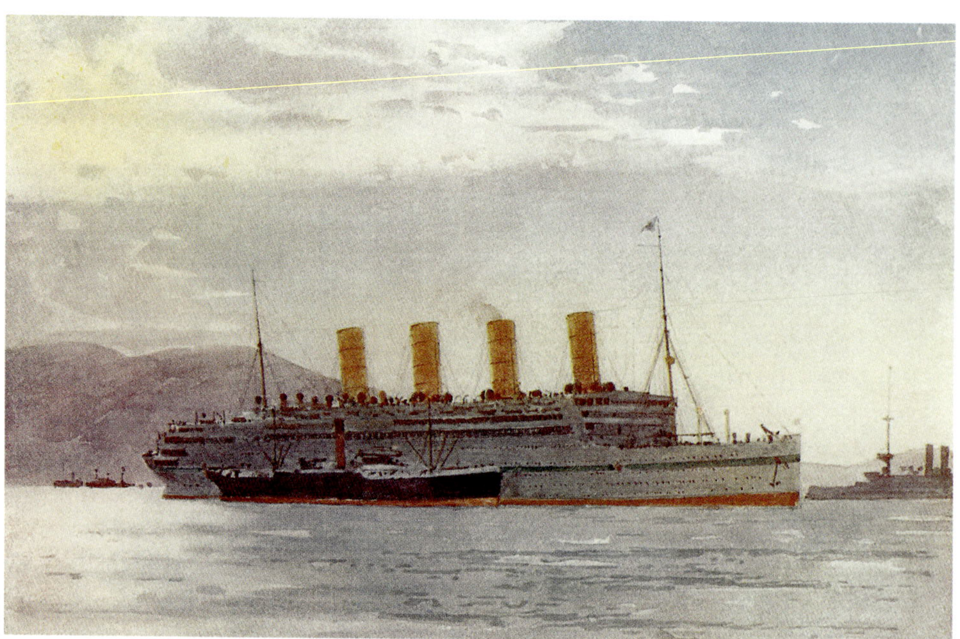

This document is filled with fascinating historical details. It was dated 9 September 1914, and was an estimate of what it would cost to restore the *Aquitania* to civilian service. The total of £175,310 is estimated; of that, £55,000 would go for joiners, £11,000 for painters, £26,000 for the main public rooms and £15,000 for work on the electric lighting. These conversions were extremely expensive affairs.
(Authors' collection)

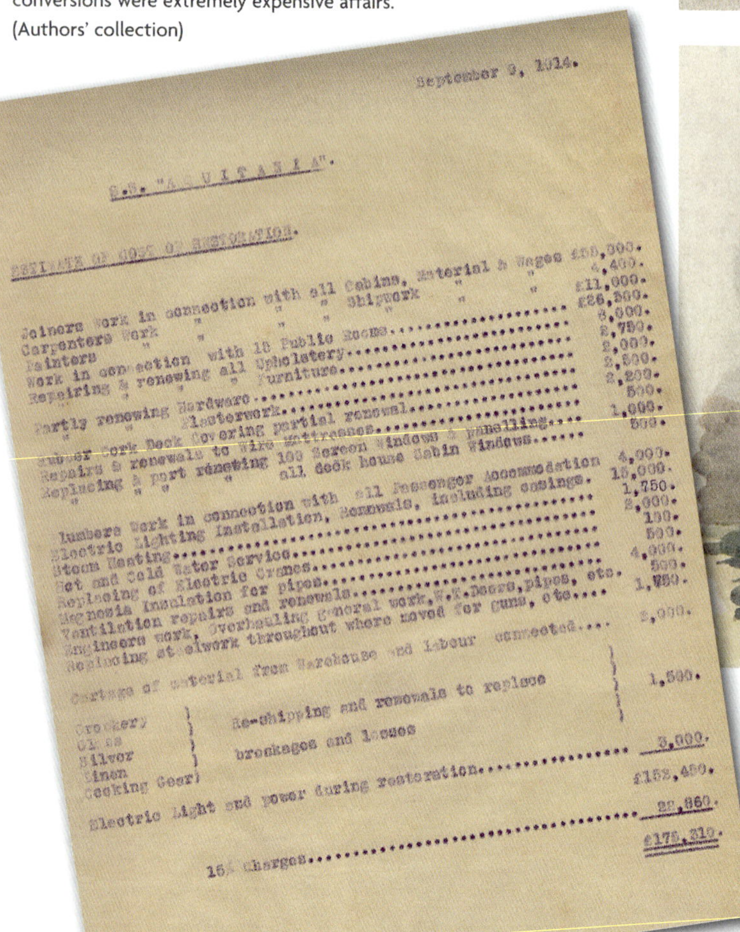

September 9, 1914.

S.S. "AQUITANIA".

ESTIMATE OF COST OF RESTORATION.

Joiners Work in connection with all Cabins, Material & Wages £55,000.
Carpenters Work " " " " " " 4,400.
Painters " " " " Shipwork " " £11,000.
Work in connection with 18 Public Rooms £26,300.
Repairing & renewing all Upholstery 8,000.
" " " Furniture 2,790.
Partly renewing Hardware 2,000.
" " Plasterwork 2,500.
Rubber Cork Deck Covering partial renewal 2,200.
Repairs & renewals to Wire Mattresses 500.
Replacing & part renewing 100 Screen Windows & Panelling 1,000.
" " " all deck house Cabin Windows 500.
Plumbers Work in connection with all Passenger Accommodation 4,000.
Electric Lighting Installation, Renewals, including casings. 15,000.
Steam Heating 1,750.
Hot and Cold Water Service 2,000.
Replacing of Electric Cranes 100.
Magnesia Insulation for pipes 500.
Ventilation repairs and renewals 4,000.
Engineers work, Overhauling General work, W.T. Doors, pipes, etc. 500.
Replacing steelwork throughout where moved for Guns, etc. 1,750.
2,000.
Cartage of material from Warehouse and Labour connected 1,500.

Crockery)
Glass) Re-shipping and renewals to replace
Silver) breakages and losses 3,000.
Linen)
Cooking Gear)

£152,400.
Electric Light and power during restoration 22,860.

£175,310.
15% Charges

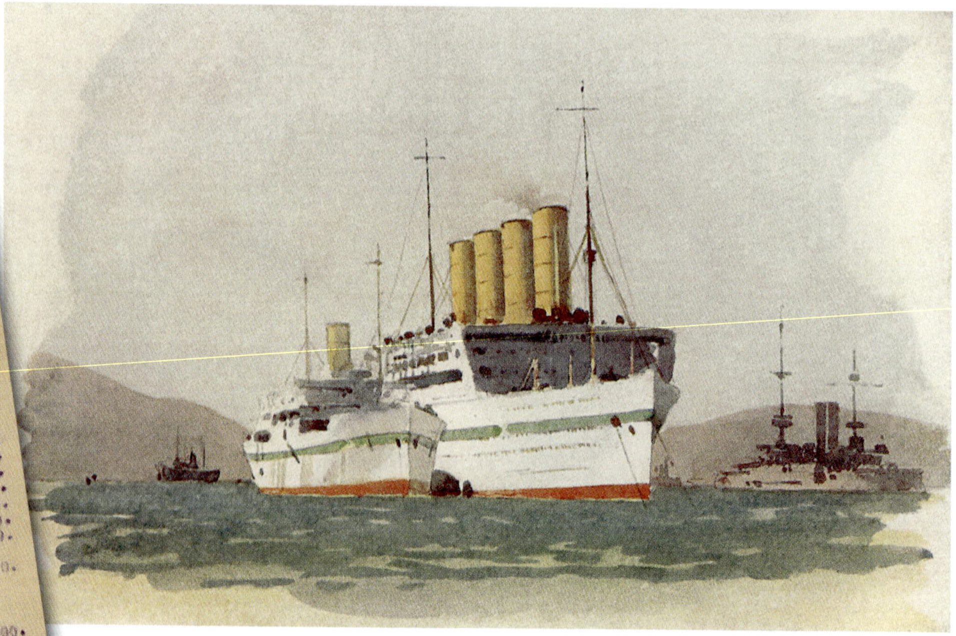

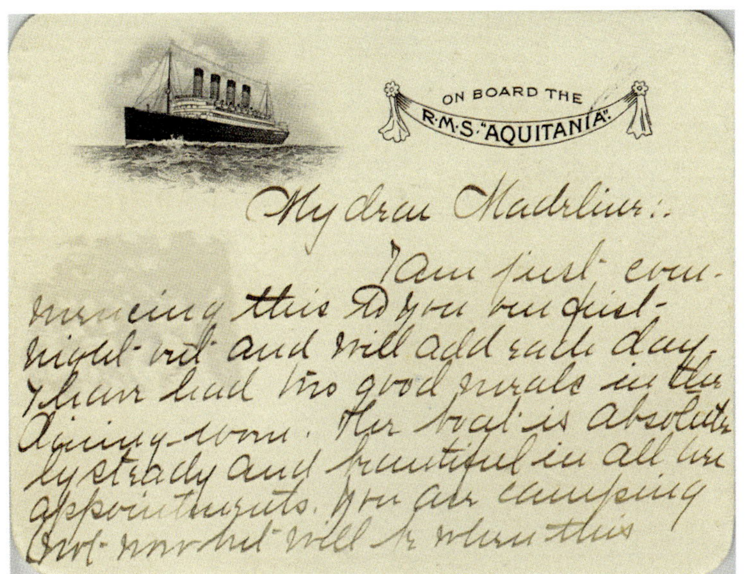

Right: This Saloon Accommodation card from 1 October 1918 includes the cabin and berth assignment for the holder, 2nd Dr Chester Blaubaugh; it also shows the meal sitting, as well as the table and seat assignment. (Mike Poirier Collection)

Left: The first page of a letter written on *Aquitania* stationary; the exact date is difficult to pin down, but it seems possible that it was written in 1917 or 1918. It is all but illegible – perhaps to everyone except the original author. What fragments can be read are fascinating. 'The boat is absolutely steady and beautiful in all the appointments,' the author claims. While bemoaning a complete absence of 'delightful girls', he says that it was 'a wonderful trip. Perfect weather and moonlight nights and the boat is so wonderfully steady.' He goes on to describe a hysterical fancy dress ball with prizes for the most original costumes. One man was dressed as a baby, complete with nightgown, boudoir cap and nursing bottle – he was even in a baby carriage. Another man was his nurse, even dressed in an English nurse's uniform, while someone else had done something very strange with his 'false hair'. The author concluded that it was all 'so funny'. On the previous night, apparently 4 July, the Dining Saloon had been decorated with flags all around. (Authors' collection)

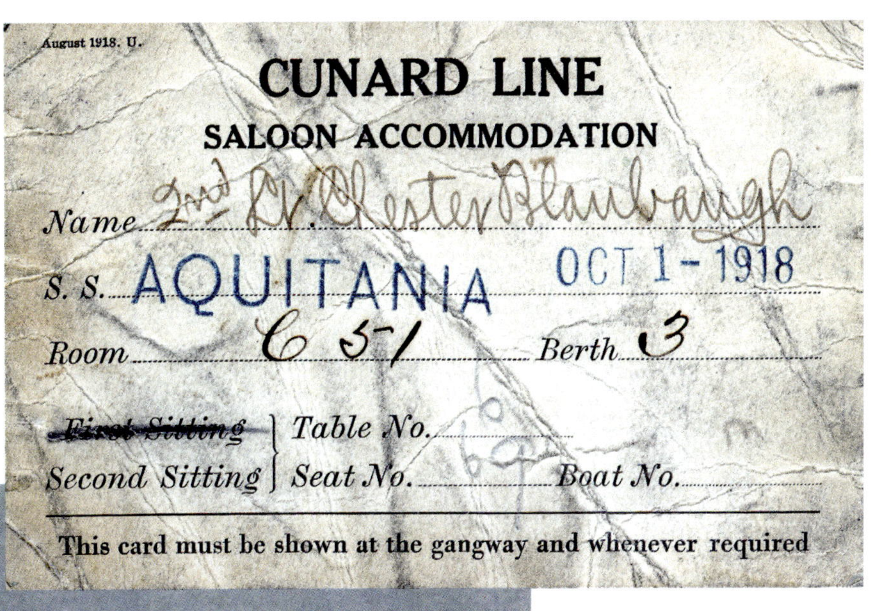

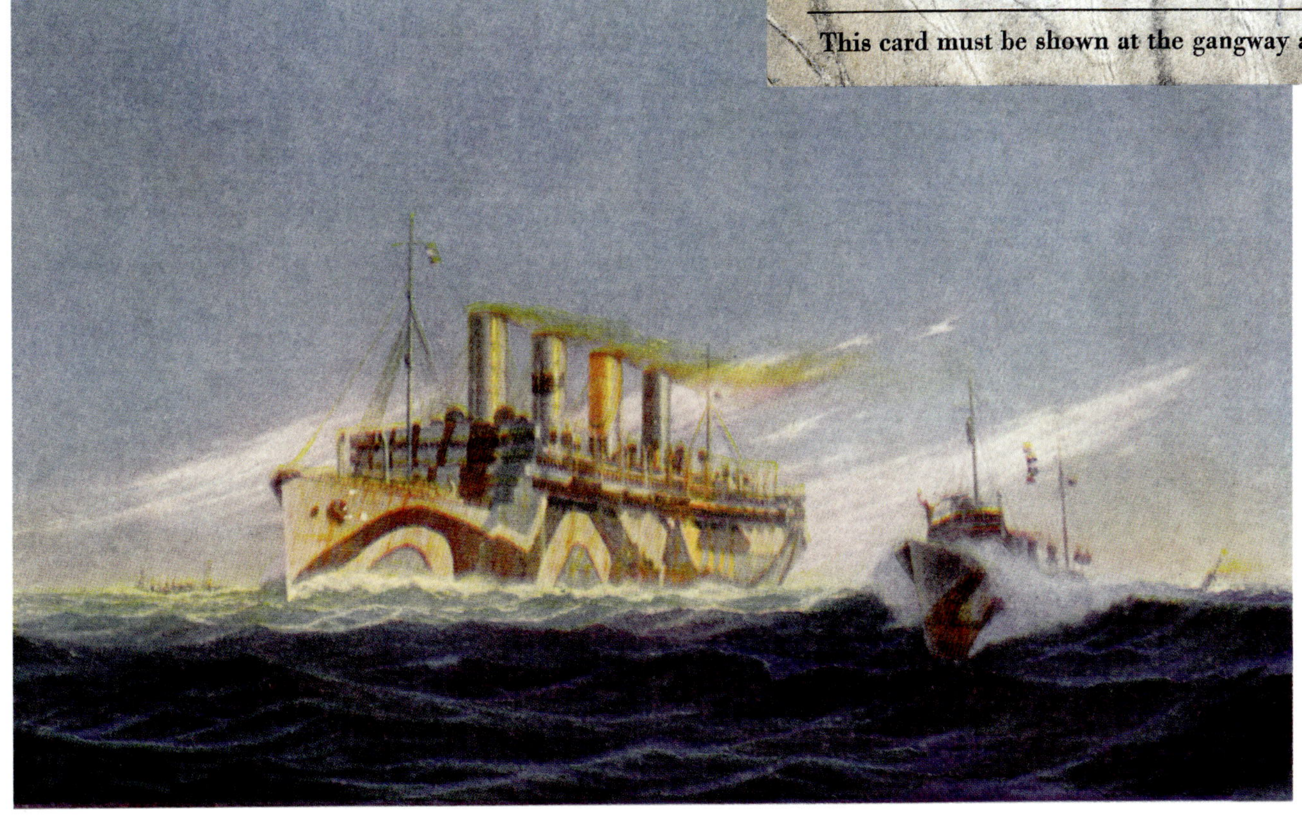

Left: A splendid portrait of the *Aquitania* while serving as a troop transport during the closing months of the war. She is here being convoyed by a small warship. While the little vessel crashes through the seas, the *Aquitania* rides atop them serenely. (Authors' collection)

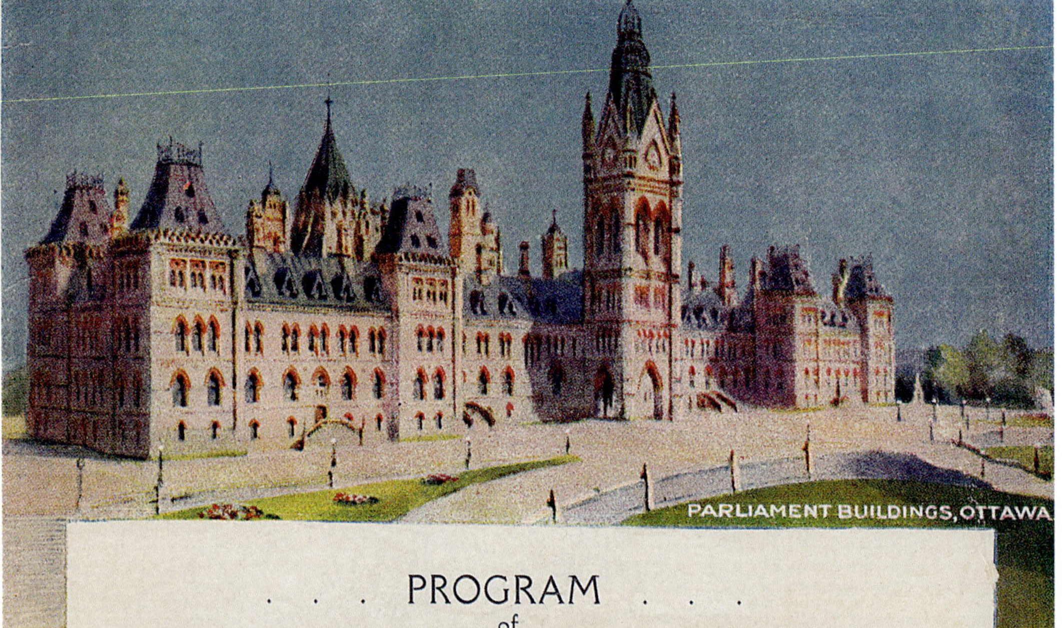

. . . PROGRAM . . .
of
ENTERTAINMENT,
held on board
H.M.S. "AQUITANIA,"
(Capt. J. T. W. Charles, C.B., R.D., R.N.R.)

March	... ...	"The Diplomat."	... ...	*J. P. Sousa*
		BAND		
Overture	... ...	"Poet and Peasant."	... ...	*Suppee*
		BAND		
Tenor Solo	...	"The Sunshine of your Smile."	... ...	——
		Private EDWARD FLANNERY		
Piano Solo	...	"Perfect Day" (variations) ...	... ...	*Lampi*
		Sgt. R. A. SMITH		
Vocal	...	"There's a long, long Trail."	...	...*Zo Elliot*
		BAND		
Guitar Special	... ...	...Selected	... ...	——
		Private PHILLIPS		
Tenor Solo	...	"A Baby's Prayer at Twilight."	... ...	——
		Private EDWARD FLANNERY		
Violin Solo	... ...	...Selected	... ...	——
		Sgt. J. F. SMOLKA		
Popular	... ...	..."Aida."	... ...	*Verdi*
		BAND		
Selection	... ...	"We're going over."	...	... *A Lange*
		BAND		

"AMERICA."   **"GOD SAVE THE KING."**

President - - Brigadier-General C. Crawford, Sixth Brigade, N.A.

Chaplain Committee—Lieuts. Earl H. Weed, S. P. Stapp and W. S. Sewell

Proceeds in aid of Seamen's Charities in Liverpool and New York.

This programme of entertainment from the HMS *Aquitania* shows the Parliament Buildings in Ottawa, Canada. The selections range from the overture from *Poet and Peasant* to patriotic airs from both the United States and England. (Steven B. Anderson Collection)

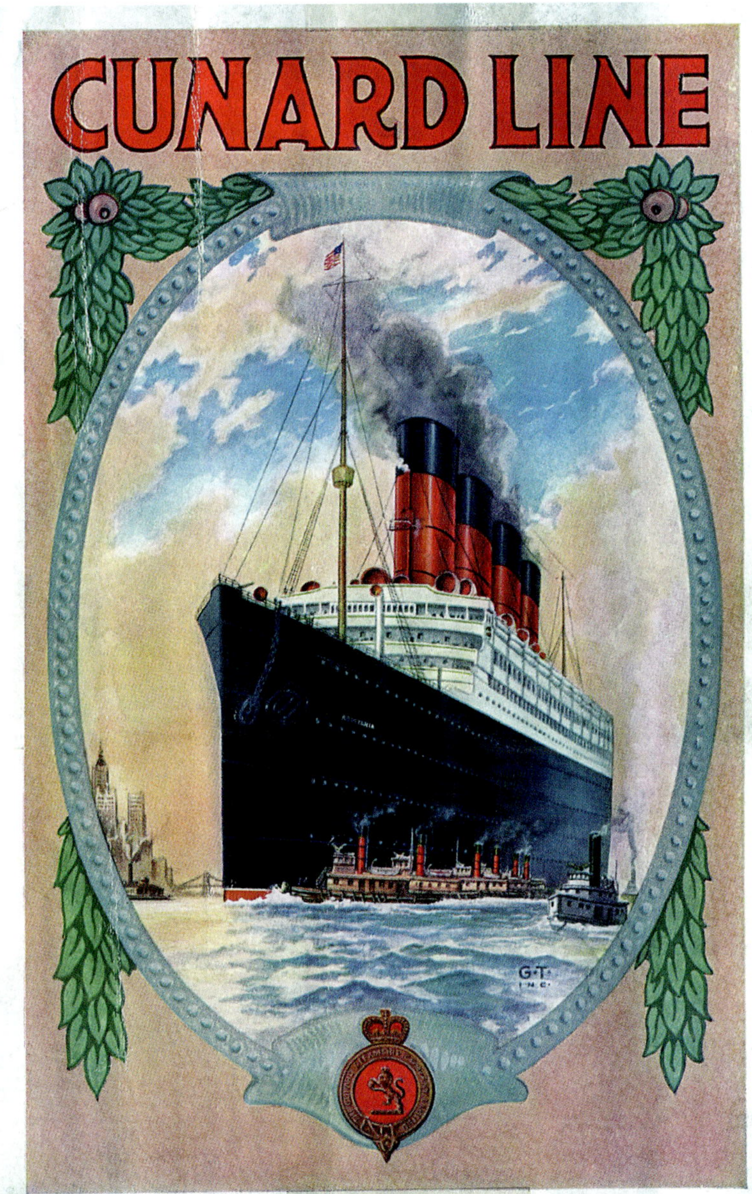

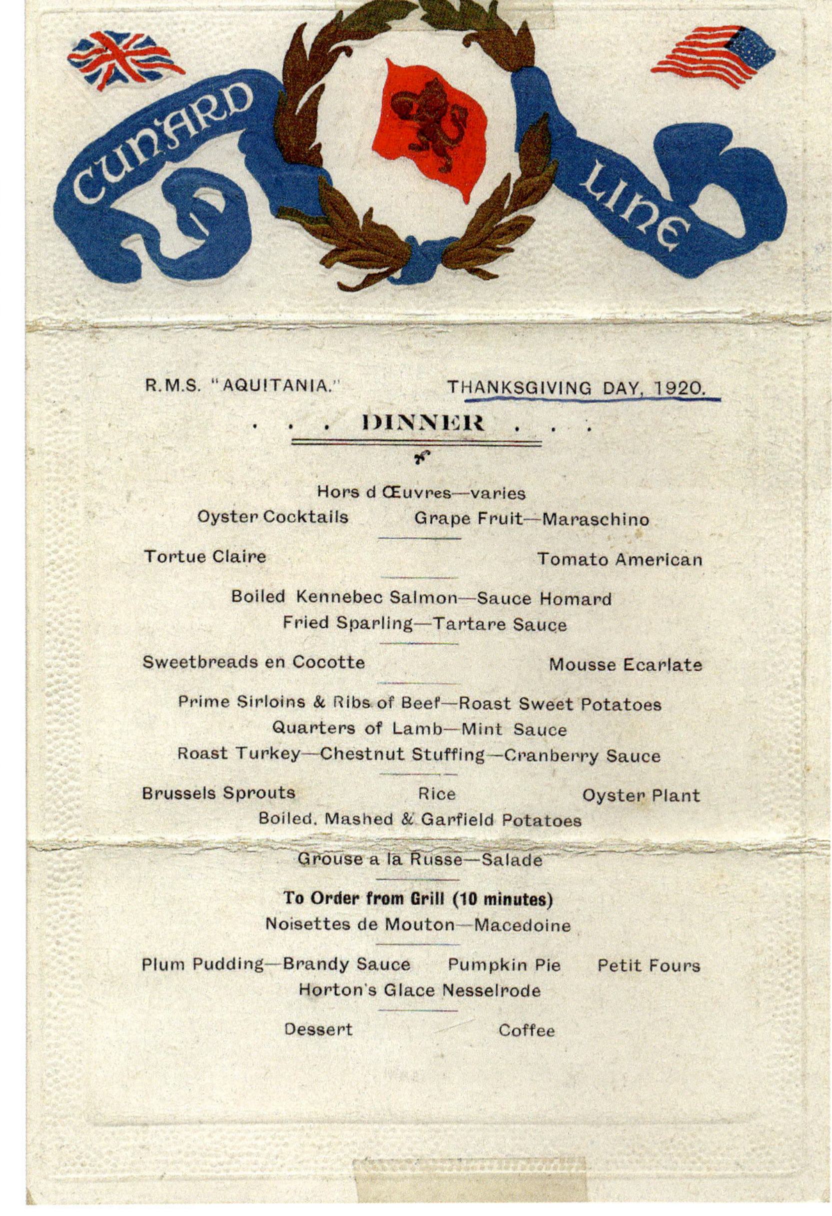

The cover of the passenger list from 23 November 1920 for an eastbound crossing. The crew listed included Captain Sir James T.W. Charles, Staff Captain H.S. Horsbrough, Chief Officer F.W. Robinson, Chief Engineer G. Paterson, Staff Chief Enginer R. Shortridge, as well as Chief Steward Fred Jones and his friend, Purser Charles Spedding. Among the prominent passengers was actor Roscoe 'Fatty' Arbuckle. (Mike Poirier Collection)

This dinner menu comes from the same crossing as the passenger list, and dates to Thursday, 25 November – Thanksgiving Day. Selections included oyster cocktails, boiled Kennebec salmon, sweetbreads, roast turkey and pumpkin pie. (Mike Poirier Collection)

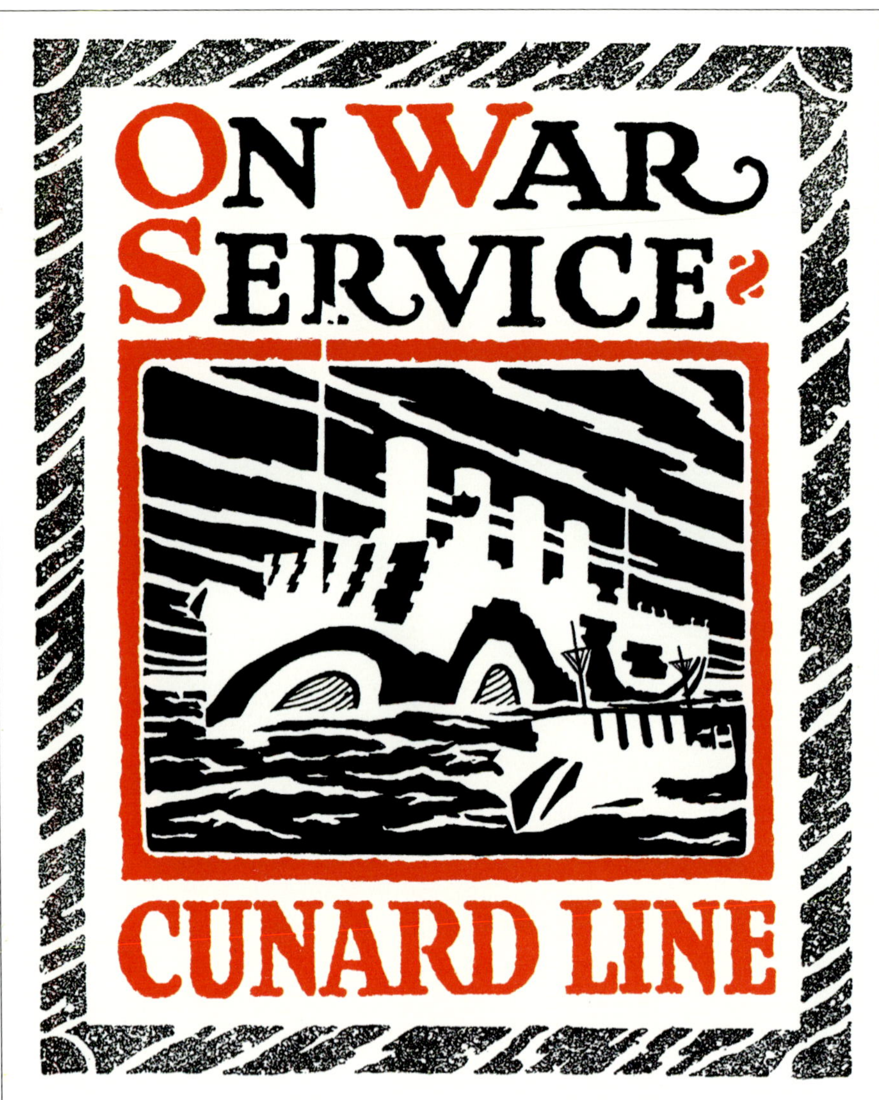

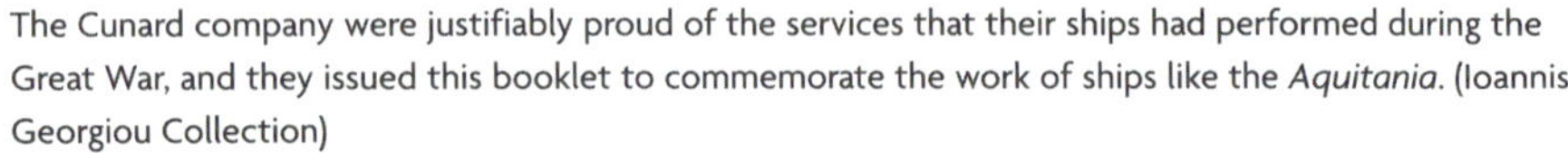

The Cunard company were justifiably proud of the services that their ships had performed during the Great War, and they issued this booklet to commemorate the work of ships like the *Aquitania*. (Ioannis Georgiou Collection)

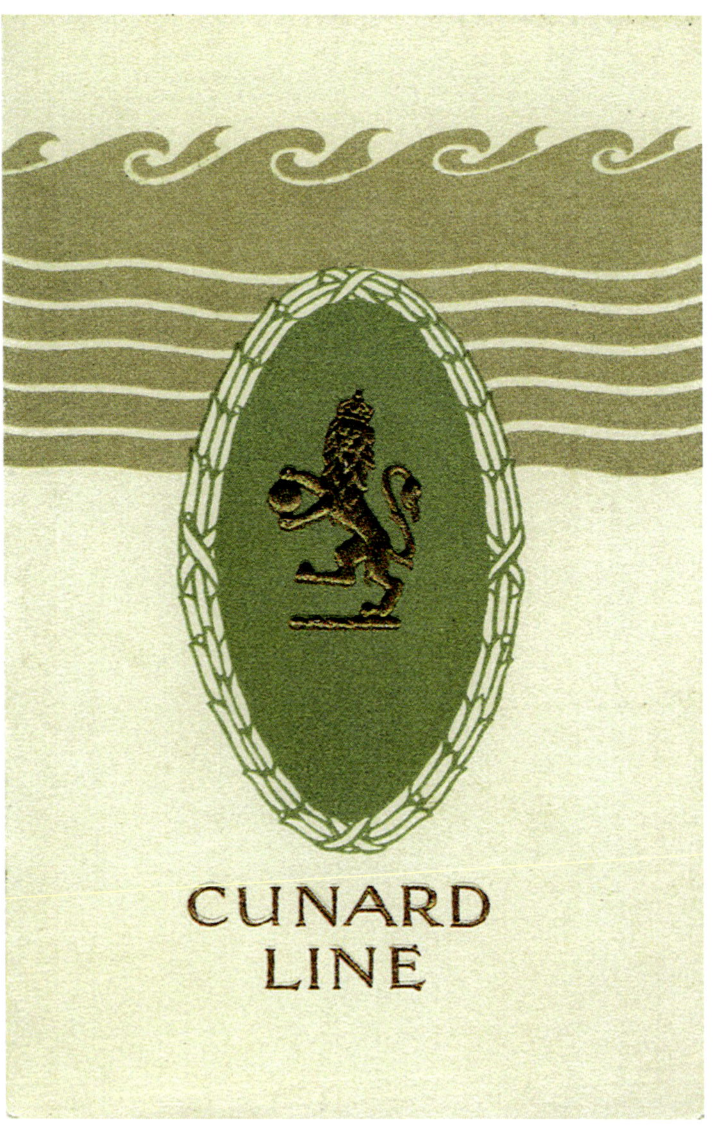

The cover of an *Aquitania* menu dating to 24 July 1922. (Mike Poirier Collection)

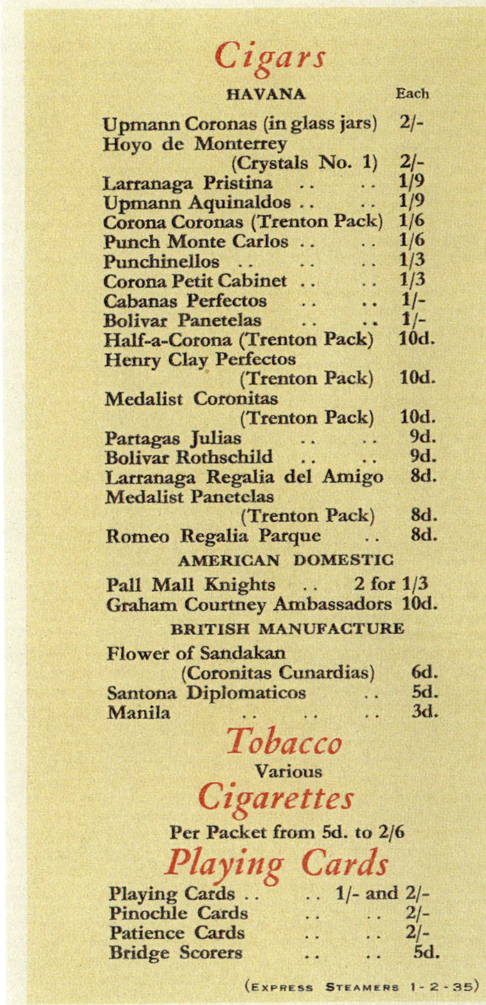

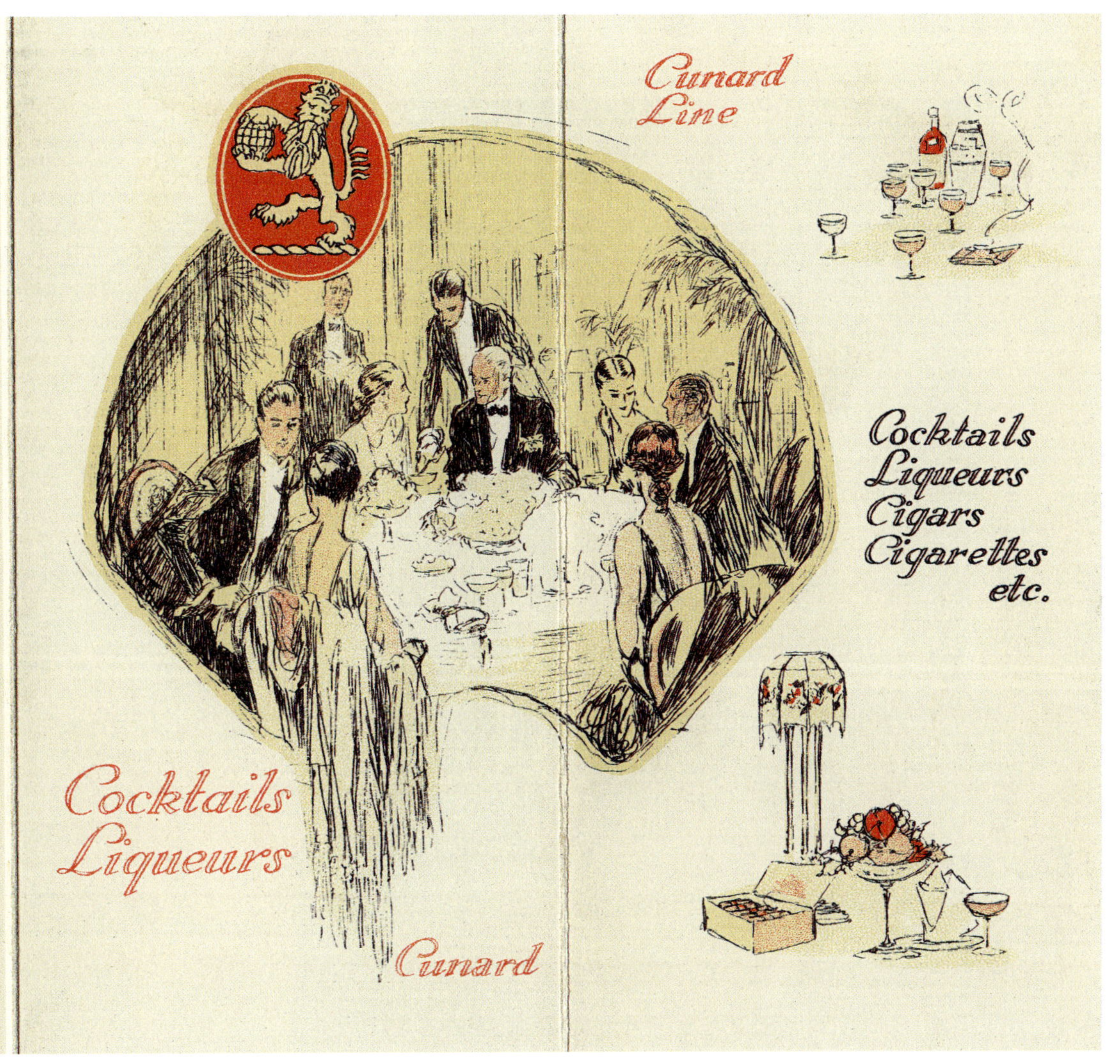

A splendid Cunard trifold cocktail list from the 1920s. On the inside, not shown, were offerings for various cocktails and aperitifs, liquors like Courvoisier Napoleon Cognac and Drambuie, and 'mixed and fancy drinks' such as various eggnogs, whisky sours, sloe gin fizz and whisky flips. On the back portion were various cigars and cigarettes from Havana, America and Britain, as well as decks of playing cards – regular, Pinochle, Patience – and Bridge scorers. (Steven B. Anderson Collection)

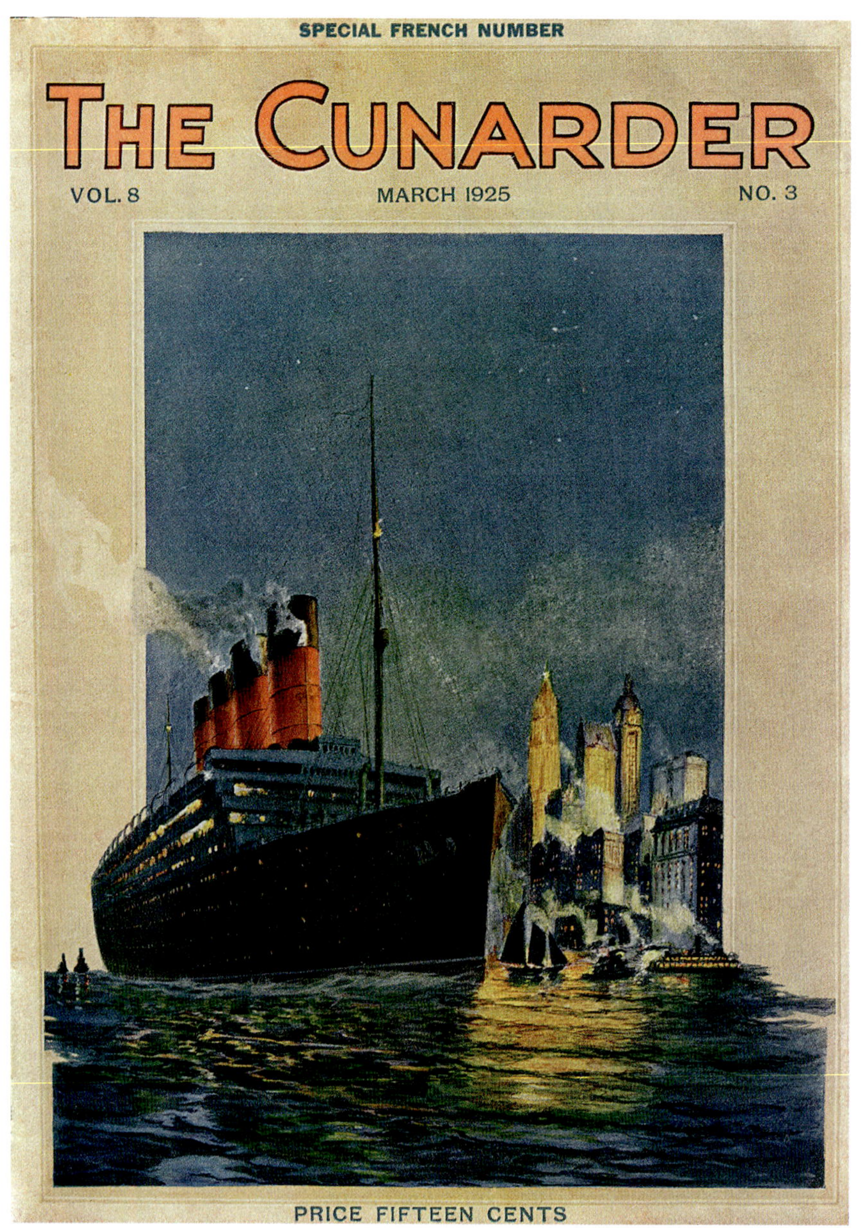

The cover of the March 1925 issue of *The Cunarder*, the company's house organ, featured this brilliant illustration of the *Aquitania* departing New York at night. (Ioannis Georgiou Collection)

ON BOARD THE CUNARD R.M.S. "AQUITANIA"

17 July / '26

My dear Dr. Ordway:

Thank you very much for your kind message. We are having a fine voyage with good weather; yesterday very cold but today perfect.

I hope you will have a pleasant summer. The only thing that worries me is that you will overwork yourself. Don't do it. Take time to get things as you want 'em. I hope soon you will have help.

My kind regards to Mrs. Ordway and my love to all my children on the mountain.

Affectionately your friend,

William Ordway, M.D.

This letter was written on official stationary from the liner on 17 July 1926. Again mostly illegible, it does point out that they were 'having a fine voyage with good weather'. (Authors' collection)

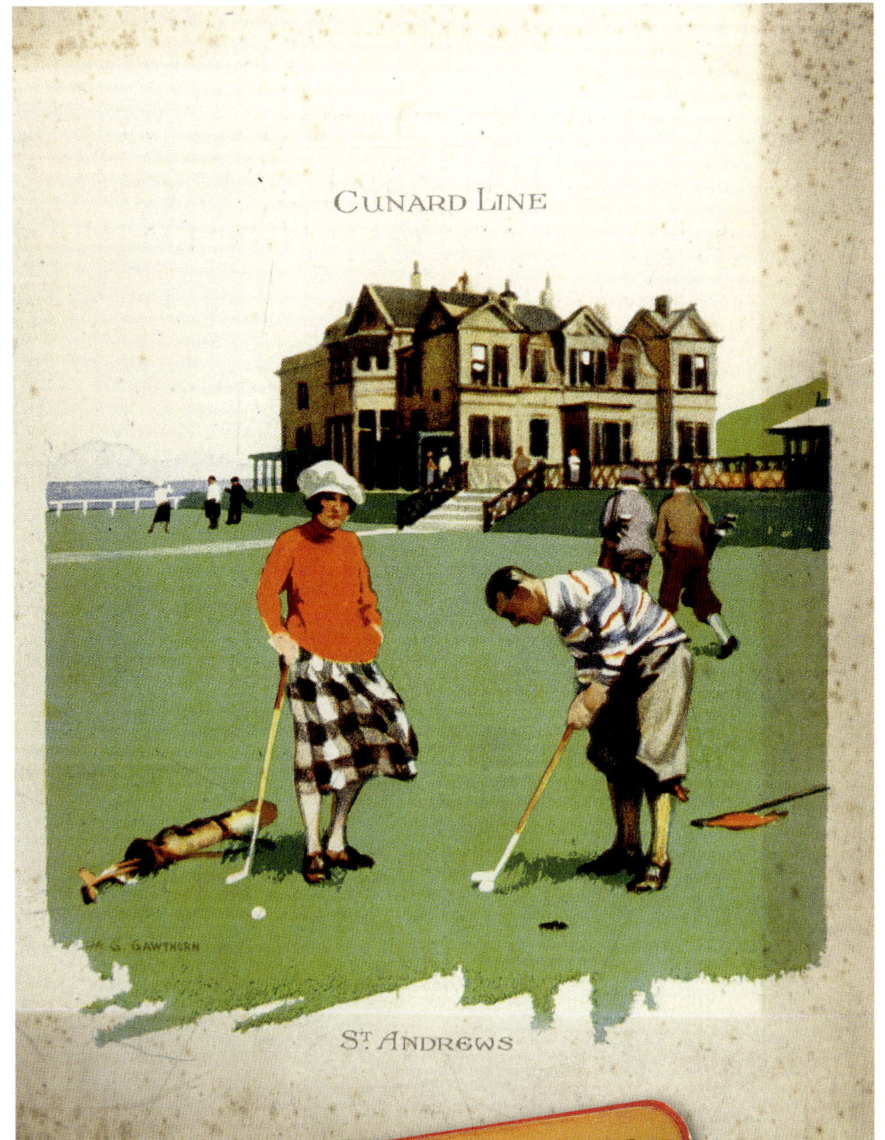

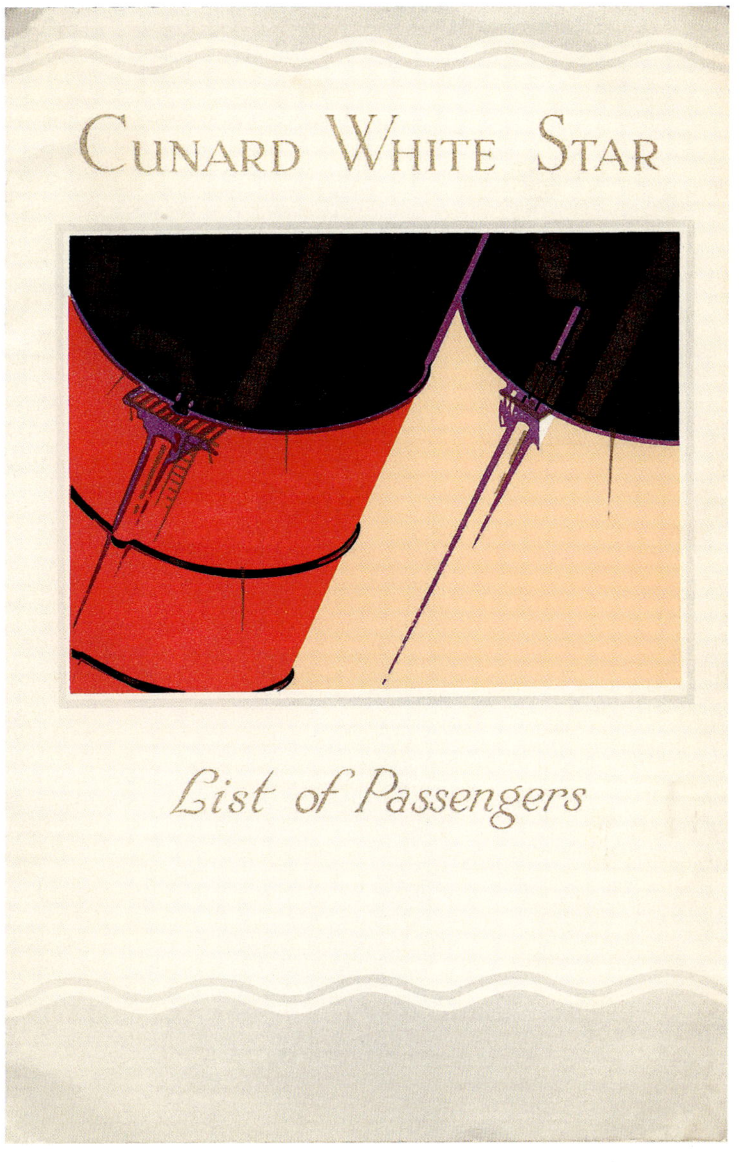

Above left: The cover of the 1928 menu seen on page 96. This colour illustration shows golfers at the famous links at St Andrews. (Beamish Museum)

Above right: After the May 1934 merger between Cunard and White Star, the *Majestic* and *Aquitania* were running mates rather than competitors. This c. 1934–5 Cunard-White Star passenger list features an *Aquitania* funnel beside a *Majestic* funnel. (Steven B. Anderson Collection)

Left: The *Aquitania* and *Majestic* appear together in this luggage tag. Since the Majestic left service in early 1936, like the menu above, this item come from a very narrow window of time. (Steven B. Anderson Collection)

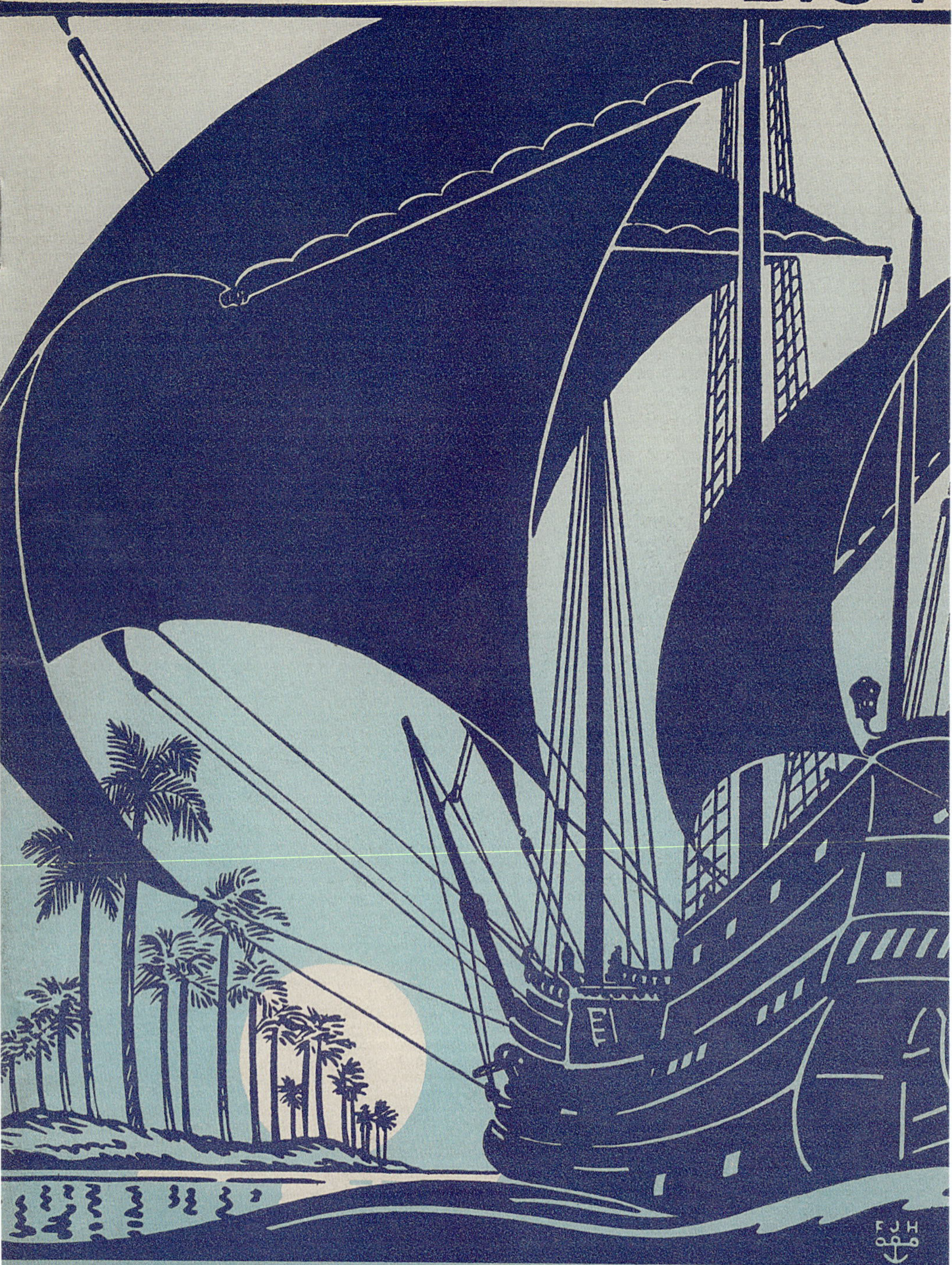

PASSENGER LIST
CUNARD WHITE STAR LINE

Opposite: This is the cover to the passenger list for the 1938 cruise to Rio de Janeiro. (Steven B. Anderson Collection)

Right: An inside page of the 1938 Rio de Janeiro cruise passenger list. (Steven B. Anderson Collection)

This postcard was sent on 5 July 1938 via the SS *Normandie*. The message on the back read: 'Helen, Having a wonderful trip encountered a gale but didn't last long, was nauseated a little ate all the meals met loads of people from coast to coast, very courteous was very glad of this. Many experiences, how is my dear one Paula give her my love and kisses, hope all is well hope to see you all soon. Love from us both, Mimi + Arlene.' (Original spelling and punctuation retained.) (Randy White Collection)

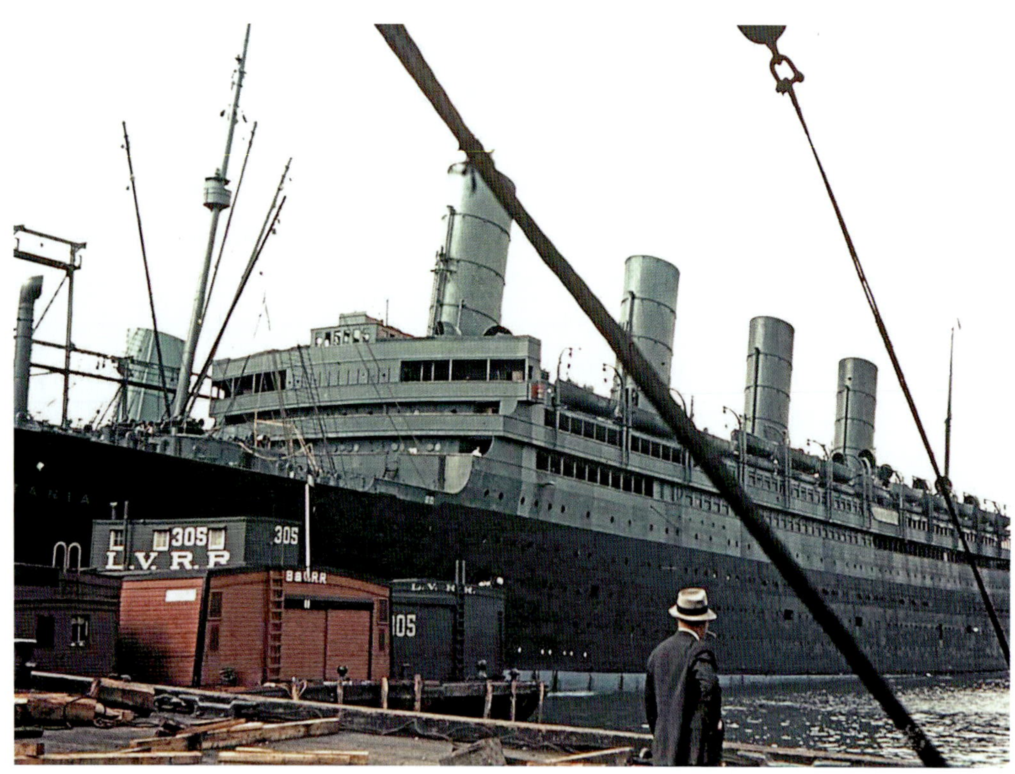

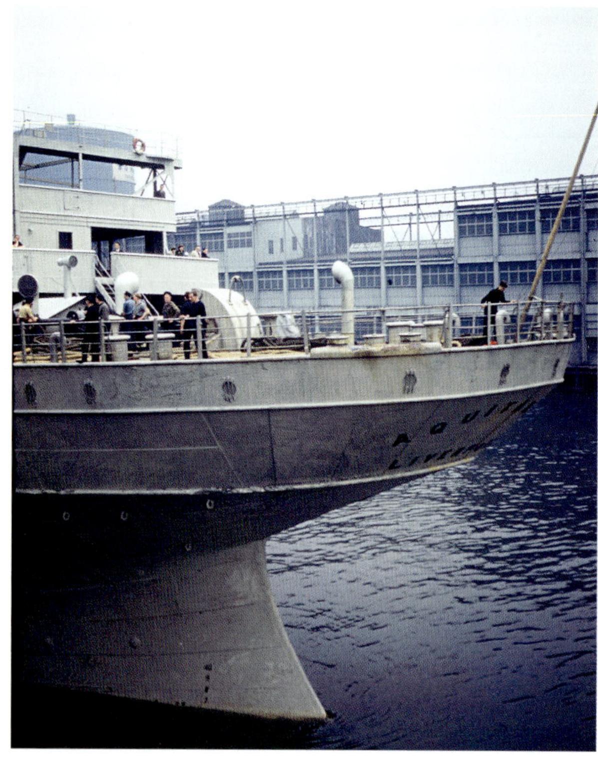

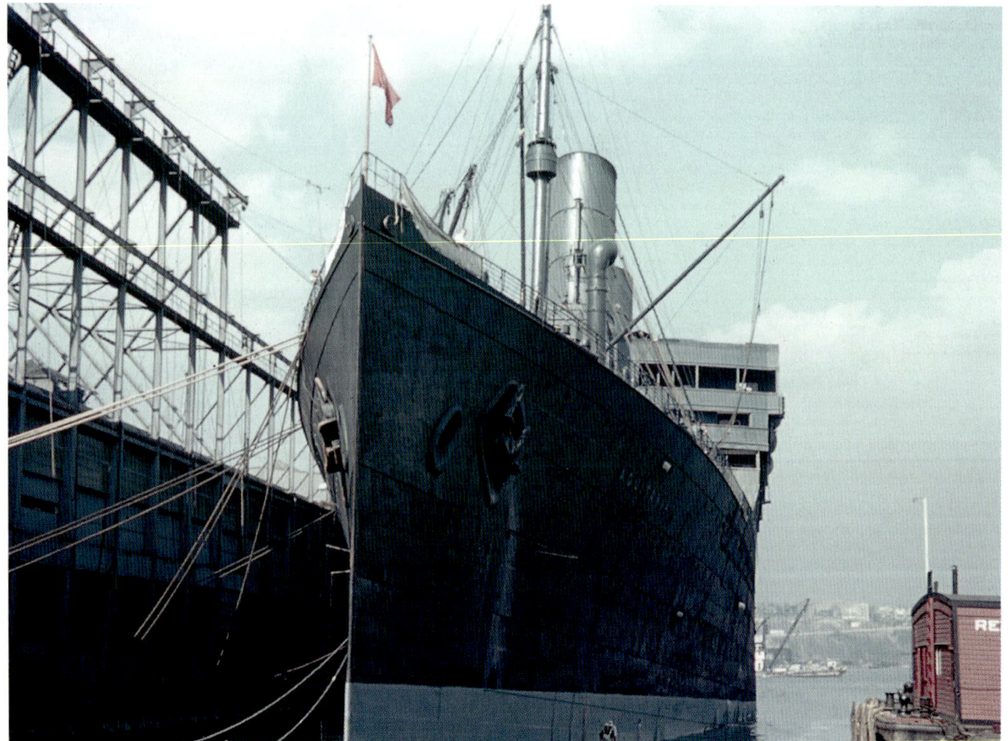

Above left: A fantastic photograph from 17 September 1939. This shows the unusual camouflage paint scheme the ship adopted right after war was declared. Notice the different shades of grey along the superstructure, as well as the blue paint of the 'boot topping' by the waterline. These and many of the photos which follow were taken by John Blake, a ship enthusiast and physician. His photographic record is a remarkable, and very rare, one. (Richard Weiss Collection)

Left: Another fantastic photograph of the *Aquitania* in New York on 17 September 1939. This shows the ship nearly bow on, and very nicely shows up the stripe of blue at the waterline. (Richard Weiss Collection)

Above right: In this 1946 photograph, the ship is in a light grey colour scheme. She is seen here at New York, and men are working on her fantail. (Richard Weiss Collection)

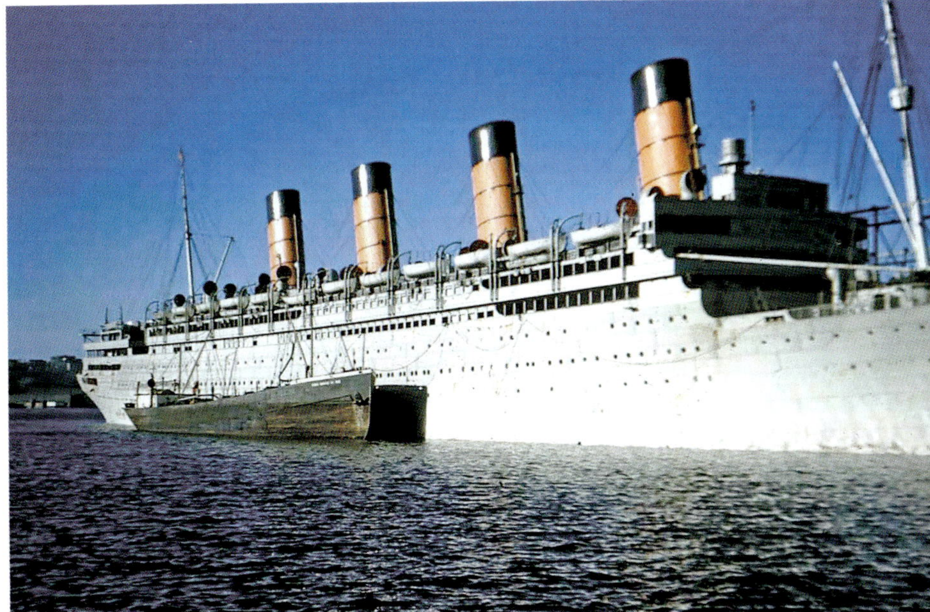

This photograph, taken in 1946, shows that the ship's funnels have been returned to their 'Cunard red' peacetime livery. This and the other photos that follow give a great idea of what 'Cunard red' looked like in the years before its more recent transition to a darker red. (Richard Weiss Collection)

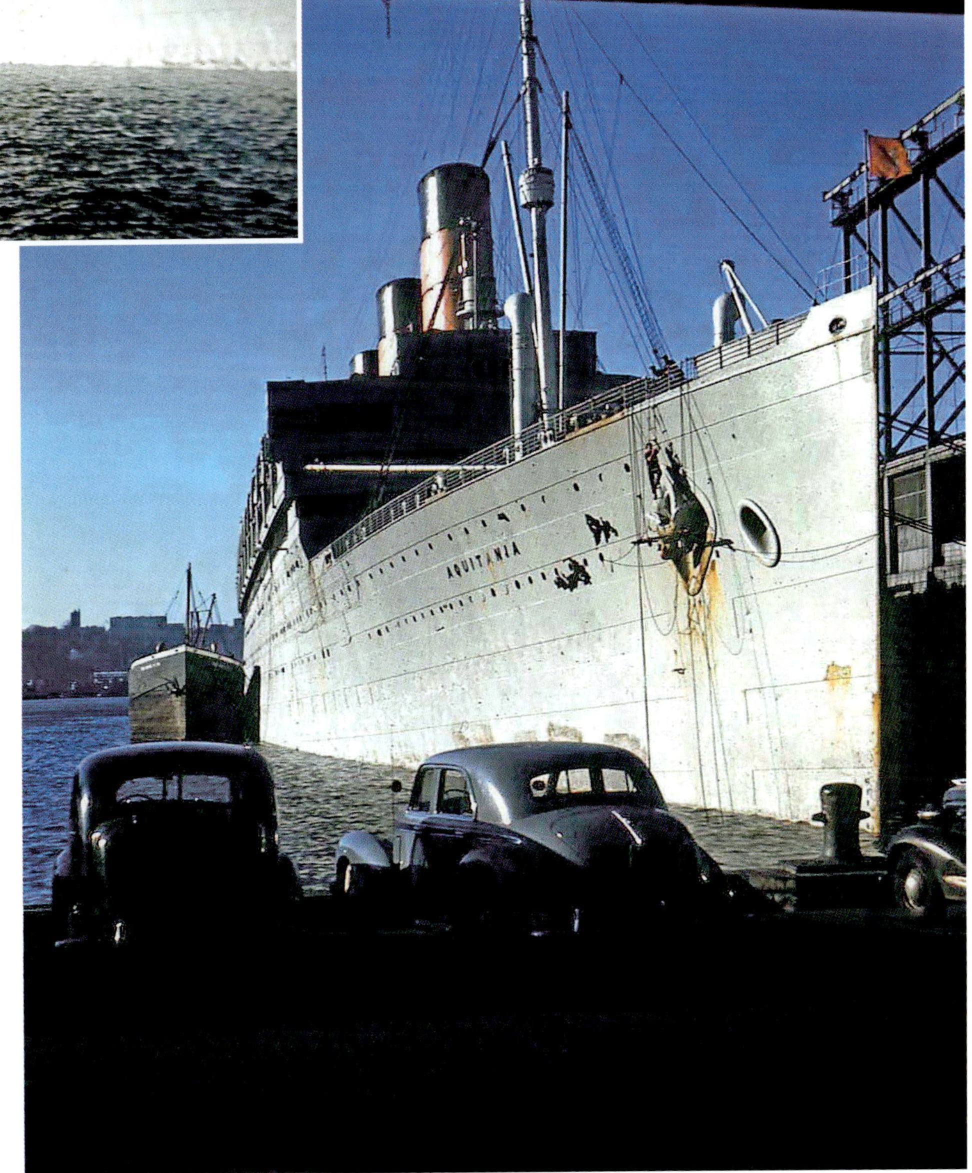

Another 1946 photo shows the *Aquitania* at her pier in New York, towering over automobiles. (Richard Weiss Collection)

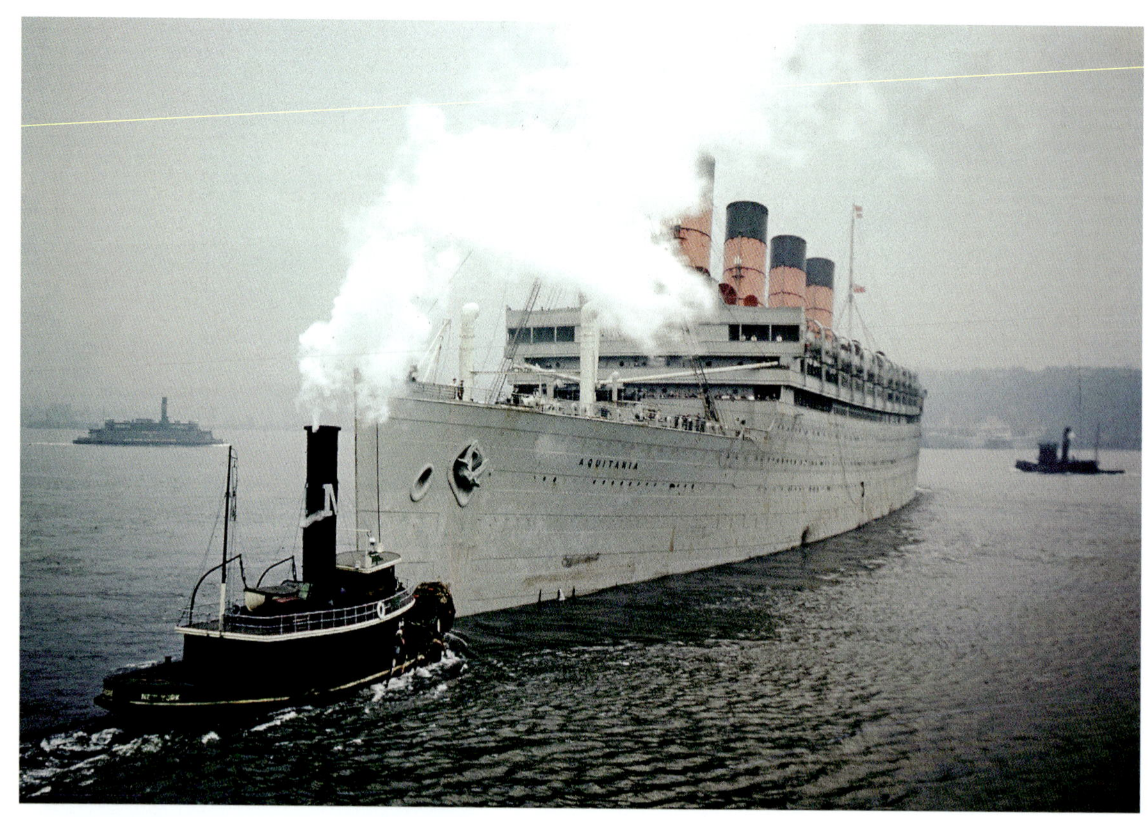

Aquitania departing port. (Richard Weiss Collection)

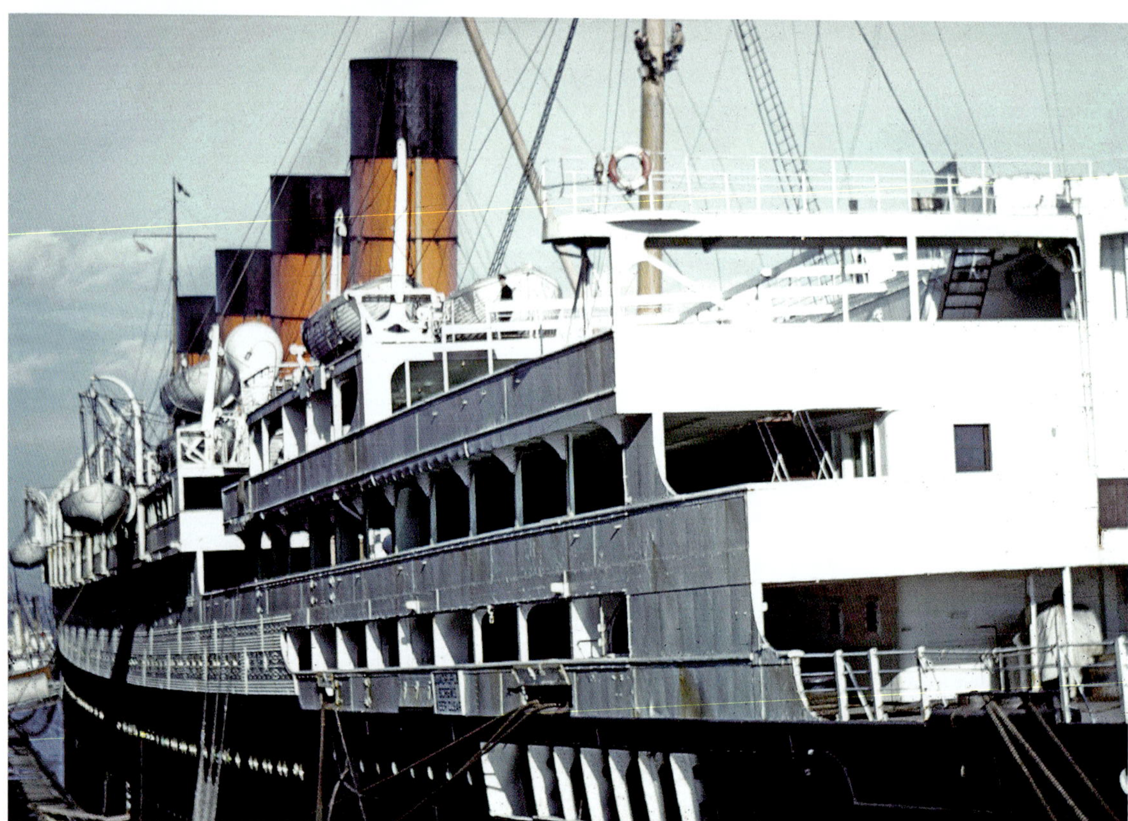

This photograph shows the *Aquitania*'s stern in Halifax, Nova Scotia, in 1949. (Richard Weiss Collection)

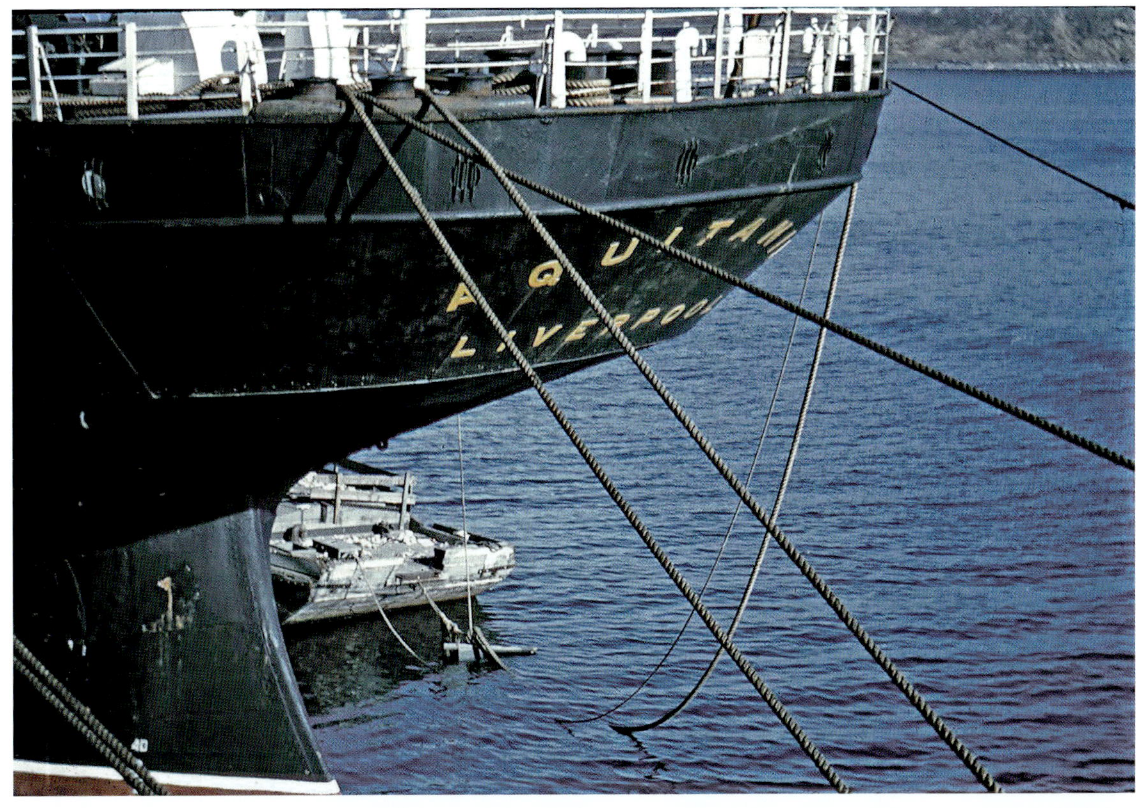

This photo, also apparently taken in Halifax in 1949, shows the ship's stern superstructure from a unique perspective. (Richard Weiss Collection)

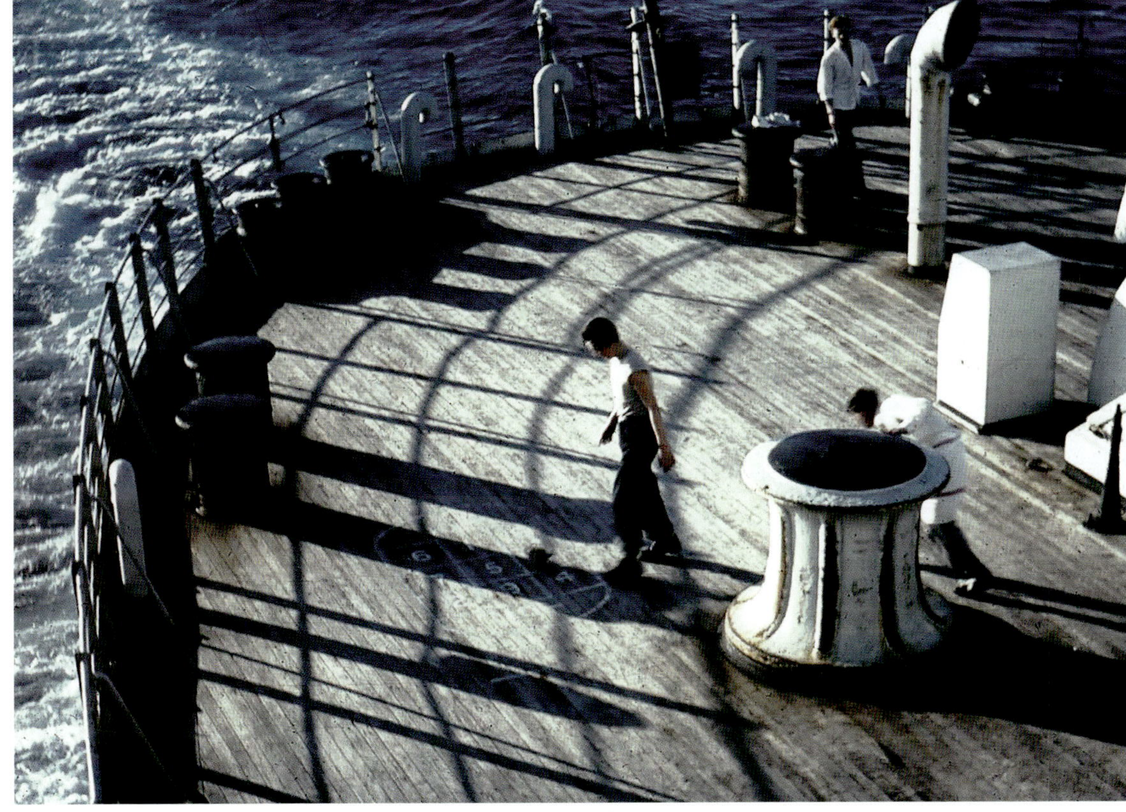

A very rare photograph of people on the *Aquitania*'s fantail during her last crossing in November 1949. (Richard Weiss Collection)

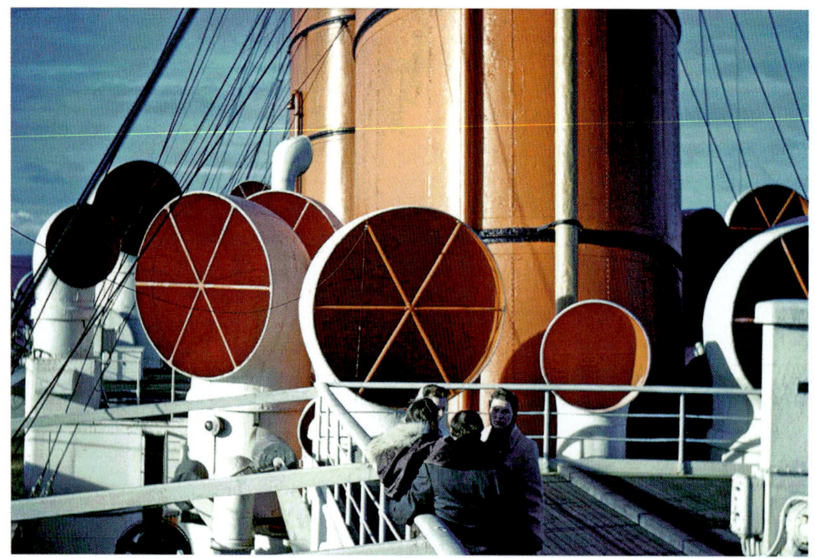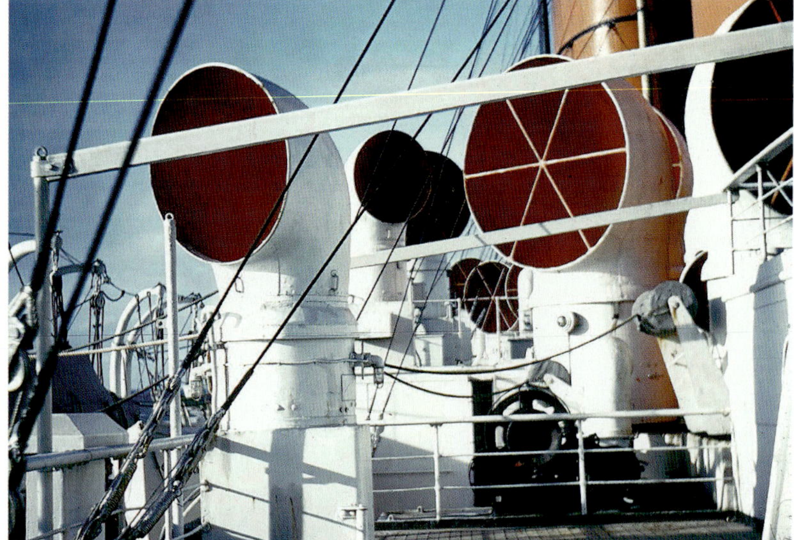

Top left: Another last-crossing photograph, this shows the funnels and vents along the *Aquitania*'s 'top of the house'. In the foreground, a quartet of girls stand not far from the bridge. (Richard Weiss Collection)

Top right: This photograph shows the vents and upper works of the *Aquitania* from a unique perspective. (Richard Weiss Collection)

Lower right: Seen from the Aft Docking Bridge looking forward, the ship's quartet of funnels gleam in the light as the *Aquitania* steams toward England for the last time. (Richard Weiss Collection)

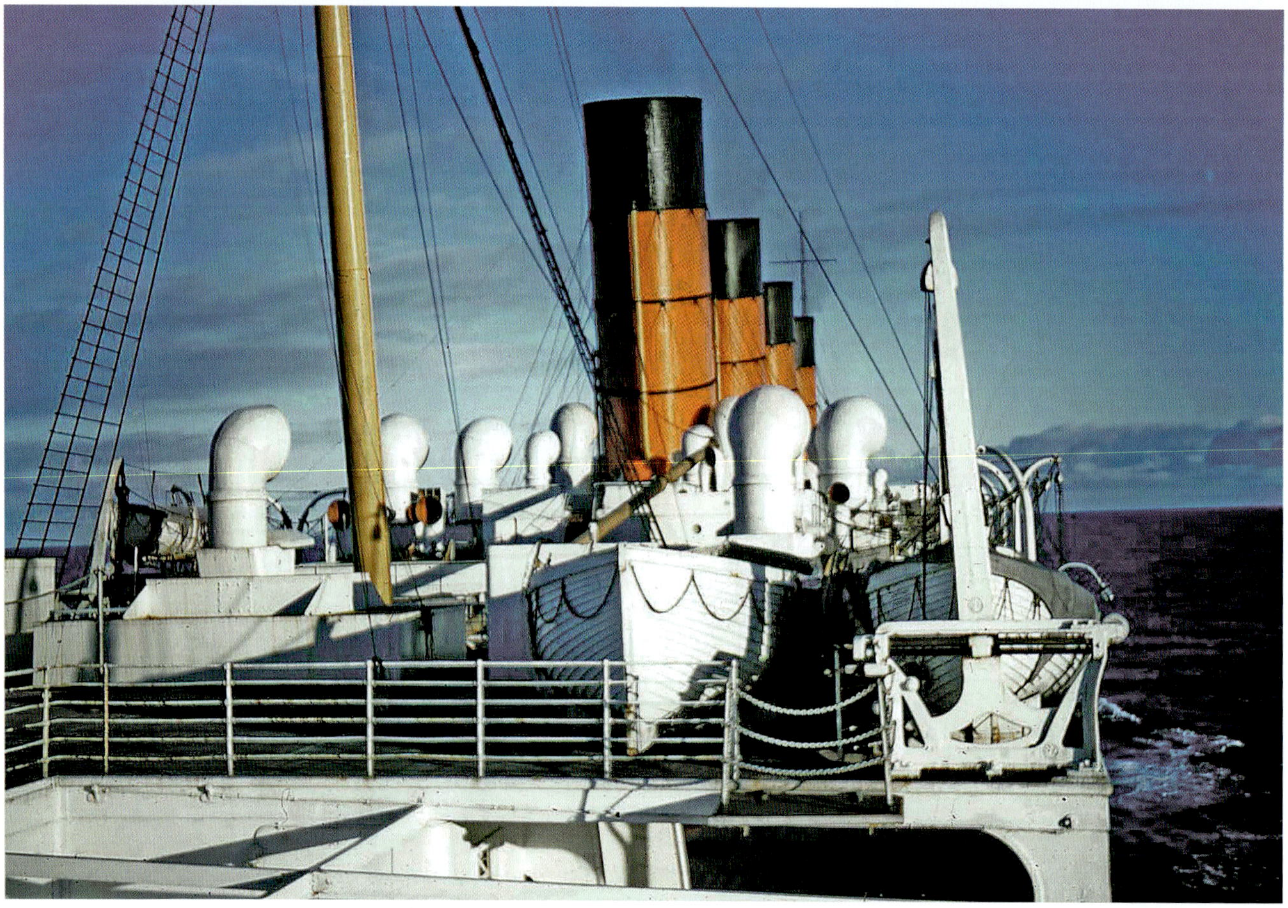

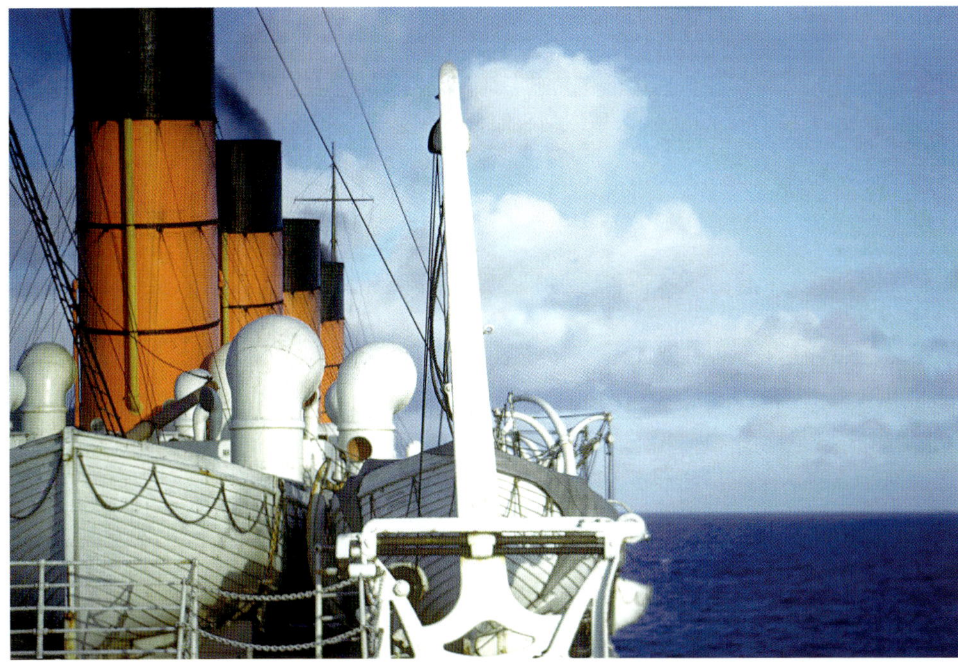

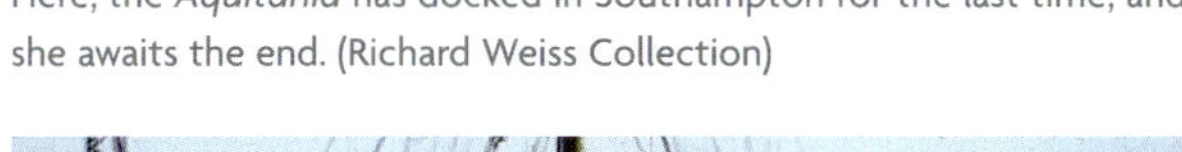

Another photo from astern looking forward as the ship steams east. Comparing this and the previous photo, notice how the subtle changes in light can make a difference in how orange or rust-tinted the funnels appear. (Richard Weiss Collection)

Here, the *Aquitania* has docked in Southampton for the last time, and she awaits the end. (Richard Weiss Collection)

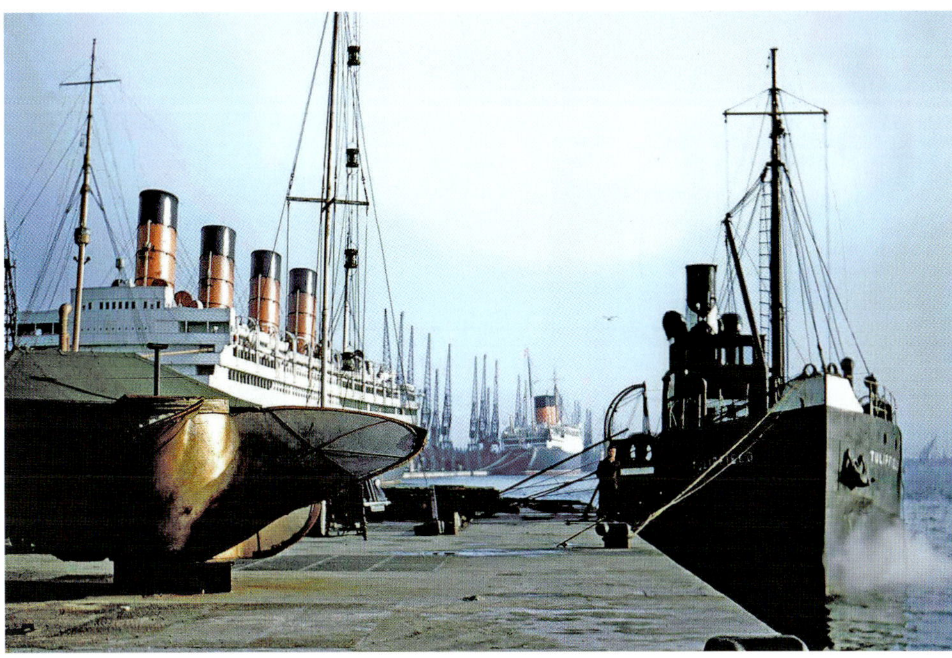

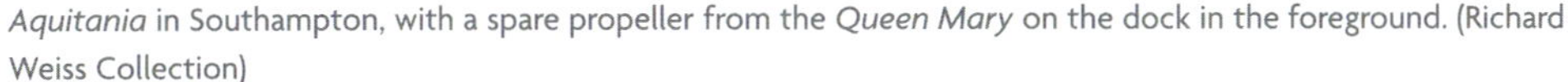

Aquitania in Southampton, with a spare propeller from the *Queen Mary* on the dock in the foreground. (Richard Weiss Collection)

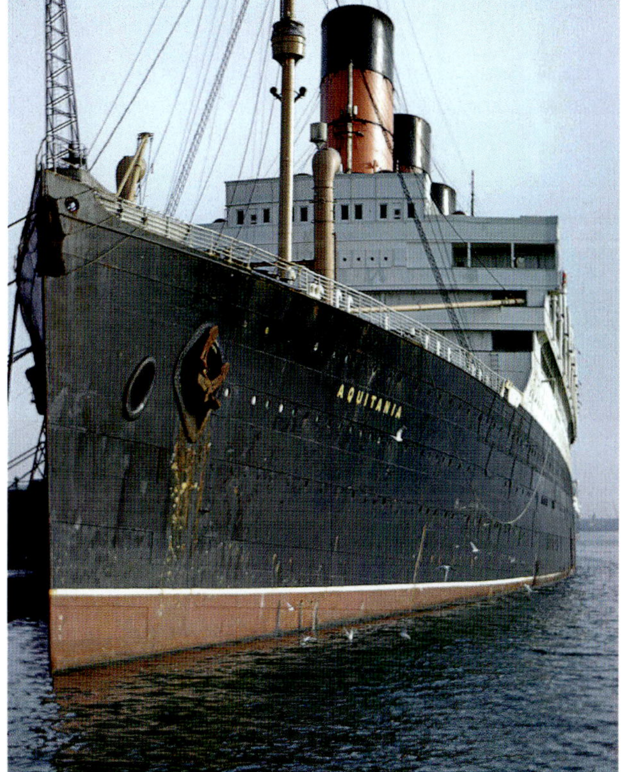

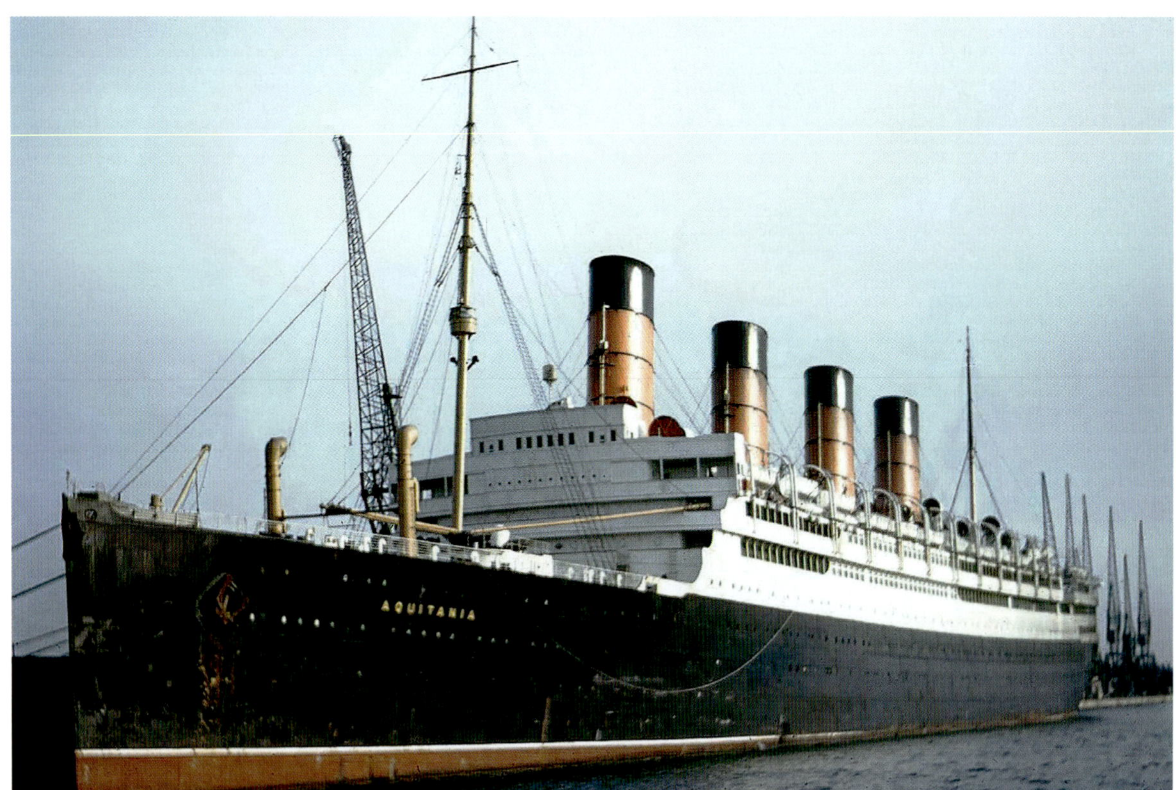

Tied up along Southampton's Western Docks, her career finished, *Aquitania*'s profile is still proud. (Richard Weiss Collection)

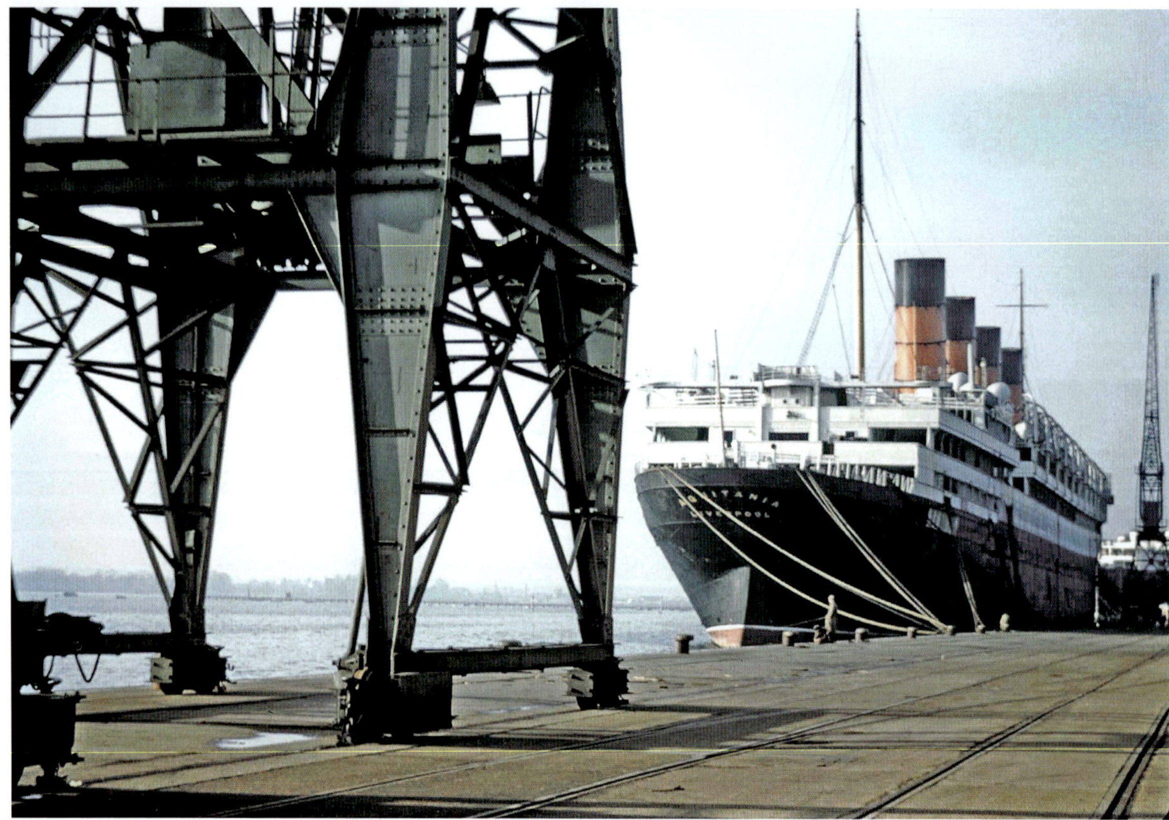

A stern view of the liner at Southampton's Western Docks, her career ended. (Richard Weiss Collection)

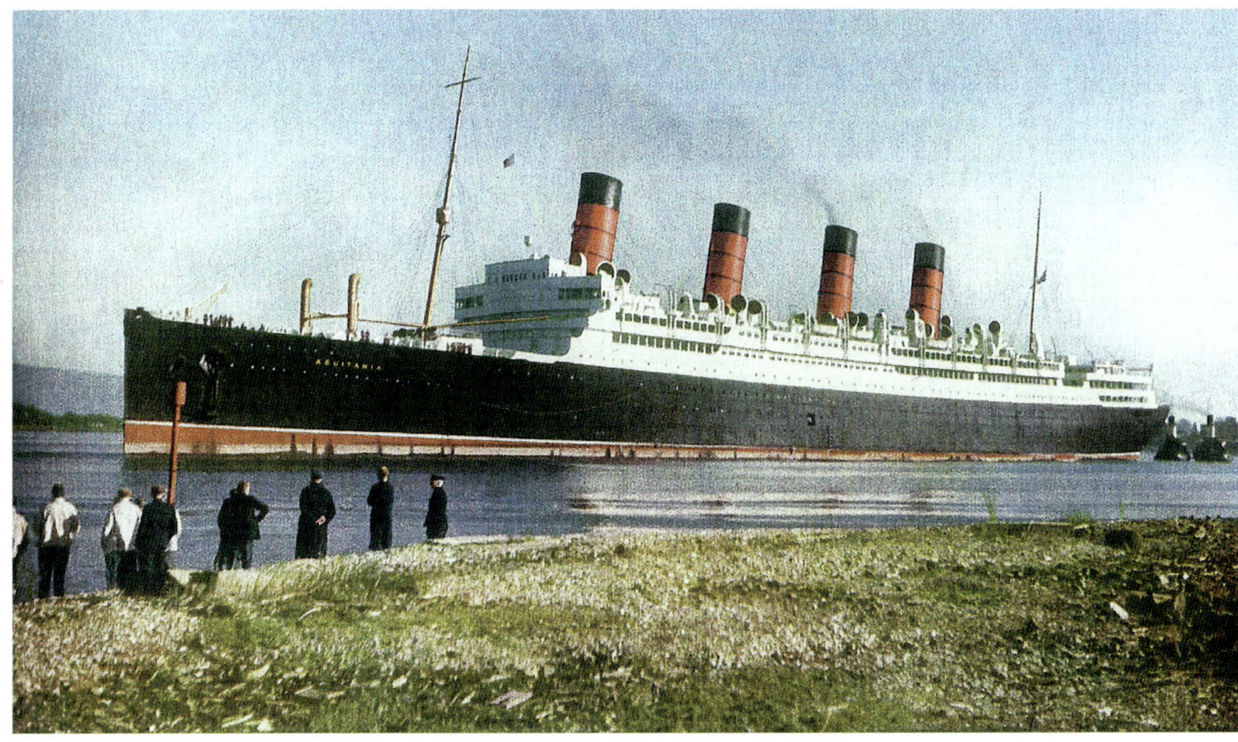

The *Aquitania* steams through the narrows of Gareloch on her way to her final destination. (Clyde George Collection)

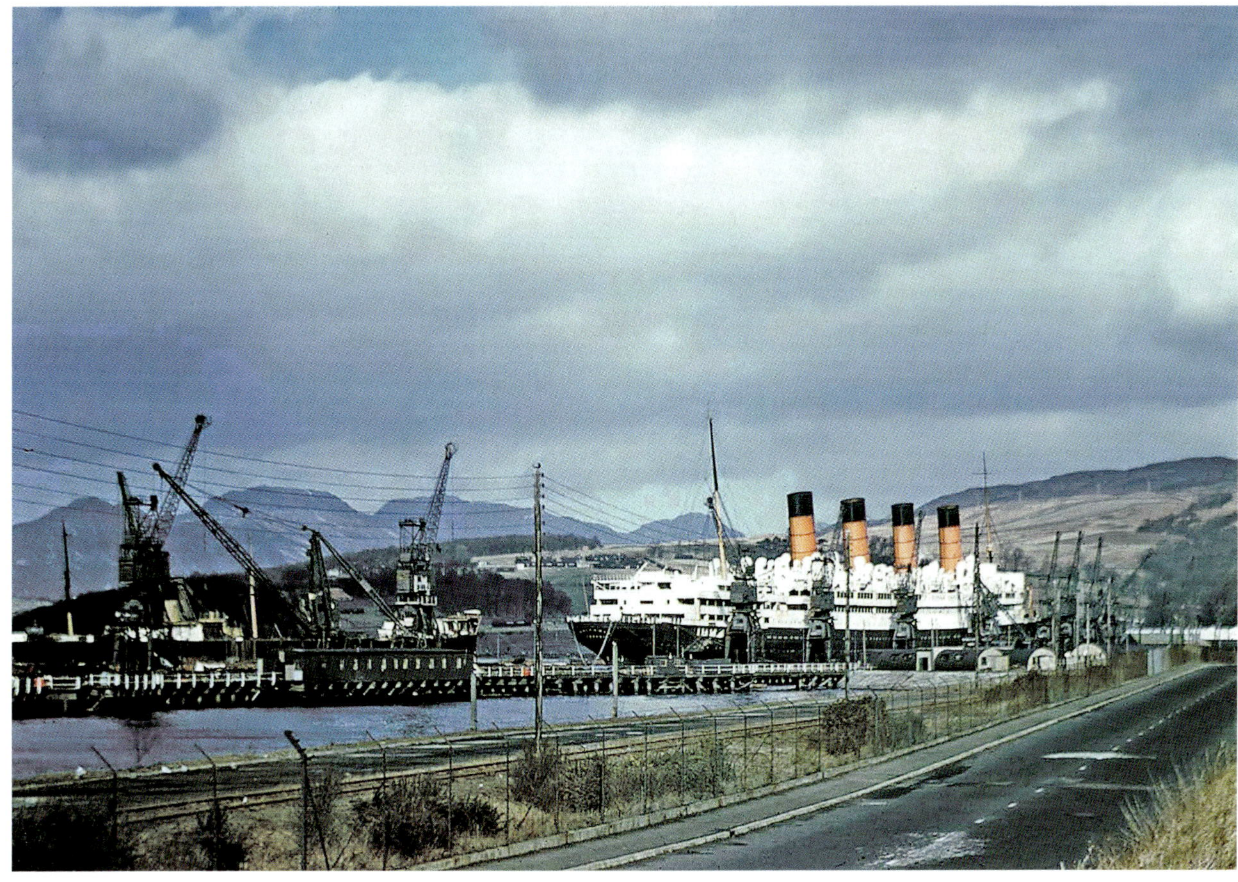

Here, the *Aquitania* has tied up for the last time, not far from where she was first 'born', and awaits demolition. (Richard Weiss Collection)

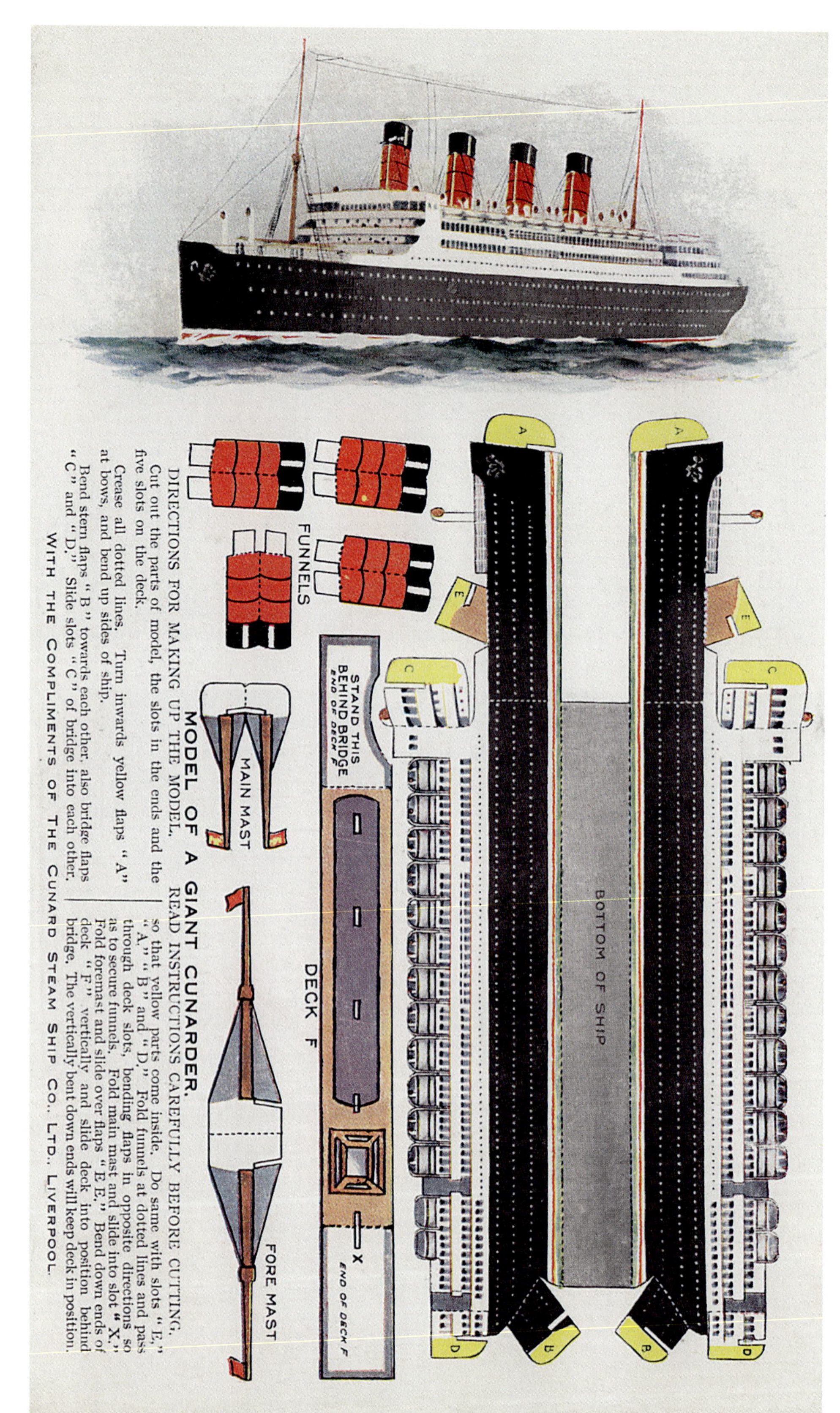

A paper model of the *Aquitania* given out by the Cunard company. (Steven B. Anderson Collection)

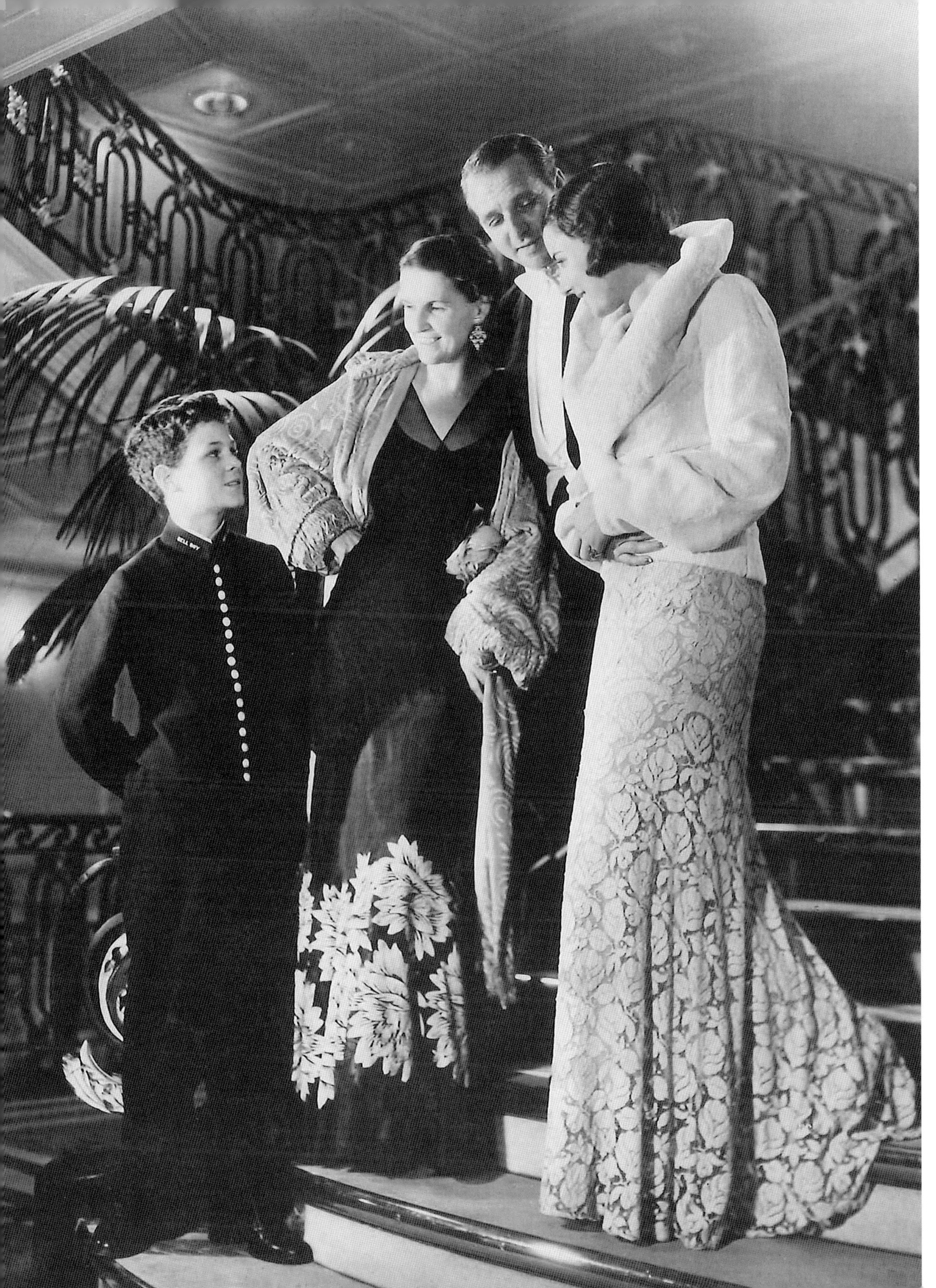

First-Class passengers on the main stairs pause to speak to a bellboy. (By courtesy of The University of Liverpool Library, Cunard Archive, D42/PR2/1/17/D133)

Not a few of the passengers left the deck carrying suspicious-looking packages, and proceeded to their cabins where, it was rumoured, 'mixtures, not of soda water, were blended in the universal language'. By the time the ship had sailed, only two of the dazzling soda concoctions had been served, the empty glasses left behind on the lounge's tables. The entire concept was an immediate failure, and the whole bar was promptly removed upon arrival at Southampton. Clearly, passengers of the day – American and otherwise – wanted liquor rather than soda. Steward Jones and Purser Spedding were openly embarrassed by the affair, and later told reporters that they were 'trying to live it down'.[123] Later on, it was permitted for British and other non-American ships to carry and serve alcohol outside of the United States' territorial 3-mile limit, as long as it was securely stored under lock and key while the ship was within American waters.

Other changes were more welcome. Accommodations were continually undergoing improvements, especially after the next generation of liners – featuring more modern interior spaces – began entering service. At the end of 1926, *Aquitania* was sent for a two-month winter overhaul. Her First-Class staterooms were redecorated, and some were enlarged into suites. In this process, the raised portion of the First-Class Promenades – which had been an innovative way, in 1914, to admit light and fresh air to inside staterooms – was removed. At the beginning of 1929, another extensive overhaul again set out to improve her passenger spaces. Over five weeks, more of her

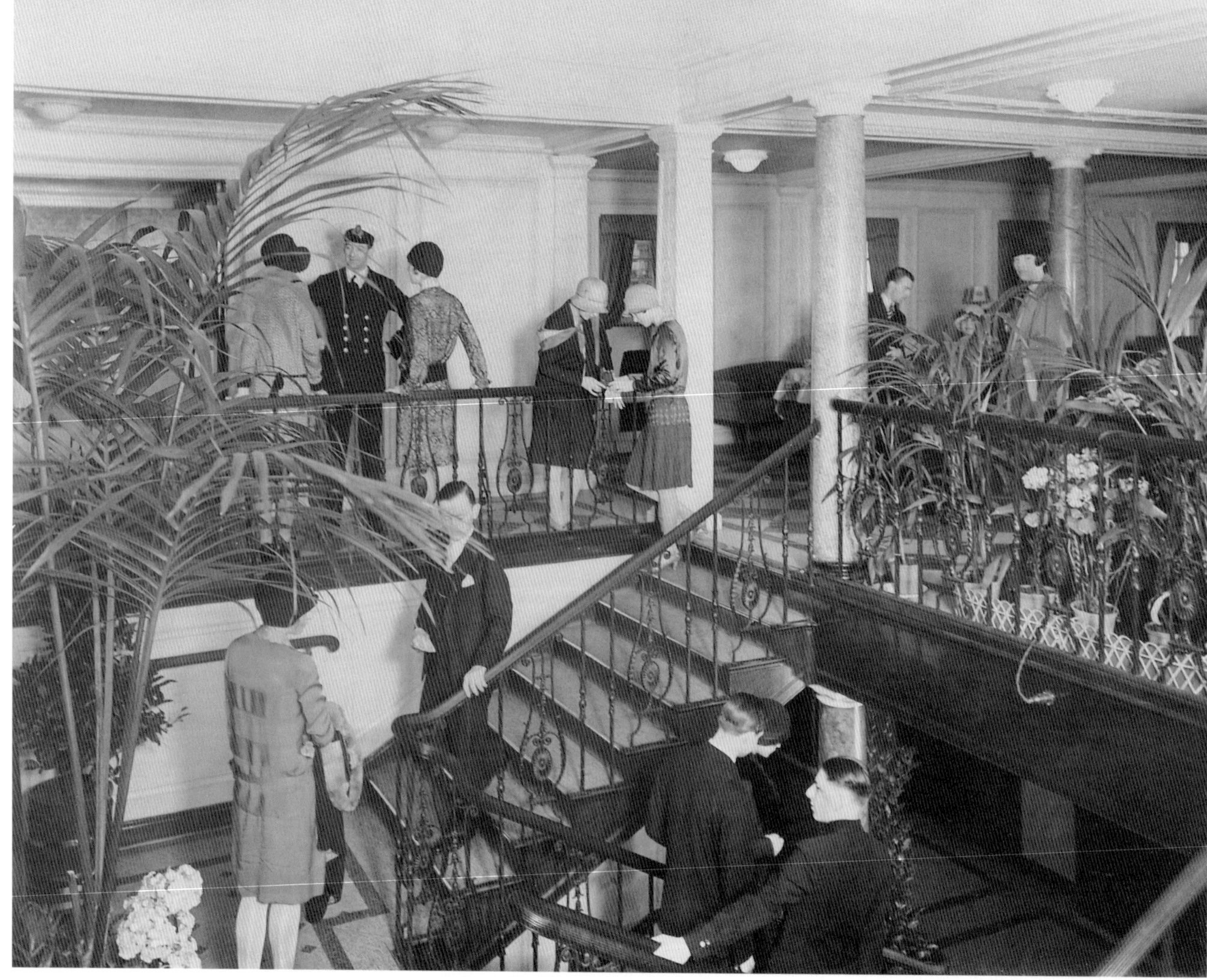

At the forward end of the Second-Class Lounge, passengers socialise with each other and with a ship's officer. At the foot of the stairs is the entry to the Second-Class Smoking Room. (By courtesy of The University of Liverpool Library, Cunard Archive, D42/PR2/1/17/D77)

cabins and staterooms in First Class areas were updated and improved; an 'American bar' was installed just off the Long Gallery; tasteful new decorating touches were given to the ship in both public and private rooms. Other passenger sections also saw improvements.

As cinema gained popularity around the world, Cunard began making arrangements for showing films aboard the *Aquitania*. Initially things were rather makeshift, but during her six-week lay-up between 1932 and 1933, the centre of the former Second-Class Dining Saloon was converted into a sound cinema with a 300sq ft stage; white panelled walls divided the cinema space from the Tourist Dining Saloon which surrounded it. The low portion of the room was used for seating, and the central well leading through to the deck above was used to make the stage appear higher. Two 'green rooms' were added behind the stage. Folding armchairs were used as seating, and could be quickly removed if passengers were going to dance instead of watch a film.[124]

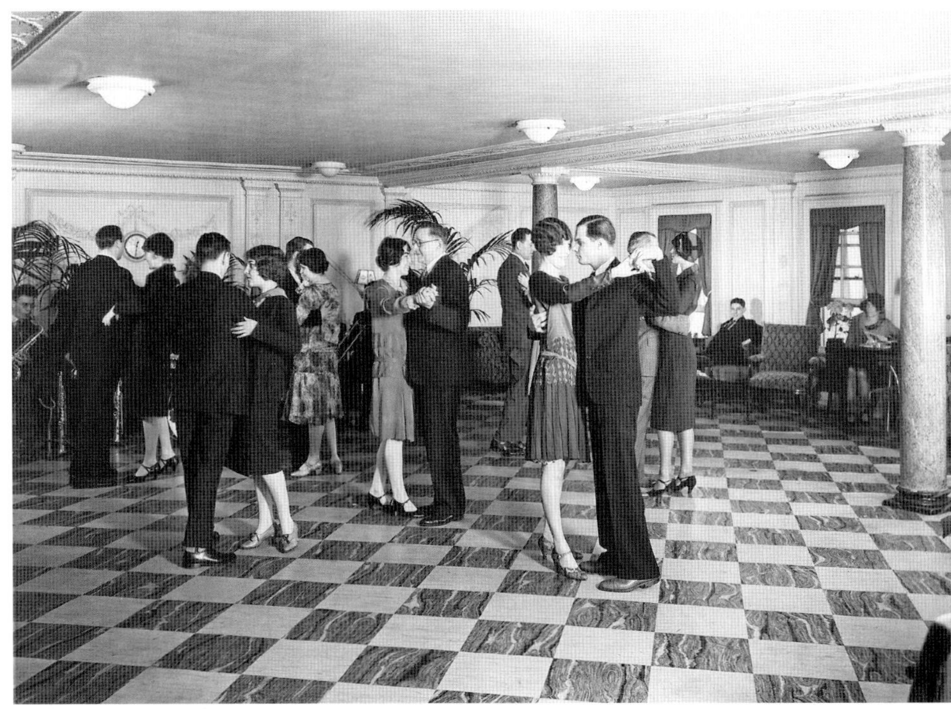

In this view, also taken in the Second-Class Lounge, the furniture has been moved in order to give passengers some room to dance. (By courtesy of The University of Liverpool Library, Cunard Archive, D42/PR2/1/17/D72)

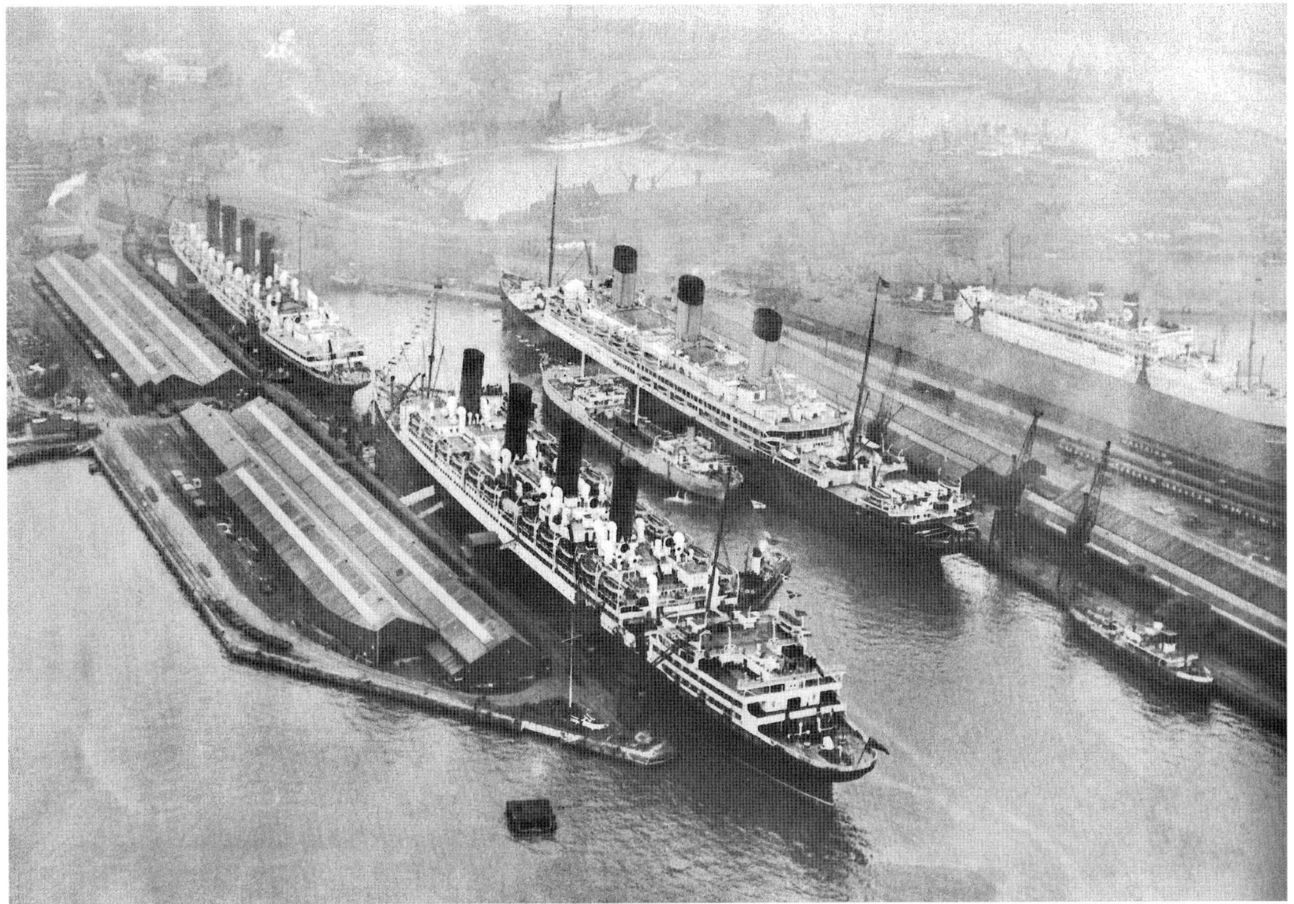

A convergence of liners. The *Mauretania*, *Aquitania*, *Majestic* and *Arandora Star* together at Southampton. (Clyde George Collection)

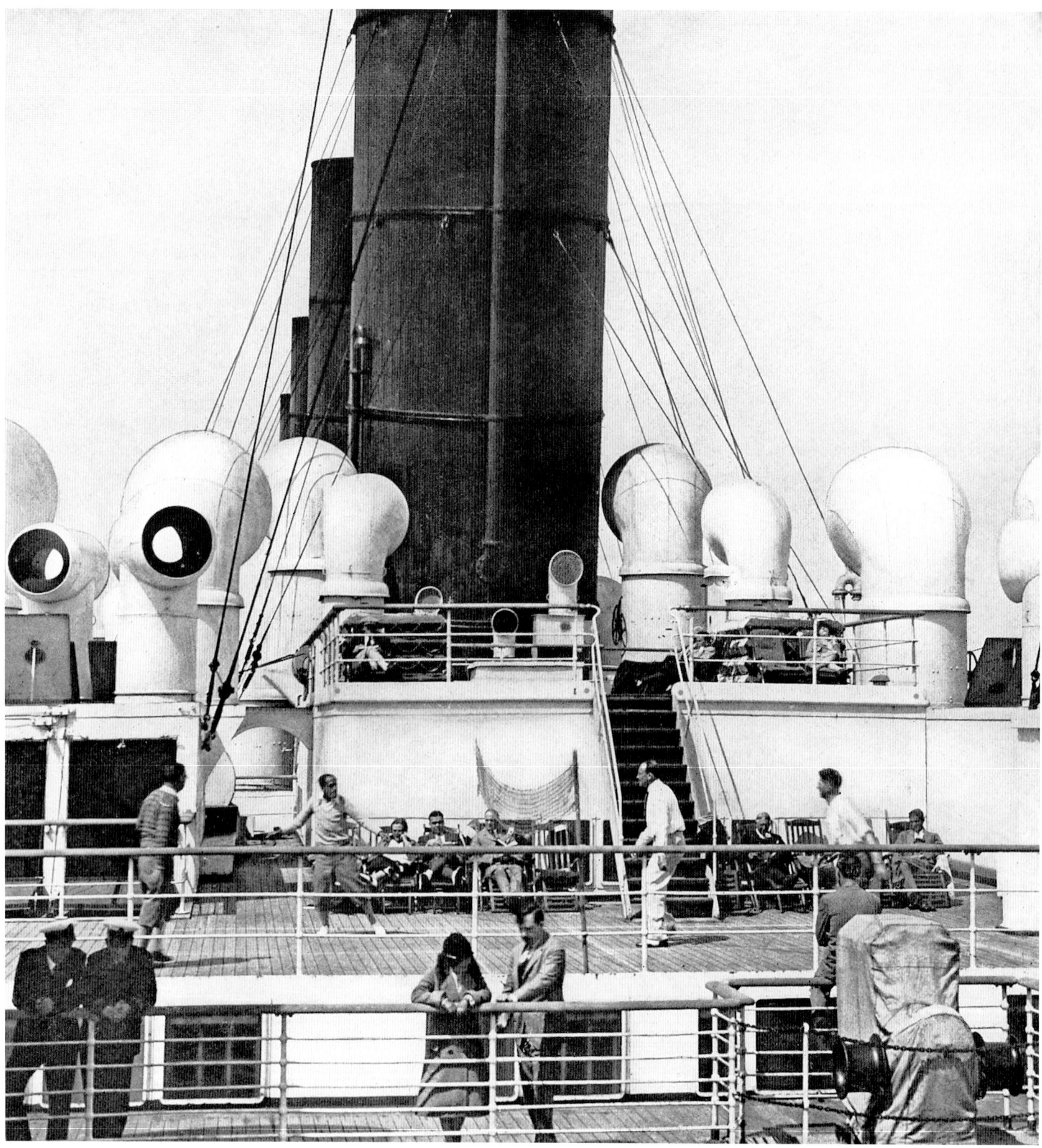

Passengers enjoy games on the after decks of the *Aquitania*. (Brian Hawley Collection)

Technical equipment was also continually upgraded. In 1930, Captain Diggle said that the *Aquitania*'s wireless was 'the most modern long-distance apparatus, capable of maintaining constant service in Great Britain, Canada and [the] United States of America, and also with vessels at sea at all times throughout the voyage'. Although her 1914 5kW Marconi set had been state of the art at the time, he said that it would 'no doubt bring a smile to the face of an amateur in these days of valve transmitters and multi-valve receivers'. Originally, the ship had carried only two wireless operators – enough to maintain a twenty-four-hour watch recognised as so important after the *Titanic* disaster; by 1930, there were eleven aboard, and Diggle believed that increasing traffic would eventually call for additional staff in the future. A central telegraph bureau had also been established on the ship.

The *Aquitania* frequently proved she was a faster ship than her original design. In June 1924, for example, she made her best day's eastbound steaming record to date: 573 miles at 25.54 knots. It was a fine performance.[125] In the opening days of August 1927, the *Aquitania*, *Olympic* and *Leviathan* left New York at the same time, bound for Southampton. Although the commanders of the three ships dismissed the idea of a race, there was keen interest in which ship would make it to England first. The ships lost sight of each other soon after departing America. The sailing was made across a sultry, unusually hot sea: the ocean's water was 85°F in the Gulf Stream, and overheated passengers who tried to use the swimming bath to cool off found it was almost as warm as the air, offering little refreshment. When all was said and done, the *Aquitania* pulled into Ocean Dock an hour ahead of either of the other vessels, proving that she could still turn in a strong performance from her turbine engines. It was the first time that three of the so-called 'Big Six' liners were seen in that port together.[126]

On 5 January 1928, while in New York, Captain Sir James Thomas Walter Charles announced that he would retire when the *Aquitania* finished that voyage and arrived back in Southampton; his place as captain of the liner was to be filled by Captain E.G. Diggle. However, Charles remained on Cunard's lists as a relief captain until he turned 63 on 2 August, and reached the mandatory retirement age. After handing the reins over to Captain Diggle, he made two 'relief' voyages, one on the *Berengaria* and another on the *Mauretania*.

Then Captain Charles made one last round-trip voyage in the *Aquitania*, signing aboard on 27 June 1928. Before leaving, he received many accolades from fellow officers; his cabin was said to have been filled with flowers sent by friends. Although he said he was looking forward to retiring, some who knew him said that he 'loathed the prospect'. As the ship departed Southampton waters, there was a strange incident: Staff Captain G.R. Dolphin noticed a buoy close to the ship's starboard bow; when the ship anchored in Cherbourg, Dolphin saw the same buoy in the same place. The French pilot said that it was an English buoy; it turned out that the ship had snagged a new buoy marking an electric cable, and had dragged it and its 100-fathom chain across the English Channel from Spithead.

In New York, at 11 a.m. on 8 July, just before leaving port, all 873 members of the crew mustered in the Louis XVI Restaurant. They gave Commodore Charles parting gifts, including a walnut leather-topped writing desk with matching chair, and a silver plaque of the ship, depicting her under full steam. Chief Engineer Llewellyn Roberts made the presentation, and spoke of the affection that the crew had for the legendary commander. Captain Charles replied that he felt he had practically become part of the ship, and that retirement would be 'a severe wrench', but that he was looking forward to a rest. He told passengers that he had not realised how difficult it would be to part from his old shipmates, and wrote a farewell letter as the ship steamed east.

During the return leg of the voyage, headed back for Cherbourg and Southampton, the weather was splendid. Captain Charles remained on the bridge almost continually for the last two days. His junior officers tried to persuade him to get some rest and let Staff Captain Dolphin bring the ship into Cherbourg; Chief Surgeon Dr Sydney Jones was called up, and joined in a growing chorus of concern. Yet Charles refused to retire.

Reaching Cherbourg, *Aquitania* dropped anchor and the harbour pilot came aboard; only then did Captain Charles leave the Bridge.[127] A few minutes later, he rang the bell to summon assistance; two officers found him

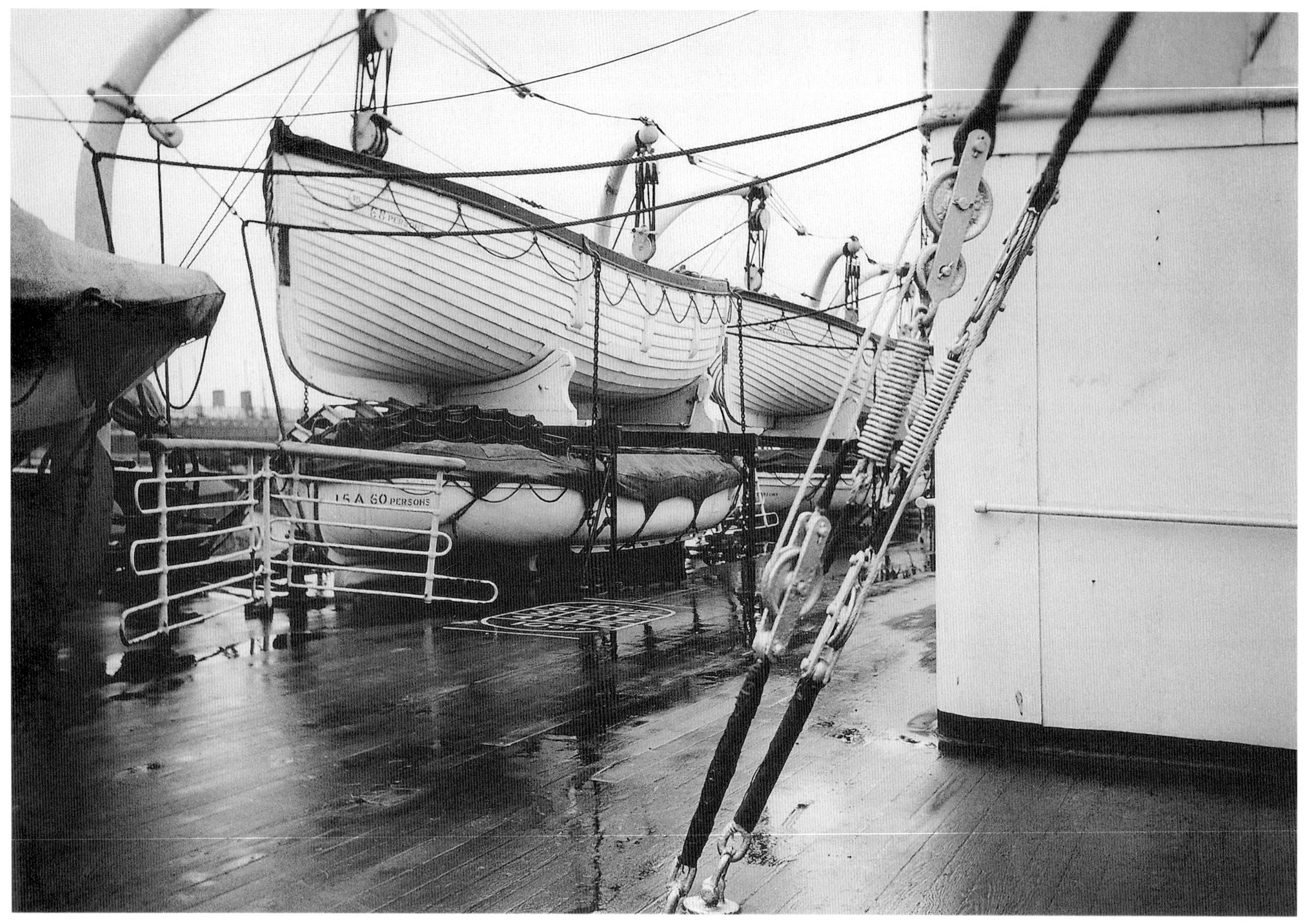

A line of lifeboats on the starboard side of the Boat Deck, leading aft from the Wheelhouse. (Ioannis Georgiou Collection)

collapsed and in great pain, apparently suffering an internal haemorrhage. Dr Jones tried to medicate and help him as the ship raced across the English Channel back to Southampton; telegrams were sent to Mrs Charles, and when the ship docked, she was waiting at the quay, but he could not recognise her. He was taken away in an ambulance, but died shortly thereafter. Back at the dock, the commodore's flag was lowered to half-mast, announcing his death. It was said that a 'hush fell over the ship and the quayside'. Commodore Sir James Thomas Walter Charles had finished his 728th voyage across the Atlantic, completing a career at sea spanning forty-eight years. It was one of the most dramatic ends to a captain's career in the history of the great liners.[128]

As 1929 was drawing to a close, the *Aquitania* had built a successful fifteen-year career, escaping from many dangers and surviving many changes. Yet some of her greatest challenges lay ahead.

STRUGGLE FOR SURVIVAL

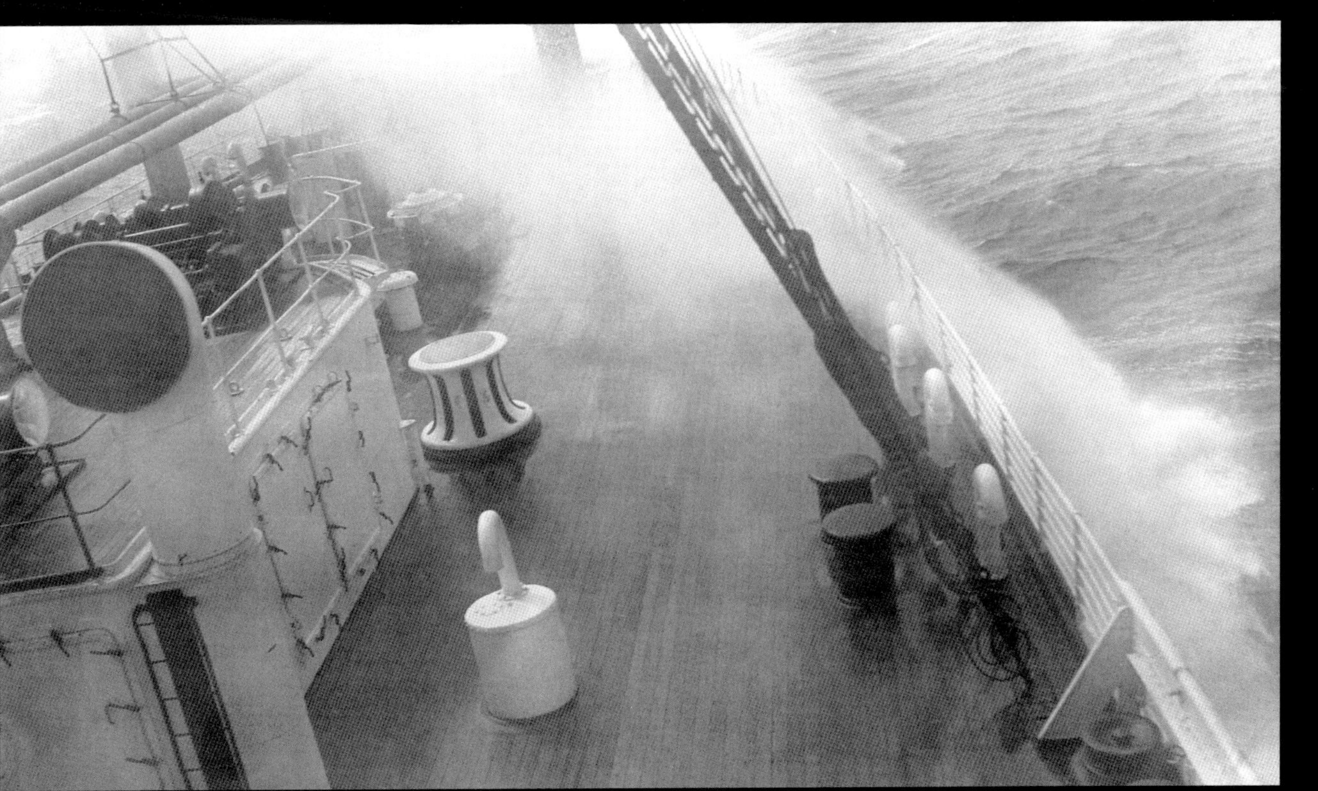

November 1930: yet again the *Aquitania* finds herself in a battle against the North Atlantic. This time, however, she is also in a fight for her life against a world economy on the brink of oblivion. (Ioannis Georgiou Collection)

The stock market crash in late 1929, and the ensuing Great Depression, could not have come at a worse time for the older liners. Newer ships were attracting passengers with greater comforts, amenities and more modern styles of decor. Some of the older ships were moving into the upper limit of their intended lifespan, and were requiring more and more expensive maintenance in order to remain in service. It was a struggle for survival that most of the Edwardian-era liners would lose: the 1930s saw the *Mauretania*, *Olympic*, *Homeric*, *Majestic* and others retired and sold for scrap.

For the foreseeable future, however, Cunard had use for the *Aquitania*; she was still a fine ship, and was an exciting assignment for youngsters eager to learn the ways of the sea. In 1936, two young 'Bridge Boys' were serving aboard. One of them was John 'Boney' Le Touzel, who was the son of *Lusitania* survivor Sidney Le Touzel. John later recalled that the First-Class Restaurant 'looked like a grand place, better than a 1st class hotel'. His duties were not quite as glorious:

My duty as a Bridge Boy ... was to keep the bridge clean and polish the brass fittings (a lot of the fittings in those days.) Get the ship's master his meals[,] also help the Pilot come on board by throwing a rope ladder over the side from the main deck and pull his bag up on a rope ... also bring the Pilot a meal from the galley (1st class) below decks.

We slept down also on the waterline, about 12 men in what was known as the 'glory hole' bunks one on top of [the] other. We did not have a mess room and got our food when in the first class kitchen/galley – standing up. As conditions where [*sic*: were] in those days, they would not be allowed today. Our pay was £2. per month, 10 shillings a week ... Work was hard but we were treated quite well by the officers and other crew.

While at sea we received instruction in order to qualify for Junior Officer Certificate when old enough. At times we were allowed to take the wheel over from the Quartermaster who steered the ship, reading the compass. If our wake wasn't straight we got told off. We also took and

lowered a life boat when in harbour, to get our life boat certificate.[129]

Although the *Aquitania*'s career continued, as the Great Depression began to draw on, Cunard needed to find new ways of earning revenue with her. They had already found great success in cruises to tropical ports – some short weekend getaways, others much longer – in the Mediterranean and West Indies with the older *Mauretania* and other ships. It was decided that the *Aquitania* would also try to turn a profit by making these cruises.

On 7 April 1931, it was announced that the *Aquitania* would make her first cruise from New York that spring. Initially, it was planned as a 'cruise to nowhere', in order to

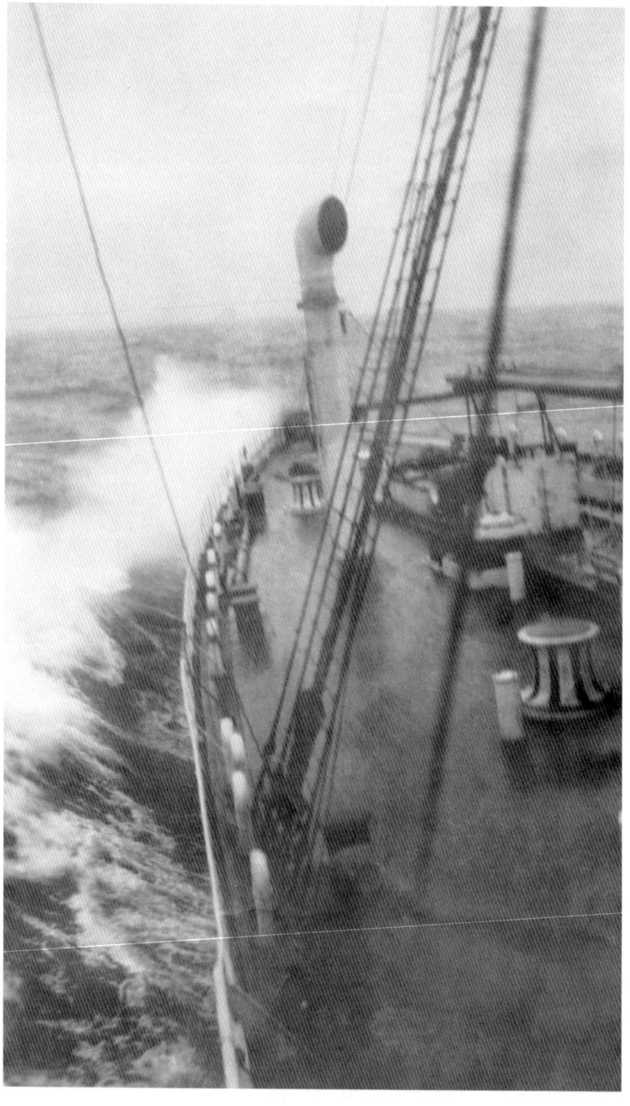

The prow of the *Aquitania* crashes through a heavy swell during a storm. (Ioannis Georgiou Collection)

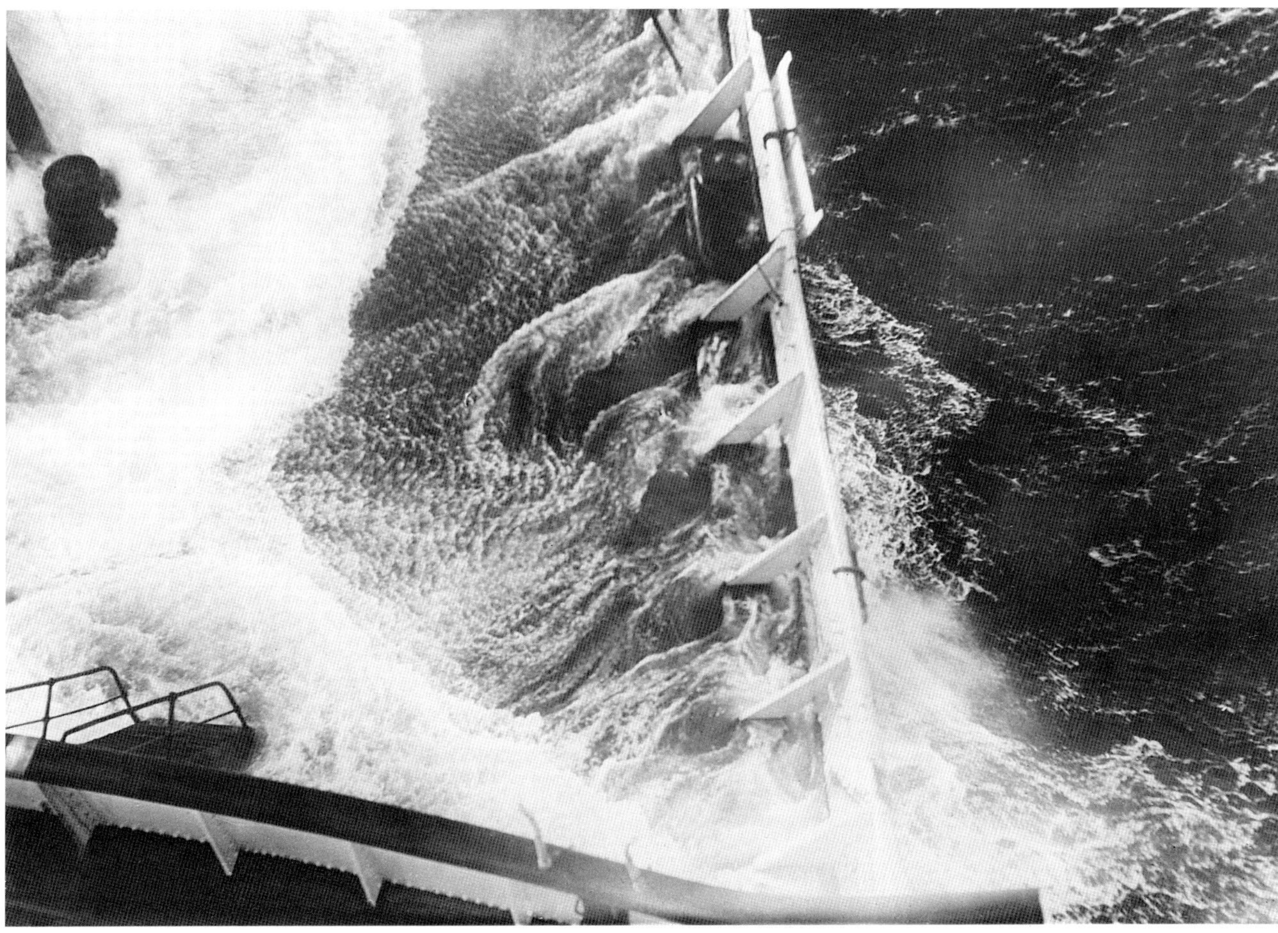

Another photo from November 1930 showing her starboard forecastle, just forward of her superstructure, awash at the peak of the storm's violence. (Ioannis Georgiou Collection)

avoid the hassles involved with customs if she entered a foreign port. However, prospective passengers were not enthusiastic about this concept. They clearly 'favored a stop for the sake of breaking up their trip and for the experience of having visited a foreign land, regardless of the length of the stay', just like many modern cruise passengers do. Thus, Cunard changed the itinerary to include a stop at St George, Bermuda. She departed on the first cruise at 10 a.m. on Saturday 2 May, and returned three days later. She carried some 600 tourists, all of them housed in former First and Second-Class accommodations. No matter what accommodations they had selected, all were given unrestricted access to spaces of both classes, and were given the same 'first class' service. The ship made a record passage from St George to the Ambrose Channel Lightship at an average speed of 23.09 knots, lowering the standing record time by over 10 hours.[130]

Fortunately, cruising proved popular for the *Aquitania*, and she also offered trips from New York to the West Indies and, in the summer months, Halifax, Nova Scotia. She also offered longer cruises of about a month, beginning in New York and visiting Mediterranean ports. However, bookings in that dark financial time were highly unpredictable: the 3 July midnight sailing for an Independence Day weekend Halifax cruise carried an astonishing 1,170; by contrast, the 23 December Christmas cruise to Bermuda attracted only 100 passengers – a figure so low that the company cancelled the trip. The good news, however, was that *Aquitania* was not alone: the *Leviathan*'s 26 December cruise to the West Indies also had to be cancelled due to a lack of interest.[131] At the beginning of 1932, it was announced:

The *Aquitania*, which has arrived at Southampton, will go into dock this week for her annual overhaul.

This photo was taken on 11 May 1931, and shows then governor of New York, Franklin Delano Roosevelt, travelling to Europe on the *Aquitania*. The man seated behind him is captioned in the original photo as Elliot May. Roosevelt was then on a hasty journey to visit his mother, Sara, who had fallen ill while travelling in Europe. It is now widely known that Roosevelt concealed his physical disability while living in public life; while not an unusual view for passengers on an ocean voyage, this view of the man in a deck chair with his legs wrapped in a blanket now provides an instantaneous reminder of the struggles he faced on a daily basis. (National Archives & Records Administration)

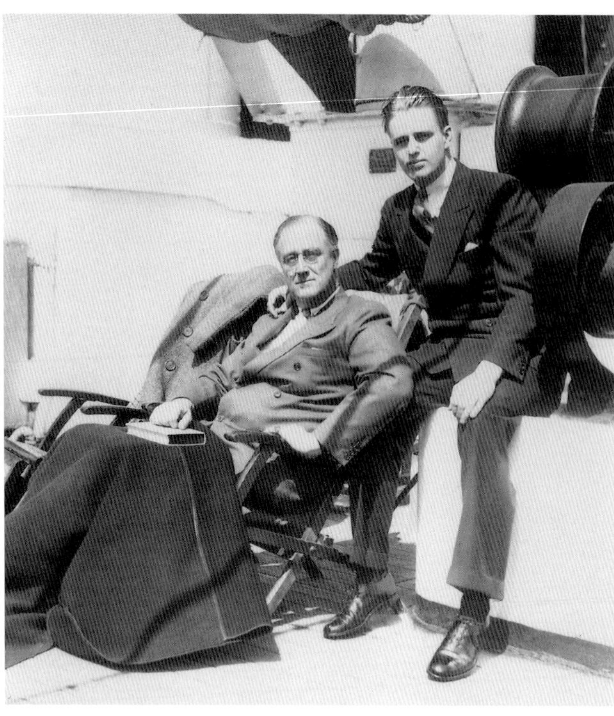

During the last twelve months she made eighteen round trips across the Atlantic, and from New York made five week-end cruises. Her total mileage for 1931 was 123,086 miles – an average of 337 miles a day.[132]

Eighteen voyages during a year was a record that she would match only once, in 1937. The series of quick turnarounds between crossings and cruises was a tremendous feat – one that would have been unthinkable when she first entered service. At nearly 20 years of age, the ship had performed remarkably.

Some idea of the actual experience of sailing aboard the *Aquitania* can be gleaned from the reminiscences of composer, actor and musician Larry Adler. In 1934, when he was working at the Palace Theatre in New York, English theatre impresario Charles B. Cochran invited Adler to join him on a trip to England on the *Aquitania*. The trip, Adler recalled, gave him an impression of the British that he never forgot:

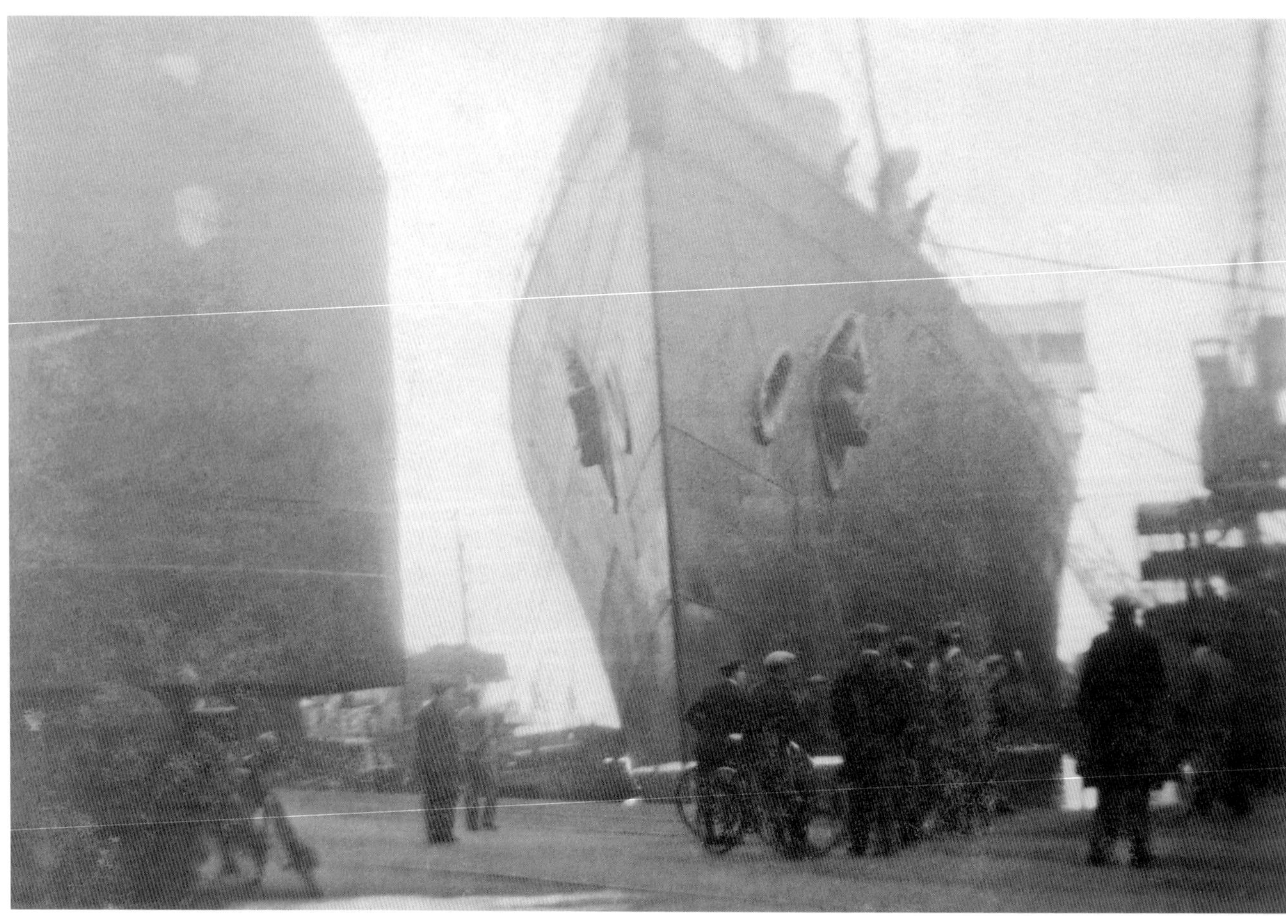

A moody, atmospheric view of the *Aquitania*'s prow in dock at Ocean Dock in Southampton during 1931. (Ioannis Georgiou Collection)

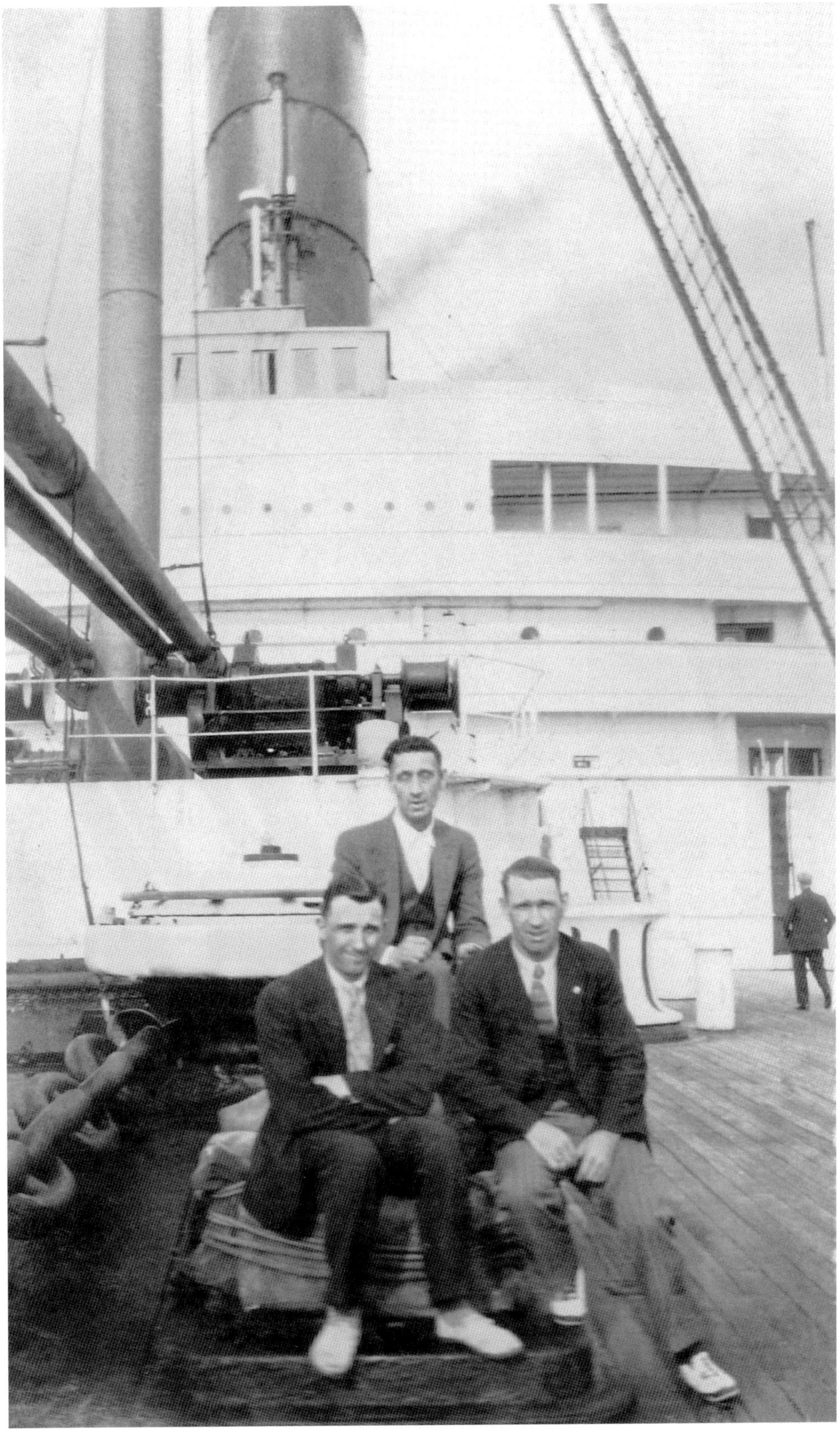

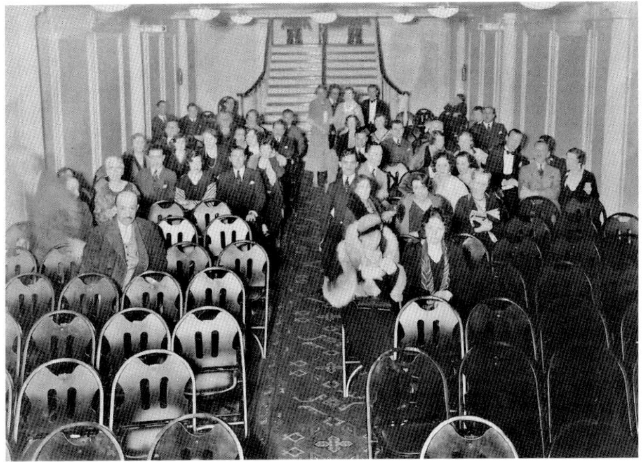

Above: This extremely rare photograph shows passengers assembling for a viewing in the newly converted cinema, sometime after the 1932 refit when it was installed. This was space formerly occupied by the Second-Class Dining Saloon; it had been closed off from the rest of that area. The view is looking forward along the centreline of the ship. The original columns visible in 1914 photographs of the Saloon are still present, but new wooden panelling has been installed between them. The photo was taken from the area of the stage and screen, and the height suggests that the camera was located in the open well which gave opening through to C Deck. (Clyde George Collection)

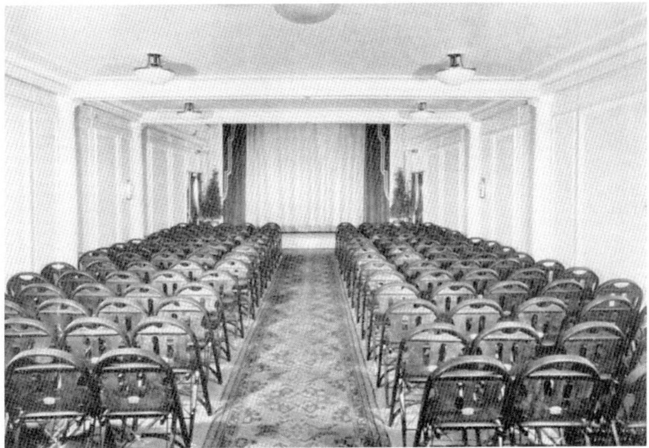

Above: This photo, taken looking aft from the staircase leading up to C Deck at the forward end of the new cinema, shows the screen and stage. Just in front of the stage is the open well. (Ioannis Georgiou Collection)

Left: This candid photo, taken in May 1932, shows passengers enjoying time in the fresh air on the *Aquitania*'s forecastle. Behind them, the foremast, superstructure and smokestacks are all visible; the face of the upper bridge has now been painted dark, a distinguishing feature of her appearance during the 1930s. (Steven B. Anderson Collection)

Right: This is the first of three photographs excerpted from a private album. It shows passengers standing on the top deck of a tender alongside the *Aquitania* at Cherbourg in 1934. (Authors' collection)

Far right: A quartet of ladies in the travelling party pose on the liner's deck with a life ring. (Authors' collection)

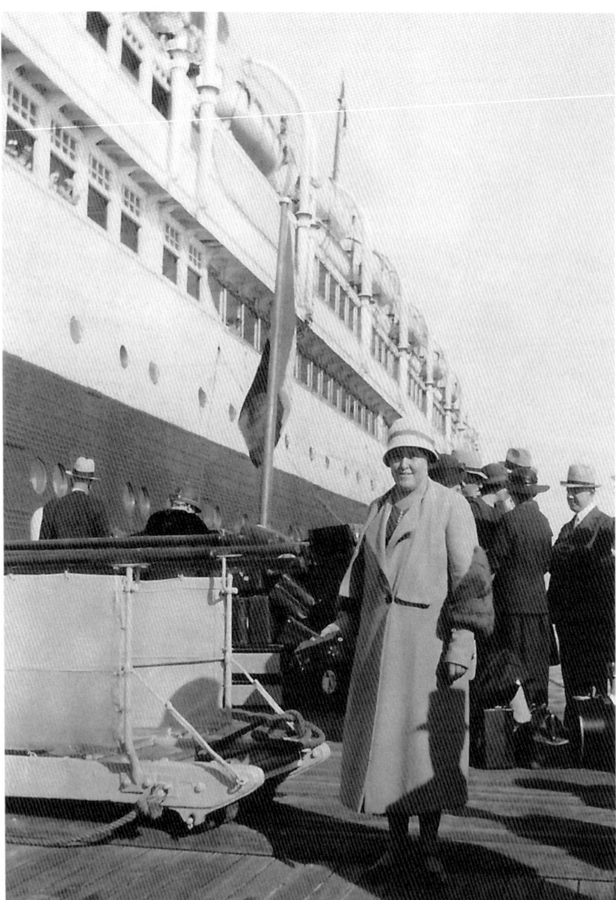

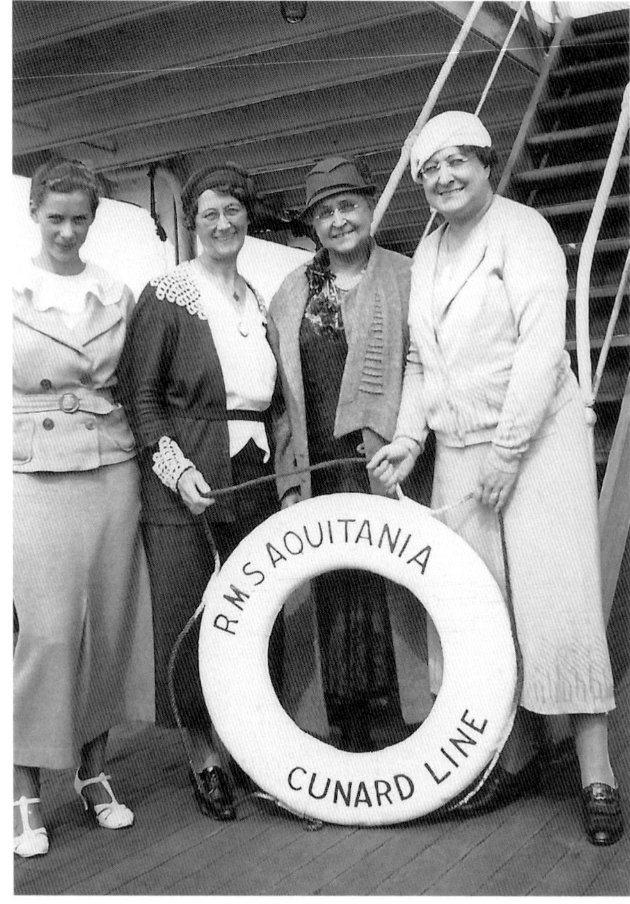

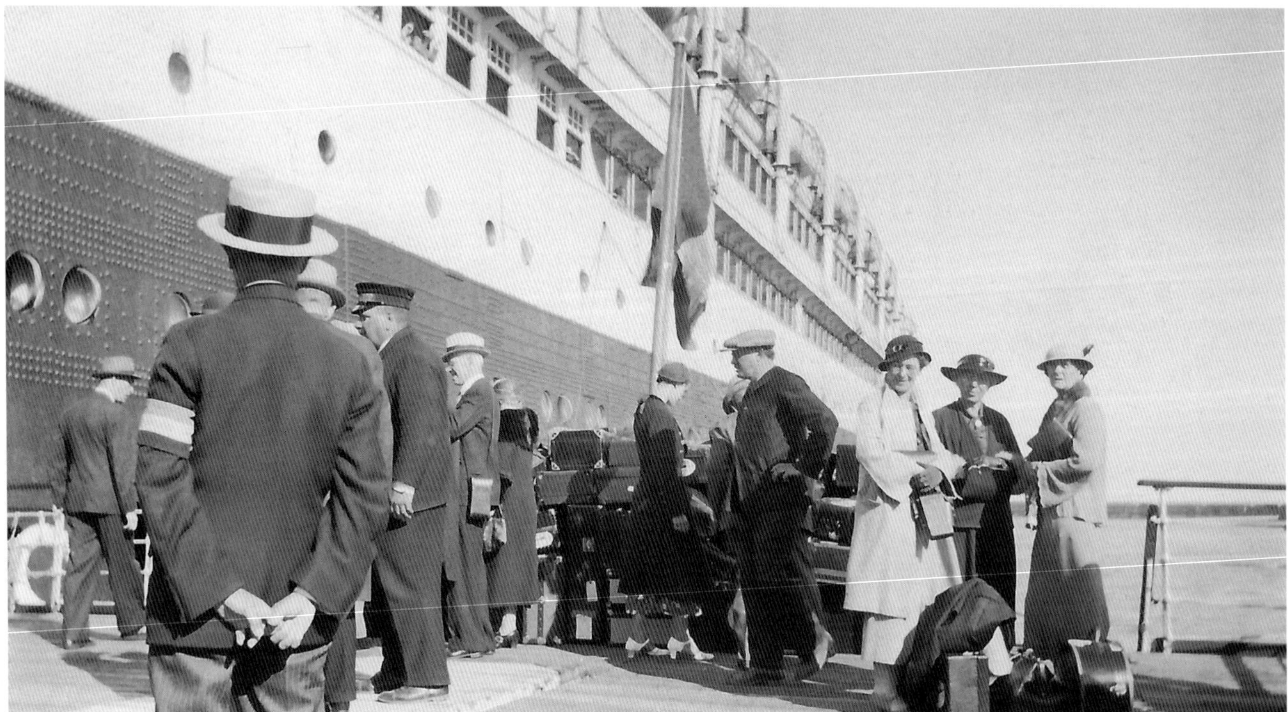

A second photo taken on the tender at Cherbourg. This wider angle shows how long the ship was compared to the tender. (Authors' collection)

The staff called me 'Sir' (I was 19). The passengers dressed for dinner and they spoke, oh boy, how they spoke! I had never encountered such respect for language. I had *read* it but I hadn't *heard* it.[133]

Apparently, Cunard advertisements from that era were not exaggerating when they trumpeted the:

special charm that differentiates the *Aquitania*, that something which wins the loyal devotion of so many people and which is difficult to define. Something that gives vitality to her days and animation to her evenings ... making the six-day voyage from New York to Cherbourg or Southampton the 'pleasantest distance' between two worlds.[134]

With their new *Queen Mary* to enter service in 1936, Cunard saw a need to improve the *Aquitania*'s performance. Accordingly, in early 1934, they made alterations to the ship's condensers – a vital component in making good speed. However, inadequate propellers prevented her from setting any records. This problem was soon remedied. In June 1936, she made her fastest westbound crossing to date, travelling from Cherbourg to New York at an average

Passengers pose on the stern of the *Aquitania* as the ship sails at high speed. (Authors' collection)

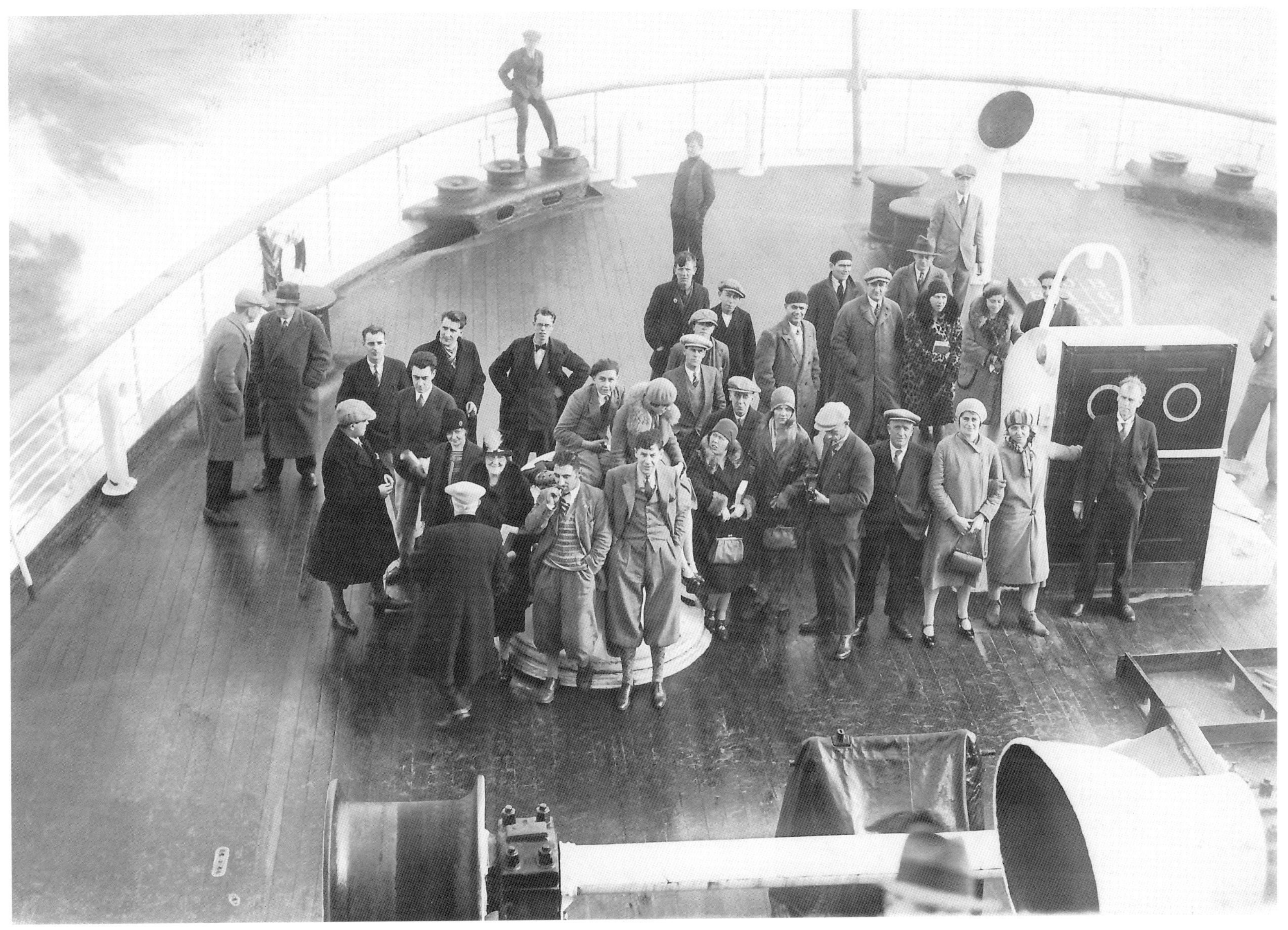

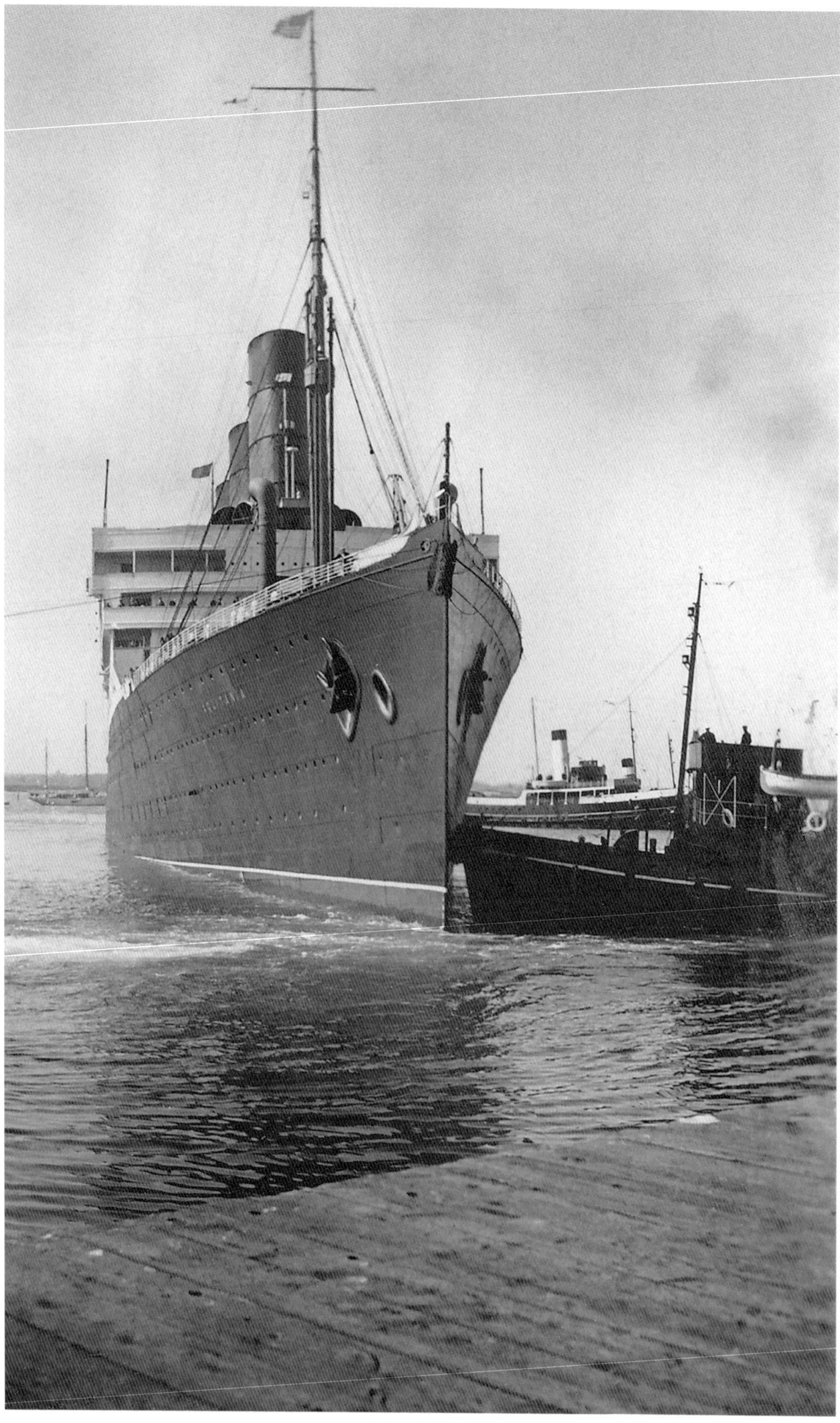

Aquitania being nosed into dock in this early 1930s view. (Russ Willoughby Collection)

speed of 24.28 knots, in five days, ten hours and six minutes. Her officers credited this feat to the new propellers.[135]

Records began stacking up. In August 1937, she averaged 24.27 knots westbound.[136] In May 1937, she averaged 24.87 knots on an eastbound crossing.[137] In April 1939, she nearly matched that average, making an eastbound crossing at 24.85 knots, arriving at Cherbourg five days, eight hours and forty-five minutes after leaving New York.[138] The following month, she made an eastbound crossing at an average of 24.99 knots, with a best day's average of 25.83 knots – a performance nearly 2 full knots better than her designed service speed. This would have been impressive for a new liner, let alone for one then 25 years of age.[139] Clearly, the *Aquitania* still had some tricks up her sleeve.

At the end of one of her Mediterranean cruises, returning to Southampton on 10 April 1935, the *Aquitania* suffered one of the most notable incidents of her career. Britain was then being swept by tremendous gales, with wind gusts of 78mph being registered in Manchester. The trouble came as the ship was attempting to navigate the ticklish 'S' bend in the channel – located in the vicinity of the Brambles Bank between Bourne Gat and Calshot Light. The *Aquitania* had trouble in this area once before: in January 1934, she had gone aground for two hours; with the assistance of six tugs she had been refloated and proceeded on her voyage. However, these waters were always a proving ground for ships and for the skills of harbour pilots.

One of the passengers was Egyptologist Howard Carter, who in 1922 had been the first to peer into the tomb of ancient Egyptian pharaoh Tutankhamun. Also aboard was Alan Muntz and his wife Lady Margaret Stewart, daughter of Lord Londonderry, who were finishing their honeymoon cruise.

The water of the channel was very rough just after 4 p.m., and at the worst moment, a gust of wind caught her and caused her nose to swing. Muntz recalled: 'We seemed to glide on the mudbank. The liner was pitching. Then suddenly the bow went down and did not come up again.' Howard Carter said: 'The liner was simply thrown on to the Bank by the wind. I felt a slight shudder and knew we were aground.'

Captain P.R. Vaughn reversed the engines, but to no avail. In short order, the signal flag indicating that the

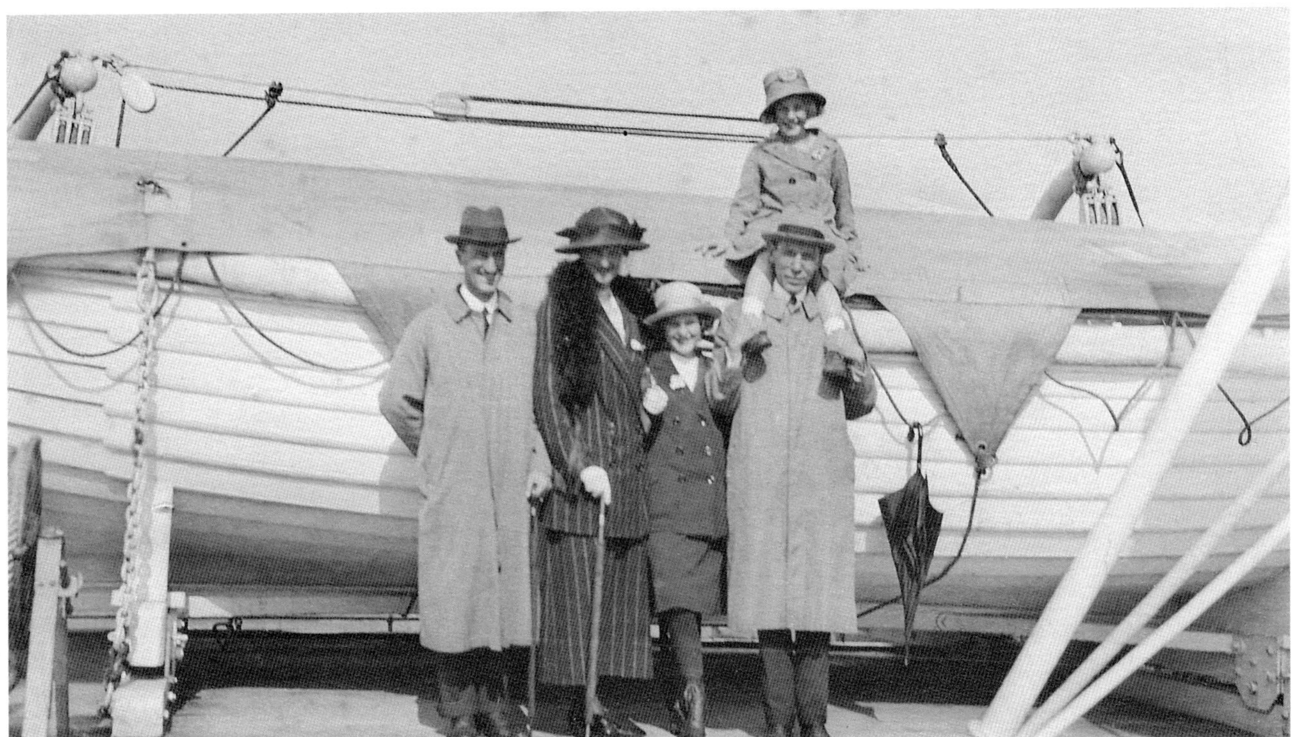

Left: A happy group poses in front of a lifeboat on the Boat Deck. (Russ Willoughby Collection)

Below: Looking aft from the starboard wing of the upper bridge, this photo captures the quartet of funnels, the numerous cowl ventilators, and below, on the Boat Deck, passengers and crew going about their daily activities. (Russ Willoughby Collection)

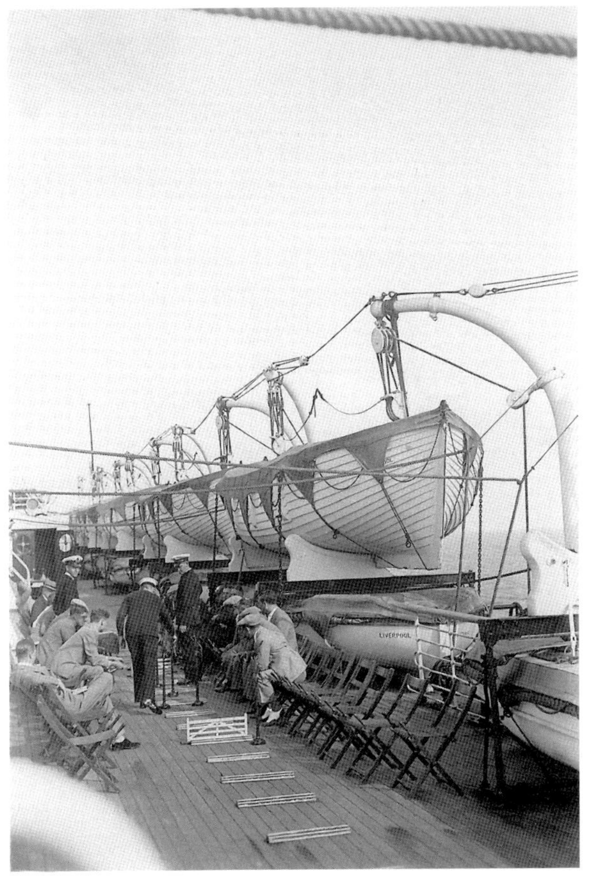

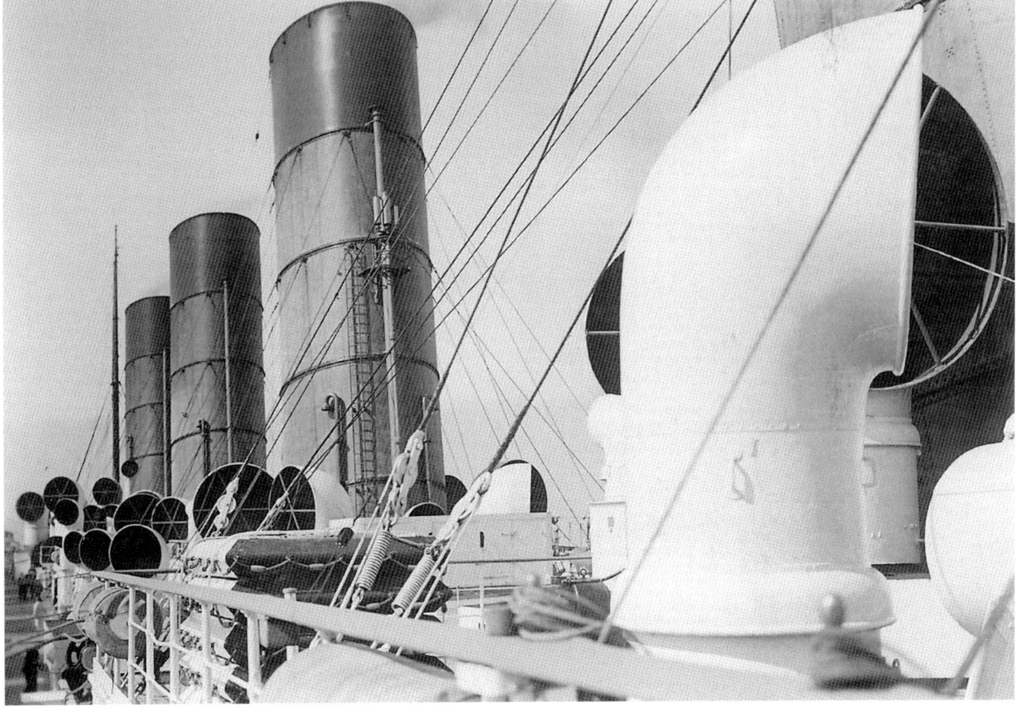

Left: Passengers play games on the *Aquitania*'s Boat Deck. (Russ Willoughby Collection)

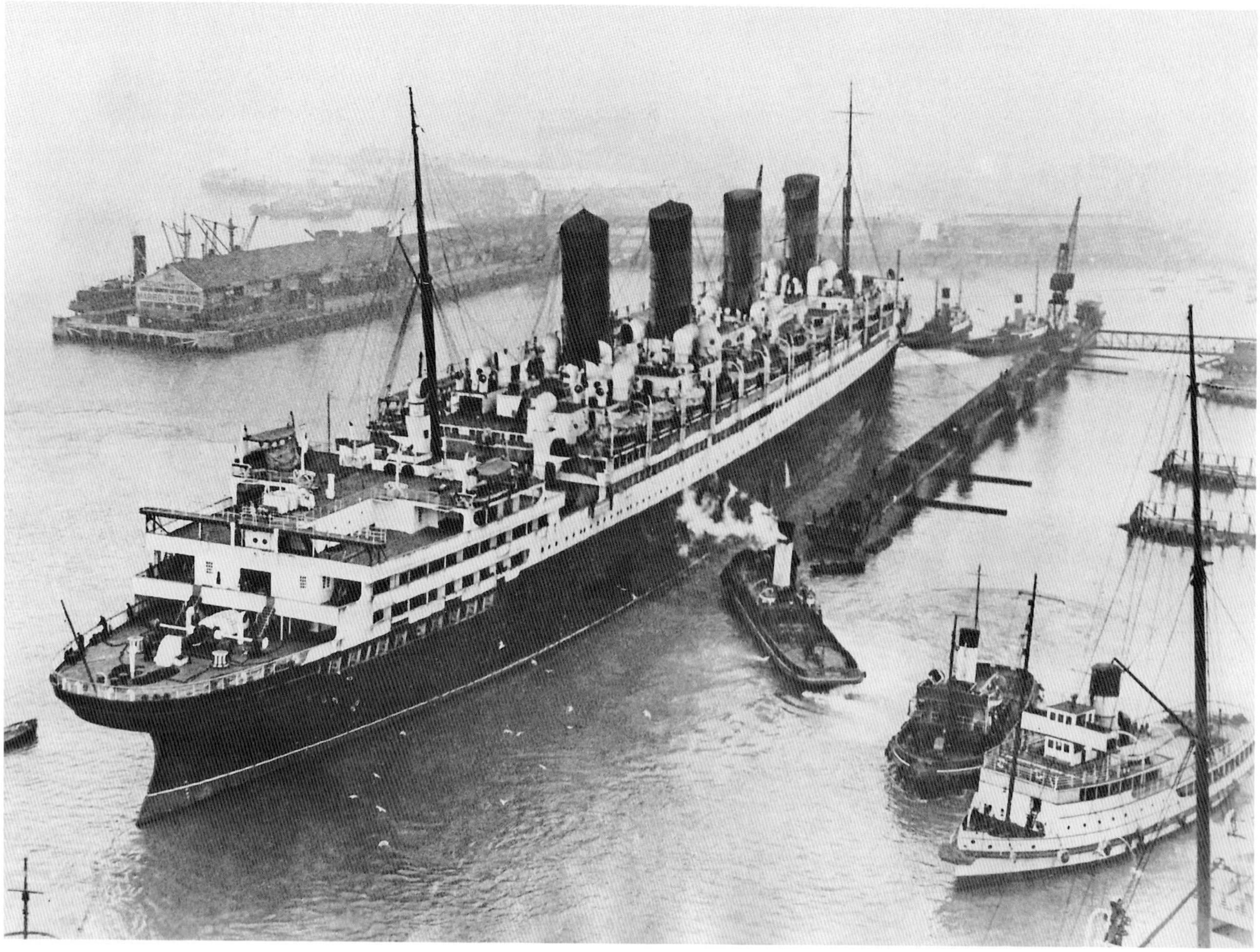

The *Aquitania* enters the floating dry dock in Southampton on 8 December 1933. (Clyde George Collection)

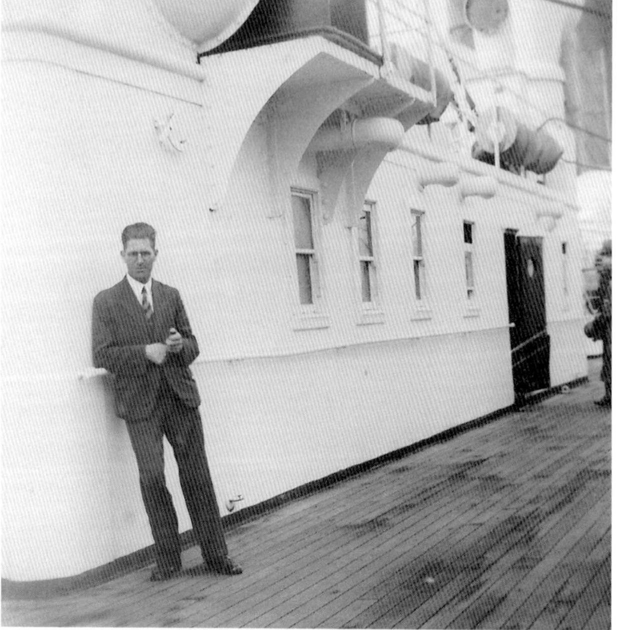

A passenger leans against the rail and bulkhead of a deckhouse on the starboard Boat Deck. (Russ Willoughby Collection)

ship was 'not under control' was hoisted, and an SOS was sent out, requesting assistance. Within an hour, eight tugs were alongside, but even with their mighty horsepower thrown into the mix, the ship remained stubbornly fast on the mud.

As the tide continued to recede, the ship's bow was raised higher than her stern, which was then still afloat. At her prow, she rose high enough to begin to show where the stem curved aft, while at the stern, water lapped up to, or in places over, the white stripe separating her boot-topping from her black hull. The next high tide was not until 4.10 a.m., and it was clear that nothing would budge the liner until then. All but twenty-seven of the liner's 300 passengers opted to go ashore via tender. Those on land and in other nearby vessels could see the remaining passengers walking the brightly lit but sheltered promenades of the ship in their evening dress.

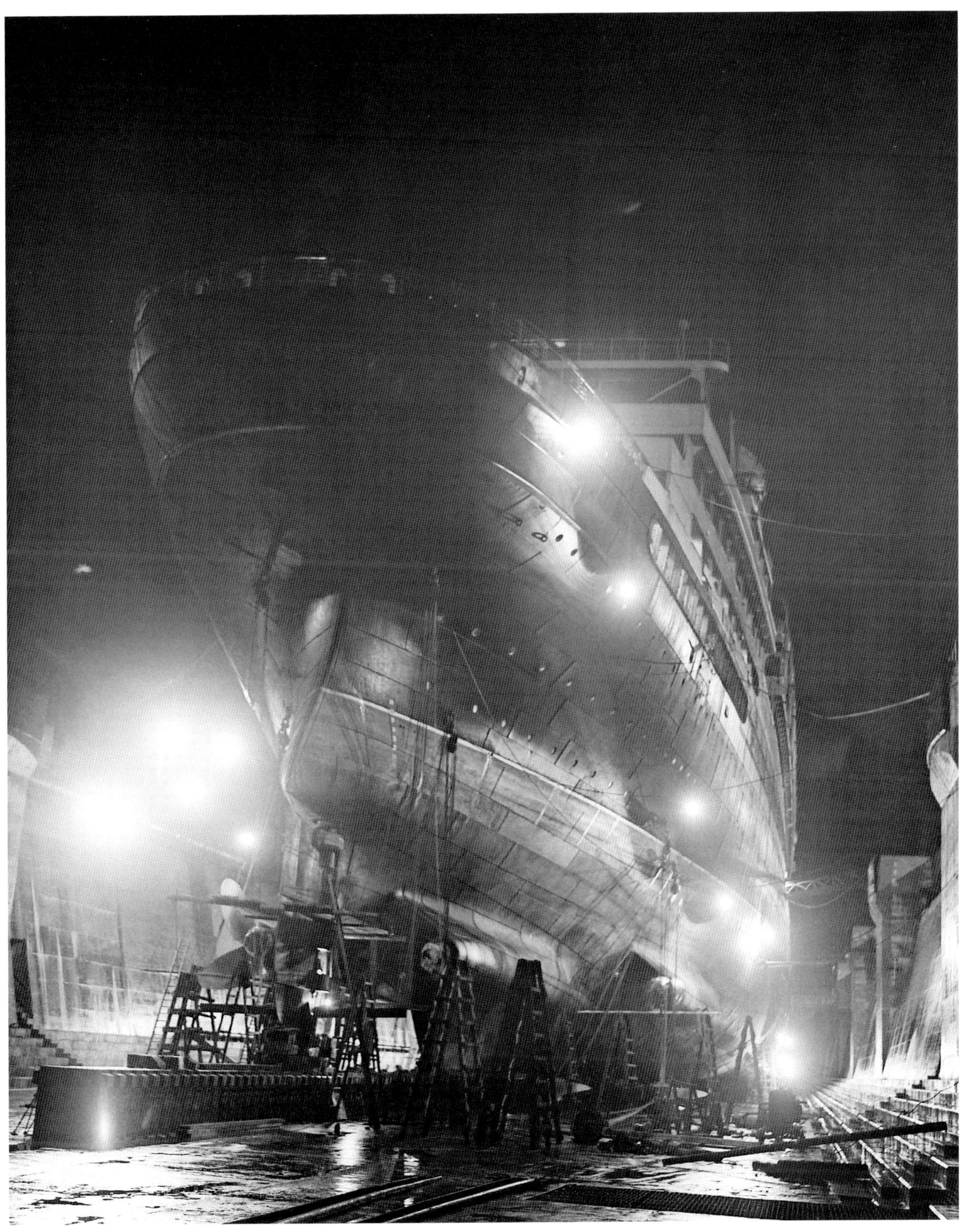

On 14 January 1935, the *Aquitania* was undergoing repairs in the new King George V Graving Dock at Southampton, after suffering damage in rough seas. Work was going on both day and night, but the night-time scene under the glare of floodlamps was truly spectacular. (TOPFOTO)

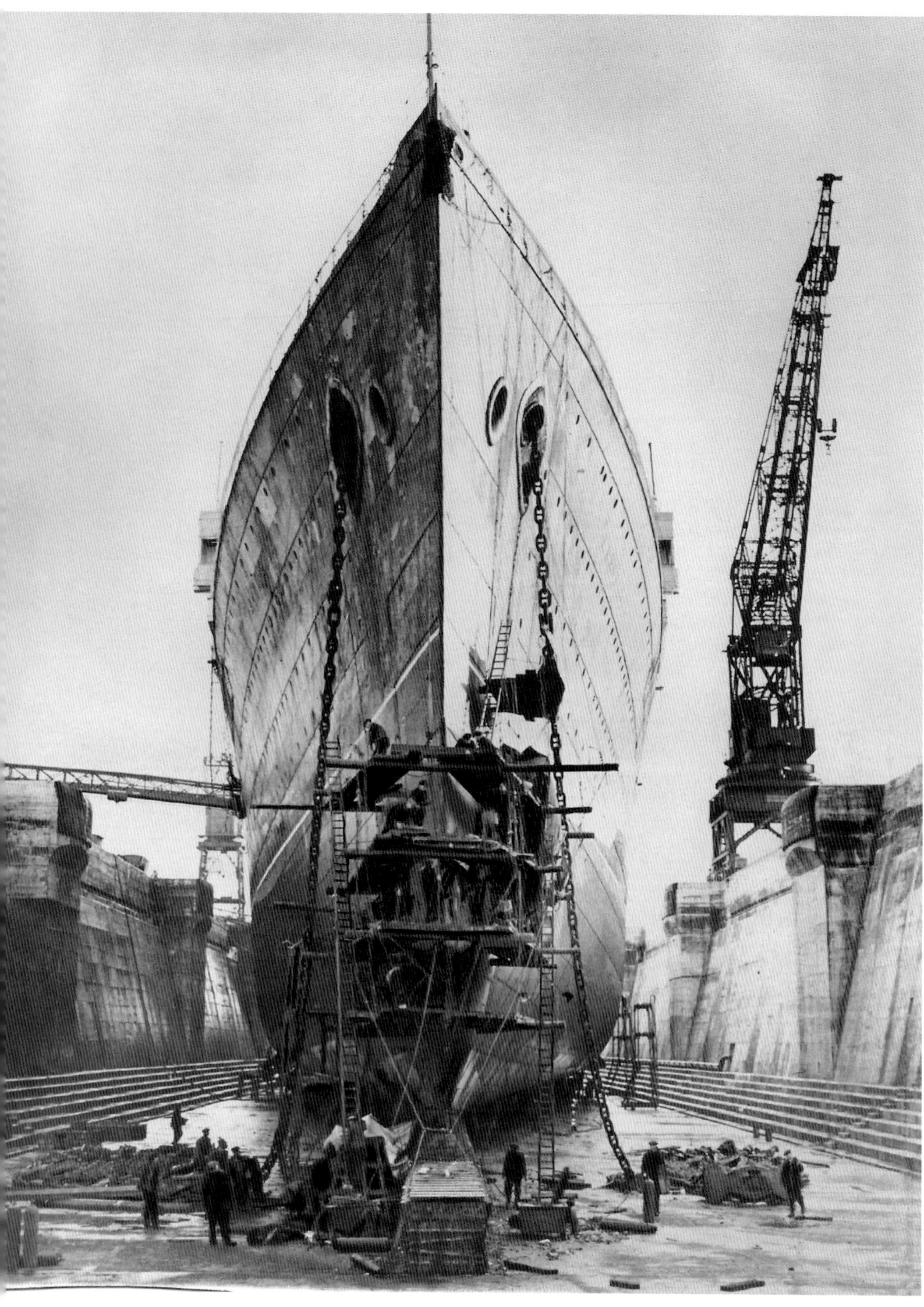

A fine portrait of the *Aquitania*'s prow while in the new King George V dry dock. The new dock had been built to service the next generation of superliners then under construction, including the *Queen Mary*. It had been formally opened – though yet incomplete – in 1933. The *Majestic* had been the first liner to use the dock, in 1934. There are some indications that this photo was taken in 1936. (Clyde George Collection)

A special dinner had been put on, and the ship's band was playing. Their music was said to have 'drifted across the angry sea' as passengers enjoyed an impromptu dance in the Palladian Lounge.

Four of the eight tugs remained connected to the *Aquitania*'s stern overnight to prevent her from swinging further onto the bank because of the wind. Chief Purser James Lawler radioed ashore that evening, stating that although it was still 'blowing a fierce gale', everyone on board was 'quite safe'. He added that Captain Vaughn had not left the bridge since the grounding. Meanwhile, the crew began to redistribute weight in order to lighten the vessel. Oil was pumped out of her forward tanks, and either placed in her aft tanks or removed from the ship altogether. The ship failed to escape on the early morning tide. That afternoon, no less than eleven tugs – five at the stern and six at the bow – made a final attempt. Shortly before four o'clock:

> [P]owerful tugs ... at the stern of the giant liner began to steam slowly. The steel hawsers tautened, and after about 20 minutes three shrill blasts sounded from the *Aquitania* and the signal was answered by the tugs. Immediately the liner's propellers began to churn the water in the hope that her nose would be loosened as the water ran under her. For 45 minutes the tugs strained and pulled, but it seemed that their efforts to refloat her would be unsuccessful. The tide was then on the turn.
>
> Then the steel hawsers were seen to slacken while the tugs backed slowly towards the liner. Two minutes later, they steamed out again, pulling hard at the stern, while other tugs at the bow slowed her round.[140]

It was said that this was 'a case of the irresistible force striving against the immovable object'.[141] At last, at 5.10 p.m., the *Aquitania* 'jumped' clear of the mudbank.[142] She had been immobilised for some twenty-five hours. An hour and a quarter later, she docked safely at her Southampton berth. A diver was sent down to examine the hull for damage; a dry-docking would have cost the company an estimated £1,000, in addition to the hundreds of pounds the company had to spend to engage the extra tugs and

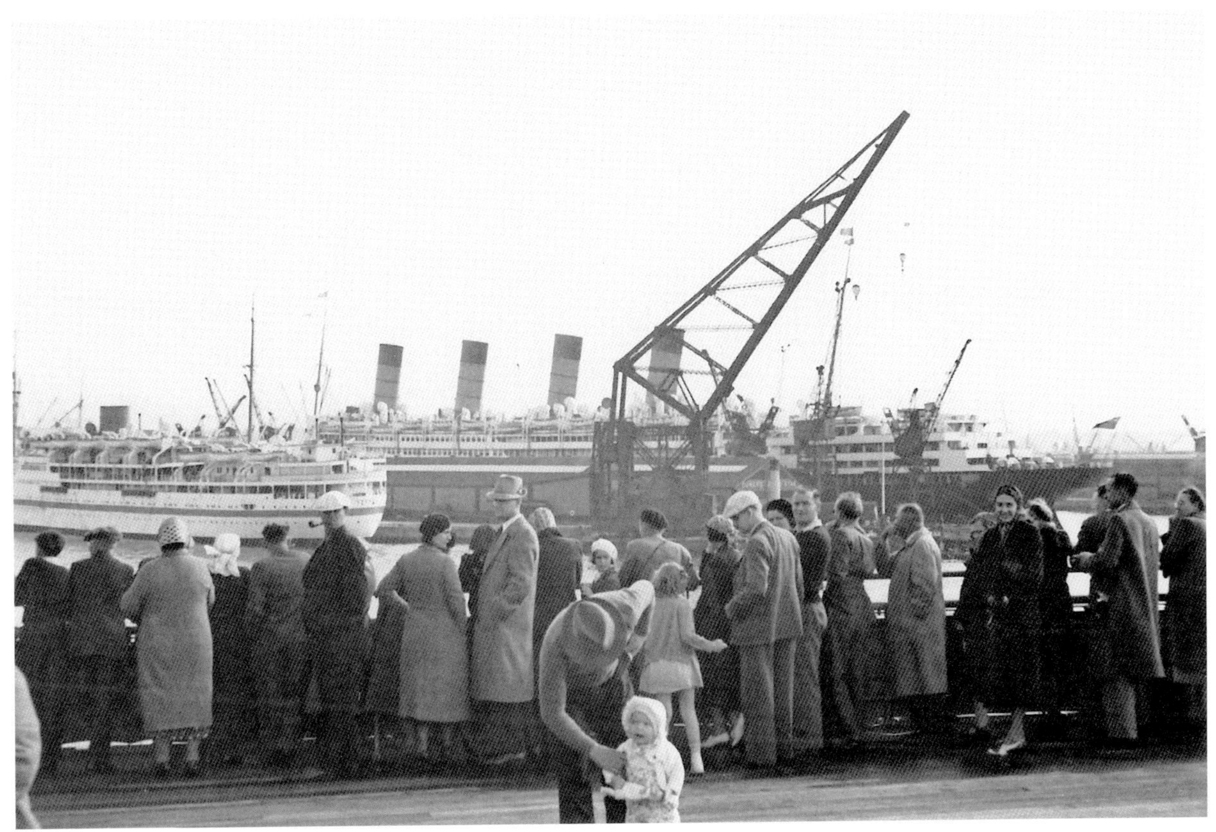

A fine photograph of the *Aquitania* in Southampton, drawing a crowd of observers on a passing vessel. A careful inspection of the pierside structure next to the *Aquitania* shows it is labelled 'Cunard White Star', indicating that the photograph was taken after the 10 May 1934 merger of the two companies. (Russ Willoughby Collection)

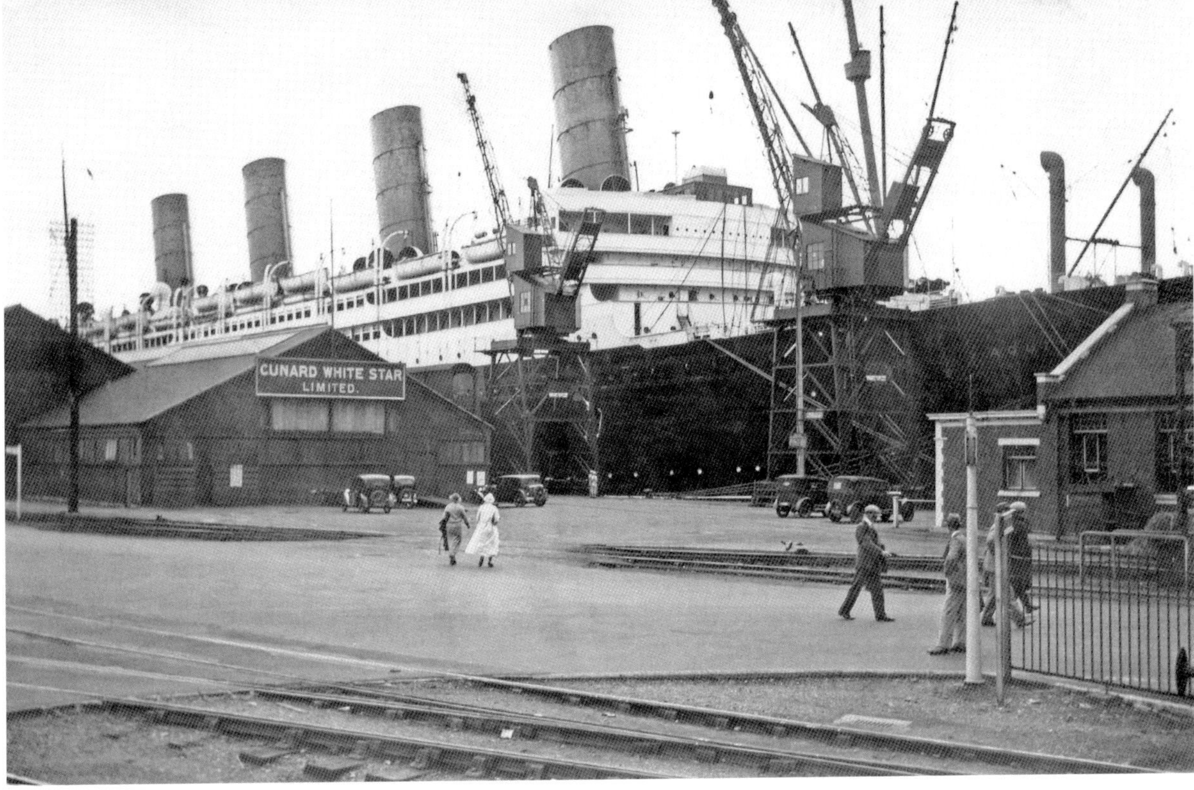

Aquitania towering over her Cunard-White Star berth in Southampton. The automobiles and people below look positively diminutive. (Ioannis Georgiou Collection)

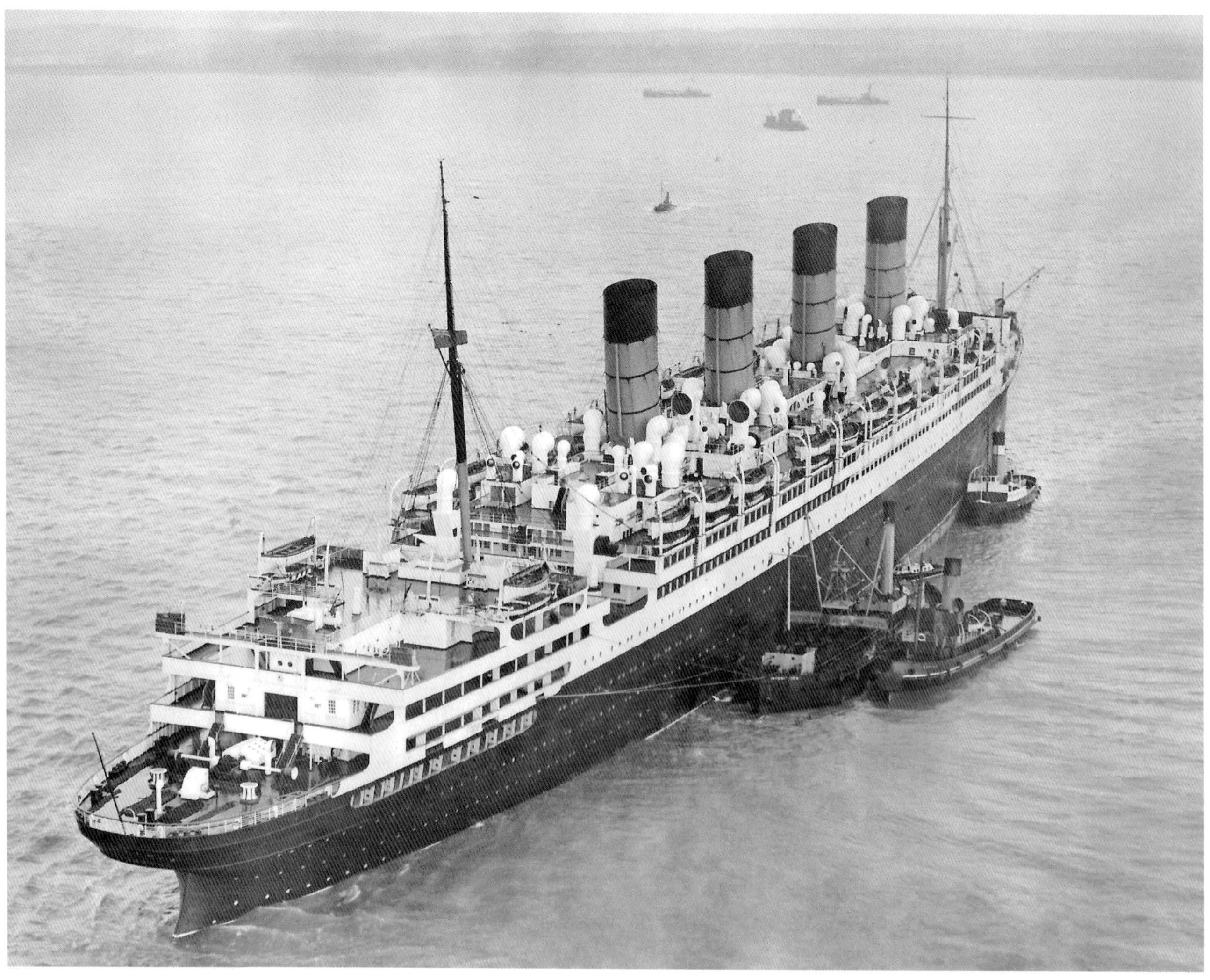

The first of three photographs showing the *Aquitania* during one of the few crises of her career. Taken either before sunset on 10 April 1935, or early in the day of 11 April, it shows the ship hard aground on Thorn Knoll (the name is spelled without the 'e' in period pilot reference manuals); this was a half-mile long hazard which extended into mid-channel between Calshot Spit and the north-western side of the Bramble Bank. A tanker lies along the *Aquitania*'s starboard flank, indicating that the process of lightening the ship by removing oil from her fuel tanks is under way. Tugs stand by, one along the liner's starboard bow and another by the fuel tanker. Two lifeboat stations on the starboard side are completely devoid of their lifesaving craft. (Clyde George Collection)

to pay the ship's crew overtime for their work. Fortunately no damage was reported, and the liner was able to resume her sailing schedule without interruption.

However, the *Aquitania* was not just combating a financial crisis and mudbanks: she was also locked in a battle with the effects of age and decay. As they aged, all ships of that era showed a degree of structural fatigue. *Olympic* had suffered some minor hull strain, but that was nothing compared to the ongoing issues that plagued the *Leviathan* and *Majestic* – both of which were nearly lost when their hulls fractured during storms, long before they had reached what might be termed 'old age' for a liner.

The *Aquitania*'s hull issues were minor by comparison, but did require consistent maintenance and attention. The ship's stern frame and rudder required attention as early as 1924, but the most consistent area of trouble was further toward the liner's bow, just forward of and below the face of the bridge; symptoms of the issue became apparent that same year. By 1933, a 12in crack on one side and a 30in crack on the other were discovered. The longer of the two had gone through the primary hull plate and its doubler plate as well. Clearly there was a weakness in the design of the hull, somewhere in the vicinity of frames 235–42. Repairs were made, but even years later the area required significant attention; at that point, it was thought that some of this damage might have been a delayed reaction caused by the enormous strain placed on that area of the ship during a grounding back in 1915.[143]

Repairs made the ship safe enough to continue active service; in fact, her workload increased. In late 1936,

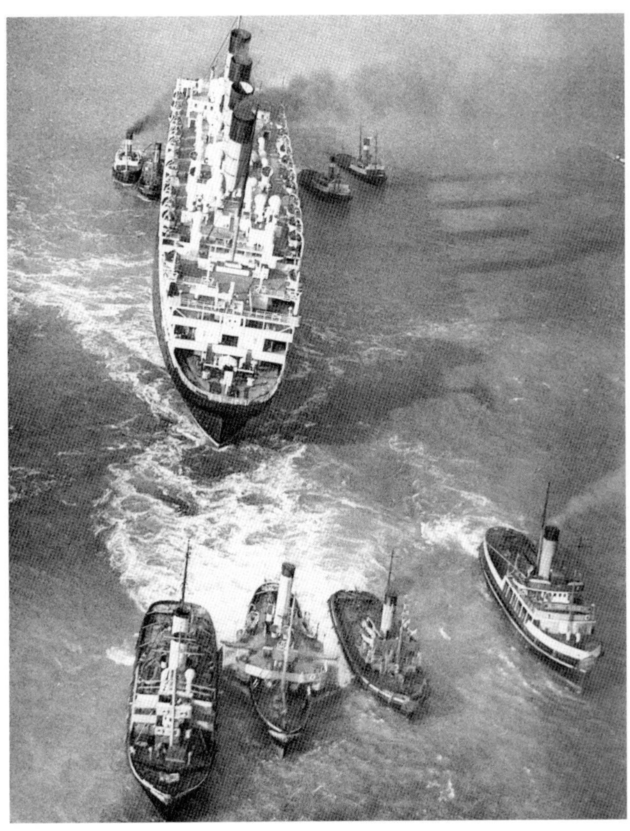

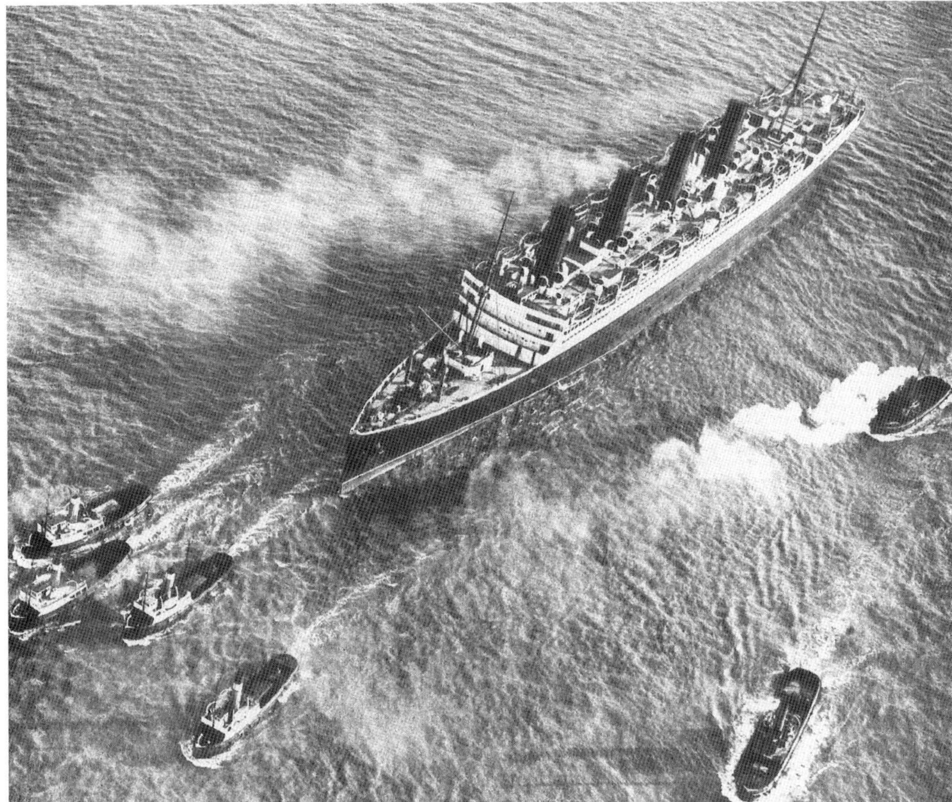

Above: A titanic effort is under way to get the *Aquitania* off the knoll on 11 April 1935. Four tugs are visible straining at her stern, and another quartet is quickly spotted at the bow. No less than eleven tugs were involved in this task, the other three not being visible in this photo. (Ioannis Geogiou Collection)

Above right: Having successfully freed the liner from Thorn Knoll, the tugs help guide the *Aquitania* up Southampton Water. (Clyde George Collection)

Right: Members of the American Naval delegation to the London Naval Conference aboard the *Aquitania*, sailing for London on 29 November 1935. The conference's purpose was naval limitation and disarmament. In the front row, from left to right: Undersecretary of State William Phillips; Chairman of the Delegation Norman H. Davis; and Chief of Naval Operations Admiral William H. Standley. In the back row, from left: Noel H. Field, Technical Assistant for the Department of State, Division of Western European Affairs; Eugene H. Dooman, Adviser to the Division of Far Eastern Affairs; David McKey, Assistant Chief, Division of Current Information; Commander R.E. Schuirmann, Technical Assistant for the Navy Department; and Lieutenant Arthur Del. Ayrault, Jr, also Technical Assistant for the Navy Department. The conference was not entirely successful: in particular, Italy and Japan refused to sign; Germany paid lip service to limiting its naval rebuilding, but in secret was beginning to create a naval force that would threaten France, Britain and the United States. The world was travelling down a one-way path towards war. (Ioannis Georgiou Collection)

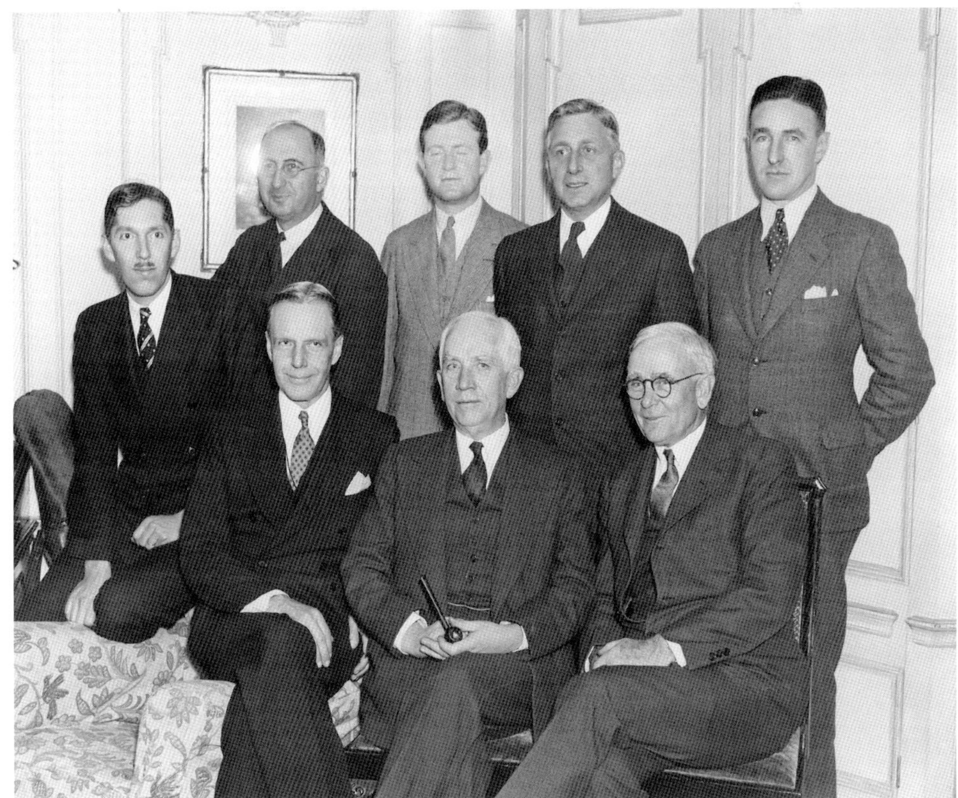

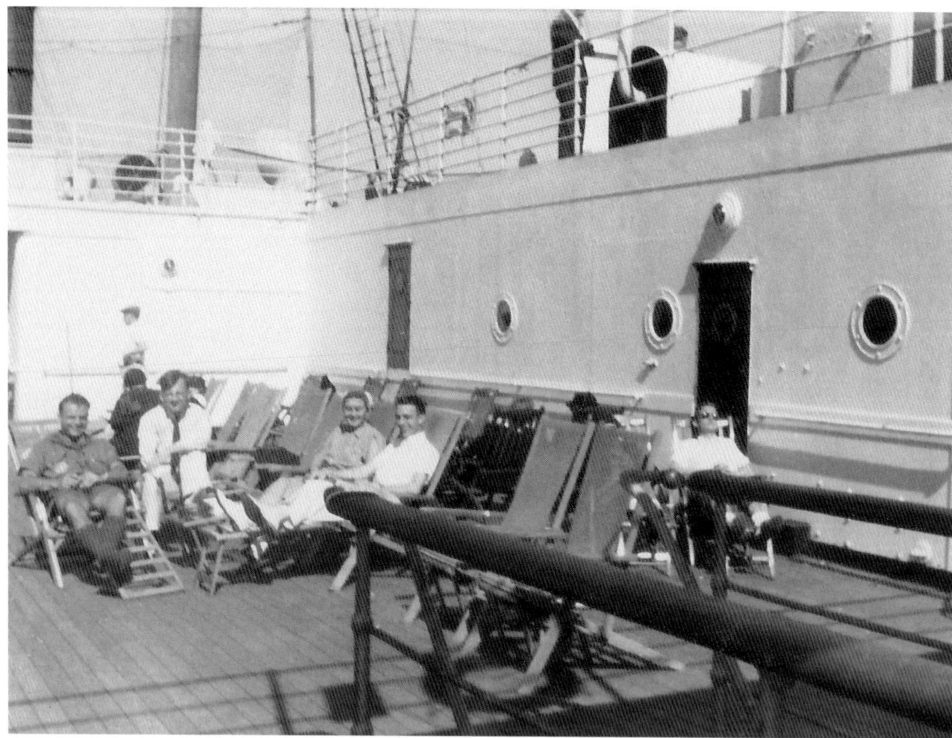

Passengers enjoy the sun in this 1937 candid photograph on the *Aquitania*'s Boat Deck. (Authors' collection)

Aquitania towers over Southampton's Ocean Dock in this 1930s view. (Authors' collection)

Cunard announced audacious plans for even longer cruises from New York. These winter pilgrimages would take the liner to Rio de Janeiro, Brazil. The Inter-American Conference for the Maintenance of Peace – held in Beunos Aires in December 1936, featuring a visit by US President Roosevelt – and all of the accompanying publicity had caused an uptake in interest for travel to Brazil and Argentina. *Aquitania*'s forty-day, 13,780-mile itinerary would begin on 17 February 1937, and started at $495 per ticket; ports visited would include Bermuda, Trinidad, Bahia, Rio de Janeiro, Montevideo, a second call at Rio, Barbados, La Guaira, Colon and Nassau. The call at Montevideo was to encompass four days, with plenty of opportunity to sightsee up the La Plata River to Buenos Aires. Some 565 passengers booked; among them was Mrs Andrew Carnegie, who turned 80 during the cruise.[144]

Aquitania arrived four hours late from her transatlantic crossing to New York because of bad weather, compressing the turnaround before she was scheduled to depart. Despite this, the cruise went very well. The biggest hitch came at Montevideo. The liner had anchored off shore, and about 430 passengers had left by tender to go on the excursion up to Buenos Aires. This trip had gone smoothly. But on 16 March, as the last batch of passengers was returning to the ship aboard a small tug, a violent gale sprang up. Captain Irving decided that it was too rough to attempt to bring passengers aboard. Indeed, waves were 'washing over the tug', forcing it to run for shelter; the stranded passengers were taken aboard a larger, safer tender and slept there overnight. As there were only fourteen bunks available, many tried to sleep on deck benches. The next day, the weather was still too rough to return to the ship, but they were able instead to dock at Montevideo and spent that night in a hotel.[145]

There was also a rather serious case of theft. During the trip, at a point when the purser's safe was open and the staff were preoccupied, 15-year-old bellboy John Cronin of Southampton, then serving in the office as a messenger boy, reached in and stole an envelope. It contained uninsured jewellery valued at $20,000 belonging to one of the passengers, mining operator R.W. Higgins. All of the ship's pursers volunteered to have their quarters searched, but after a full day, the items remained missing. Finally, Chief Purser Lawler ordered a search of the quarters of *everyone* who had access to the office; this time the packet was discovered in Cronin's locker. The passenger declined to prosecute, even offering to give Captain Irving some money for the boy, a very gregarious offer that was naturally declined. The lad continued his duties aboard the ship until she was to return to Southampton, at which time he would be turned over to the Children's Court.[146] In the end, the cruise lasted forty-one days, instead of the originally advertised forty.

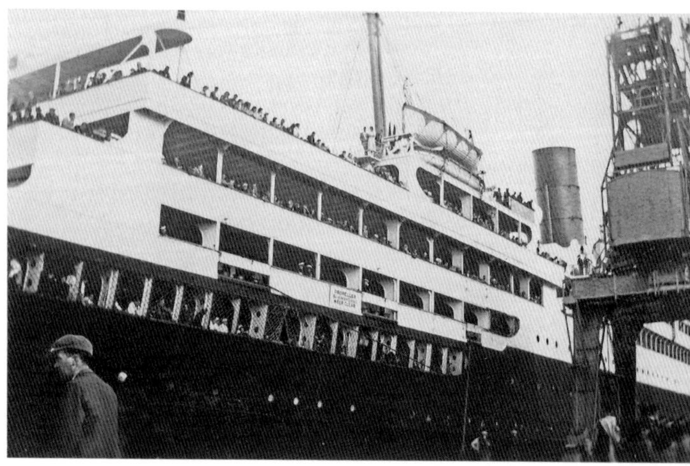

Aquitania passes the Royal Pier in Southampton, which is located just west of Ocean Dock. She had apparently just left the new Western Docks or the King George V Graving Dock. The pavilion in the foreground is now a Thai restaurant. (Authors' collection)

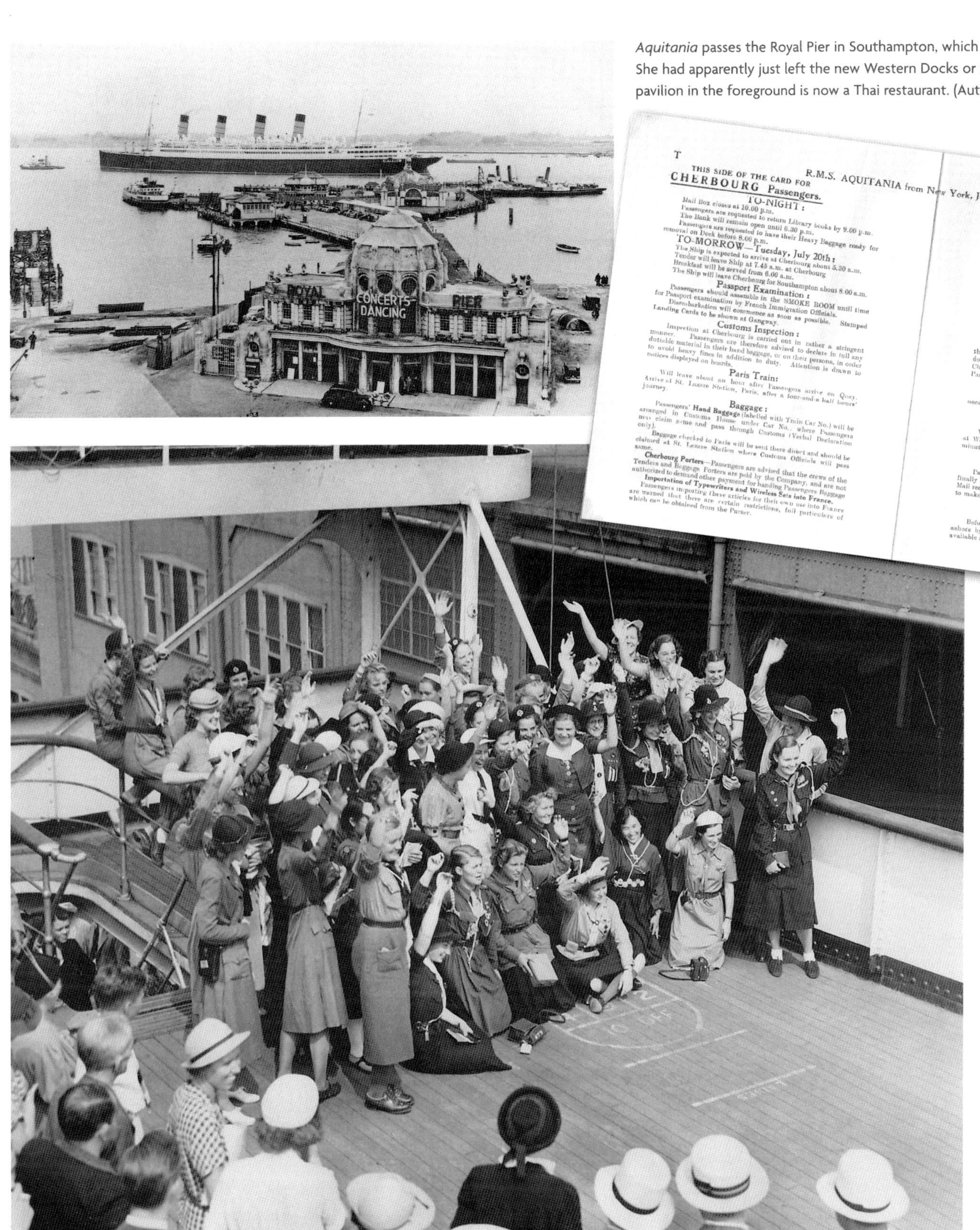

R.M.S. AQUITANIA from New York, July 14th, 1937.

THIS SIDE OF THE CARD FOR CHERBOURG Passengers.

TO-NIGHT:

Mail Box closes at 10.00 p.m.

Passengers are requested to return Library books by 9.00 p.m.

The Bank will remain open until 6.30 p.m.

Passengers are requested to have their Heavy Baggage ready for removal on Deck before 8.00 p.m.

TO-MORROW—Tuesday, July 20th:

The Ship is expected to arrive at Cherbourg about 5.30 a.m.

Tender will leave Ship at 7.45 a.m. at Cherbourg

Breakfast will be served from 6.00 a.m.

The Ship will leave Cherbourg for Southampton about 8.00 a.m.

Passport Examination:

Passengers should assemble in the SMOKE ROOM until time for Passport examination by French Immigration Officials. Disembarkation will commence as soon as possible. Stamped Landing Cards to be shown at Gangway.

Customs Inspection:

Inspection at Cherbourg is carried out in rather a stringent manner. Passengers are therefore advised to declare in full any dutiable material in their hand baggage, or on their persons, in order to avoid heavy fines in addition to duty. Attention is drawn to notices displayed on boards.

Paris Train:

Will leave about an hour after Passengers arrive on Quay. Arrive at St. Lazare Station, Paris, after a four-and-a-half hours' journey.

Baggage:

Passengers' **Hand Baggage** (labelled with Train Car No.) will be arranged in Customs House under Car No., where Passengers may claim same and pass through Customs (Verbal Declaration only).

Baggage checked to Paris will be sent there direct and should be claimed at St. Lazare Station where Customs Officials will pass same.

Cherbourg Porters—Passengers are advised that the crews of the Tenders and Baggage Porters are paid by the Company, and are not authorized to demand other payment for handing Passengers Baggage

Importation of Typewriters and Wireless Sets into France.

Passengers importing these articles for their own use into France are warned that there are certain restrictions, full particulars of which can be obtained from the Purser.

THIS SIDE OF THE CARD FOR SOUTHAMPTON Passengers.

TO-MORROW—Tuesday, July 20th:

Passengers are requested to have their Heavy Baggage ready for removal on Deck immediately after leaving Cherbourg.

Passengers are requested to return all Library books by 10 a.m.

Mail Box closes at 12 Noon.

The Bank will be open from 9.00 a.m. till 12 Noon.

Luncheon will be served at 12 Noon.

The Ship is expected to Dock at Southampton about 1.00 p.m.

Passport Examination:

All passengers should present passports and landing cards to the Immigration Authorities in the Smokeroom immediately after docking. Should the British Immigration Officer join the ship at Cherbourg a notice will be posted on C Deck showing time of Passport Examination.

Baggage will be landed **before** disembarkation commences.

Customs examination of all Baggage (only Verbal Declaration necessary) will take place on the Dock.

London Train:

Will leave about an hour after passengers disembark. Arrives at Waterloo Station, London, after a journey of one hour and forty minutes.

Important—Mail:

Passengers are specially asked to make application, before finally leaving the ship, for any Mail that may be on hand for them. Mail received later is held on the Dock and passengers are requested to make further enquiry at the Desk provided there.

Radio Services:

Before landing passengers may communicate with their friends ashore by means of telegraph or telephone. Full information is available at the Wireless Bureau.

The inside of a booklet filled with information on landing arrangements for an eastbound crossing to Cherbourg and Southampton. It is dated 14 July 1927, and contains vital information on anticipated arrival times, customs, passports and when to have luggage ready to offload. (Authors' collection)

On 1 September 1937, an international delegation of Girl Scouts left New York on the *Aquitania*. They are seen here shortly before departure, and it was said that the 'sun deck was ablaze' with the girls and their Girl Guides. They had visited the United States for the Silver Jubilee of the organisation. The girls all wore insignia representing what country they were from, but a Girl Scout from Czechoslovakia explained that they had all traded insignia, buttons, bracelets and souvenirs; she added: 'I have traded everything but my shoes and skirt.' (Authors' collection)

An atmospheric view taken from within the *Aquitania*'s New York pier. The liner is in the midst of departing, the gangway has already been pulled away and many are waving farewell. A lone passenger stands, luggage in hand, apparently having quite literally missed the boat. (Steven B. Anderson Collection)

Despite these episodes, the cruise was a success. Cunard repeated it the following year, although they shortened it to thirty-three days. Their advertising copy referred to it as the '*Aquitania* Cruise De Luxe'; stops included Nassau, Panama, La Guaira, Trinidad, Bahia, Rio de Janeiro, Barbados and Bermuda. In late 1937, Cunard announced that Colonel Charles Furlong would be brought aboard the cruise as a guest lecturer; he had spent a great deal of time in South America exploring, and had even discovered the wreck of the US frigate *Philadelphia* in the Mediterranean.[147] Total numbers of passengers cruising to Rio from New York were up through 1937; however, *Aquitania* carried fewer, some 452. The decrease may have been due to the fact that there were more, and newer, ships on this route: the new French Line flagship *Normandie* took about

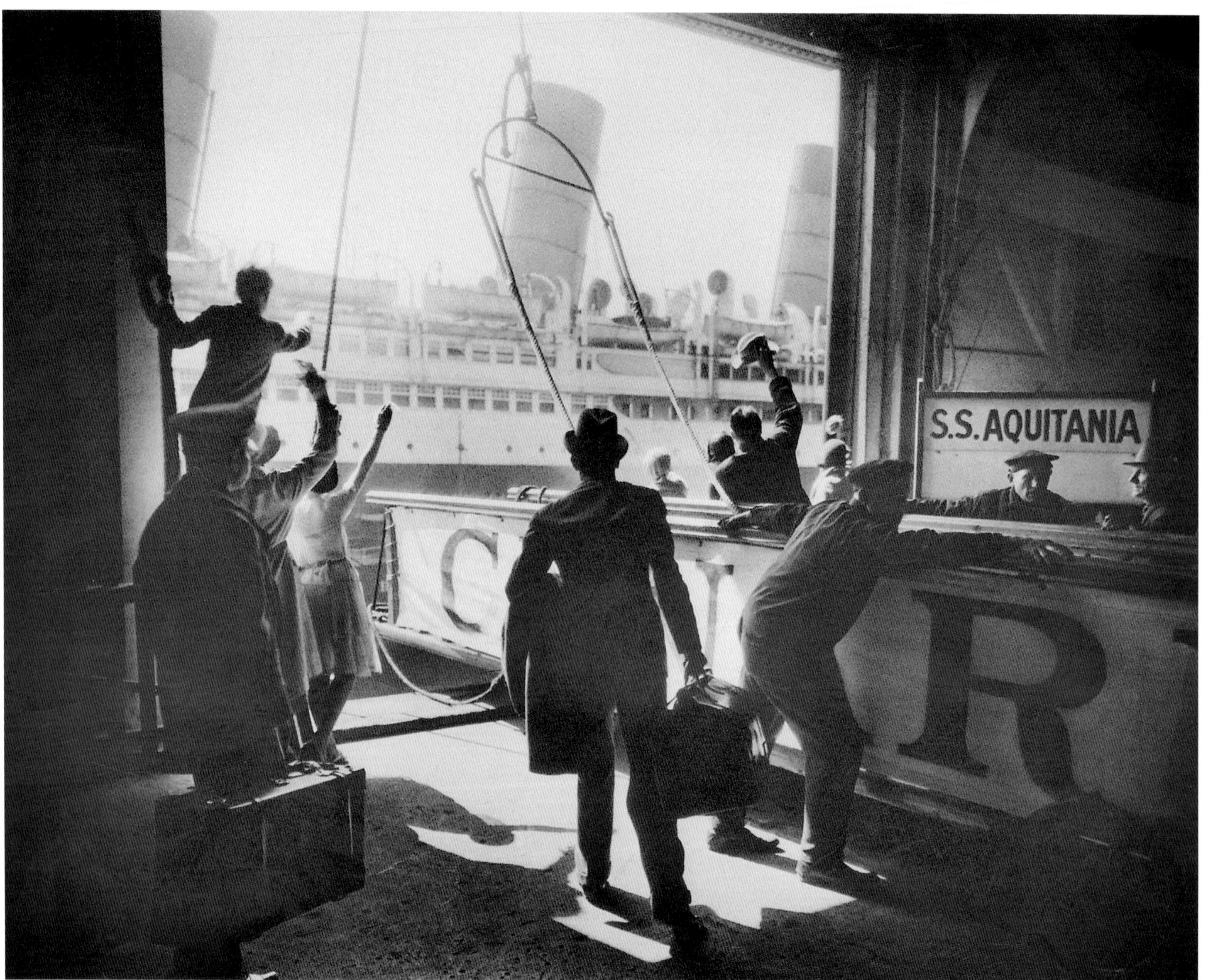

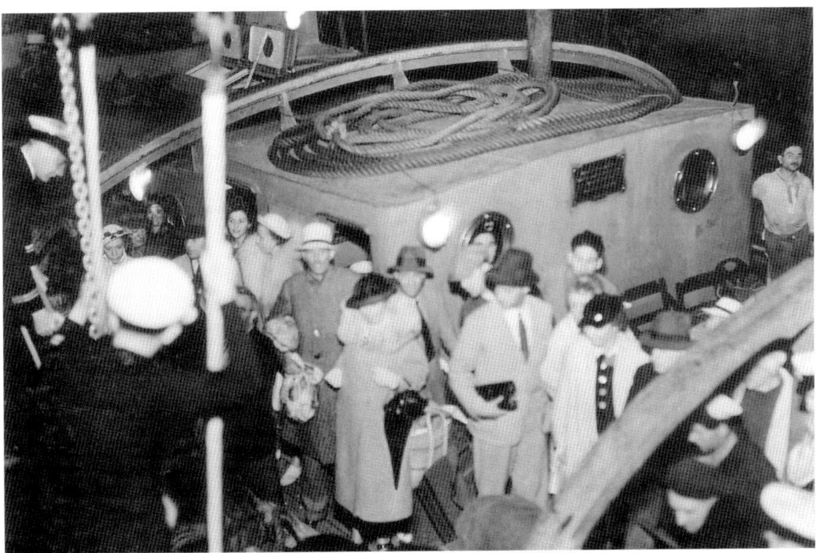

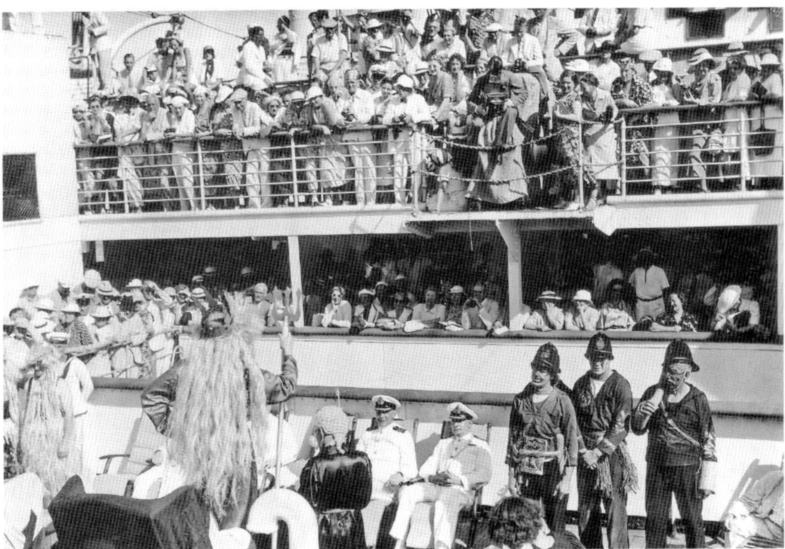

1,000 passengers; the Italian liner *Rex* and the *Empress of Australia* were making similar cruises. And while the *Aquitania* held the distinction of being the largest liner to cross below the equator in 1937, the *Normandie* took that record in 1938.

On 21 March 1938, the Cunarder *Berengaria* was suddenly removed from service after a spate of on-board fires. With the *Olympic*, *Homeric*, *Mauretania* and *Majestic* all gone, and the new *Queen Elizabeth* not yet in service, the *Aquitania* was retained on the North Atlantic service through the winter of 1938–9. It was planned to keep her in service until the new *Elizabeth* entered service in 1940. By late August, she had carried over half a million passengers since the Great War, over the course of 582 crossings or 291 round-trip voyages. Yet as the month drew to a close, the political scene in Europe had again reached a crisis that would engulf the whole world.

Above left: The first in a series of twelve photographs depicting one of the *Aquitania*'s two cruises to Rio de Janeiro. There seems to be very little way of determining whether the photos were taken in 1937 or 1938, but they would clearly represent what it was like either year. In this view, passengers are disembarking a ferry, returning to the *Aquitania*, after dark. Many of them are carrying souvenirs. (Ioannis Georgiou Collection)

Above right: The ancient tradition of the 'crossing the line' ceremony was carried out aboard the *Aquitania* as she crossed the equator. It had been performed on the ship once before, when the liner was steaming from the River Tyne after her post-war refit, and before she had re-entered service; this time, however, it was performed at the proper latitude. Here an enormous crowd of passengers has gathered to watch the proceedings. Even the captain and staff captain – seated, dressed in their summer whites – have come down to watch it all. (Ioannis Georgiou Collection)

Below: A portion of the ceremony is carried out in the exterior temporary swimming pool. (Ioannis Georgiou Collection)

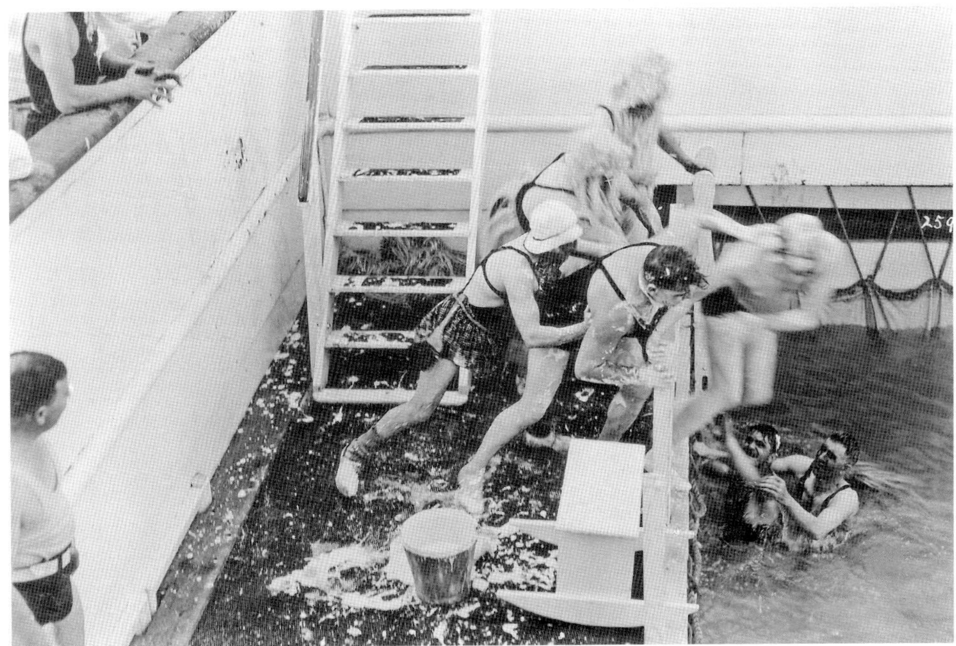

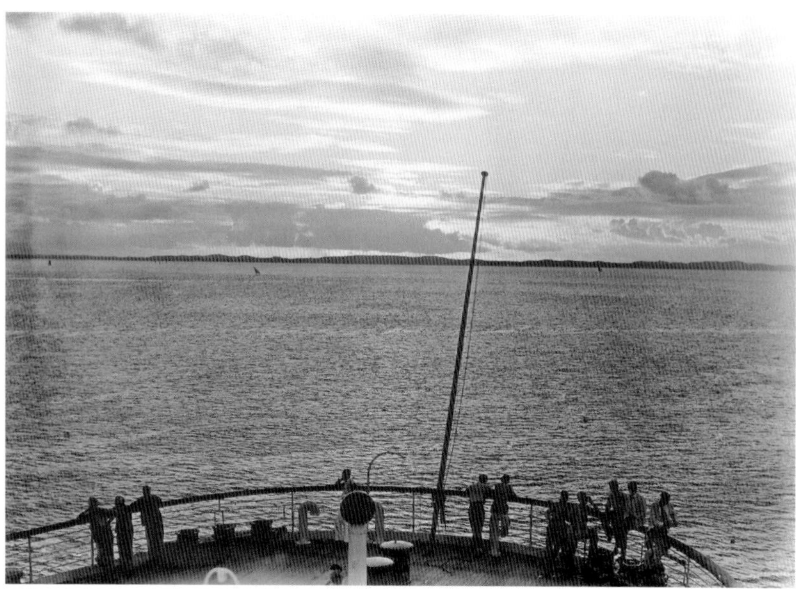

Aquitania's passengers relax on the fantail during sunset along the coast of Bahia, Brazil. (Ioannis Georgiou Collection)

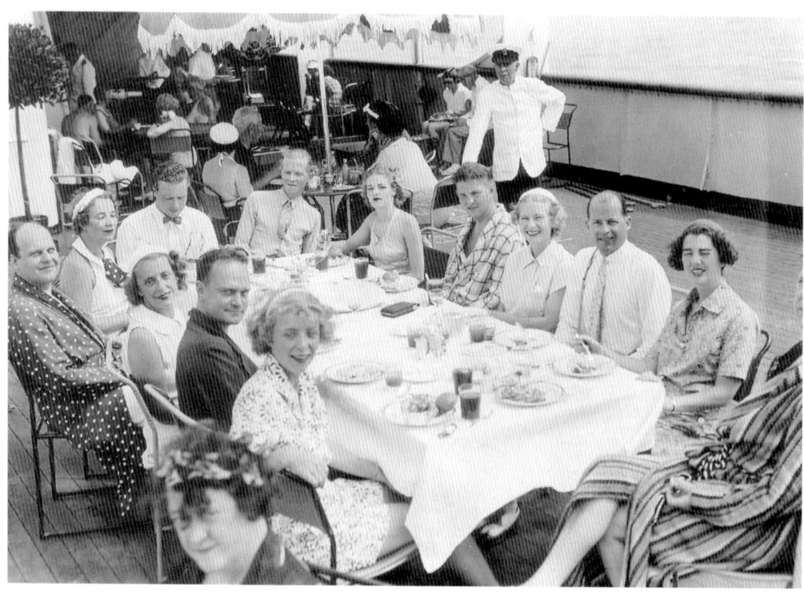

Breakfast on deck. In this photo, a group of passengers enjoys a very informal meal; many have not even bothered to dress and are still in their robes, while others seem to have made more of an effort. An attentive steward stands nearby. (Ioannis Georgiou Collection)

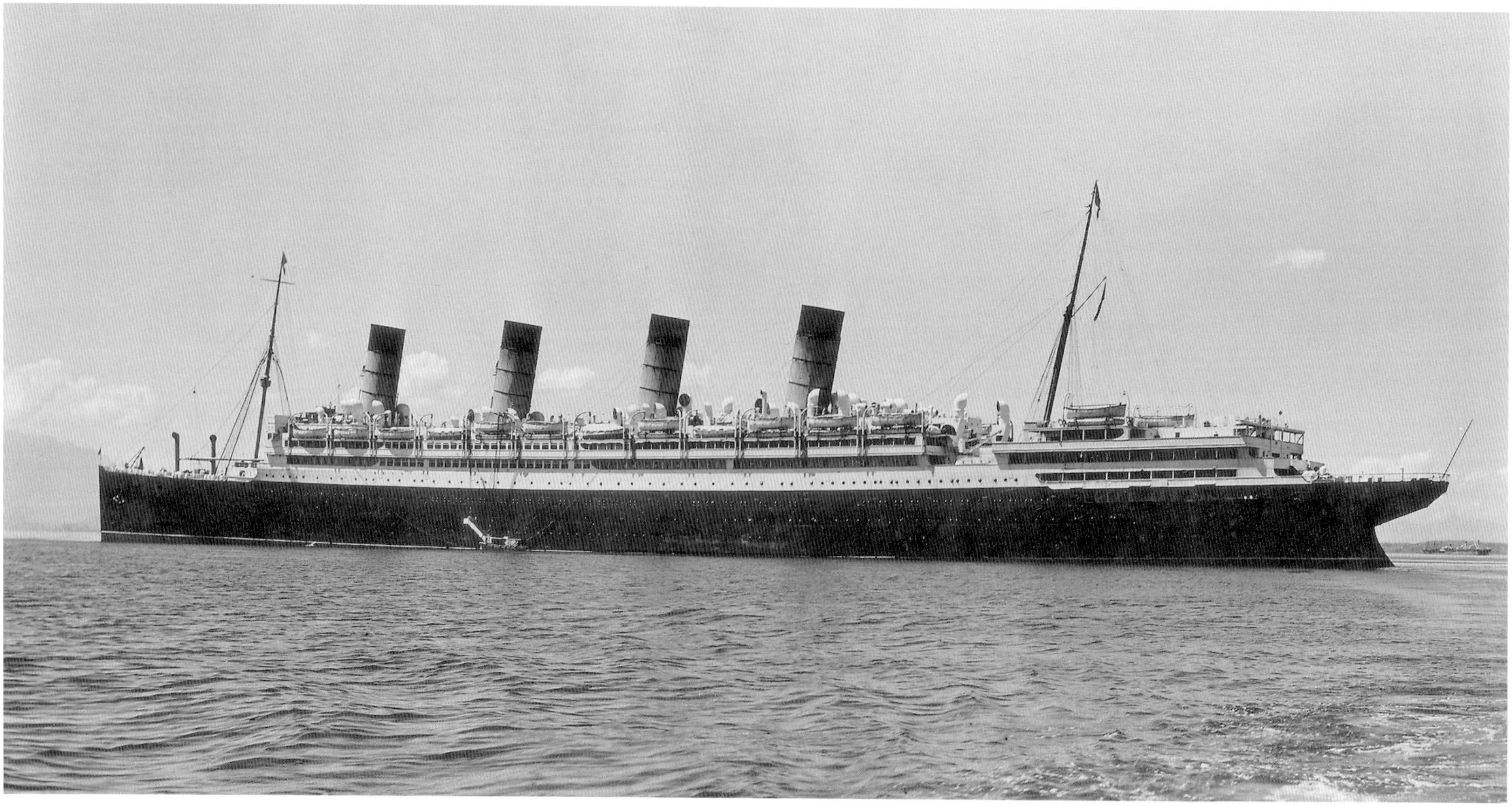

A splendid view of the *Aquitania* at Rio de Janeiro, Brazil. (Ioannis Georgiou Collection)

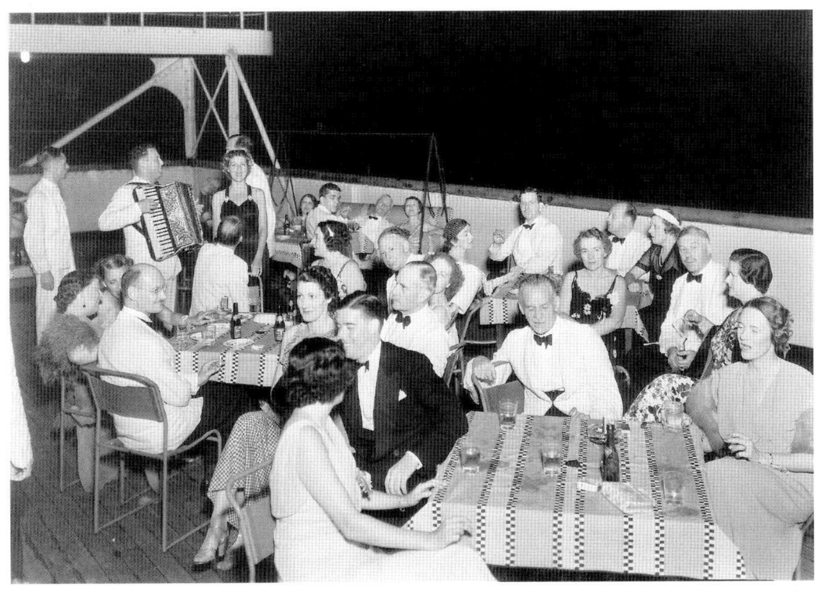

A woman serenades passengers enjoying drinks and snacks by the Stern Docking Bridge during a night of the cruise. Some of the passengers are singing along, others seem somewhat amused, and others – particularly those on the swing – look to be on the verge of falling asleep. (Ioannis Georgiou Collection)

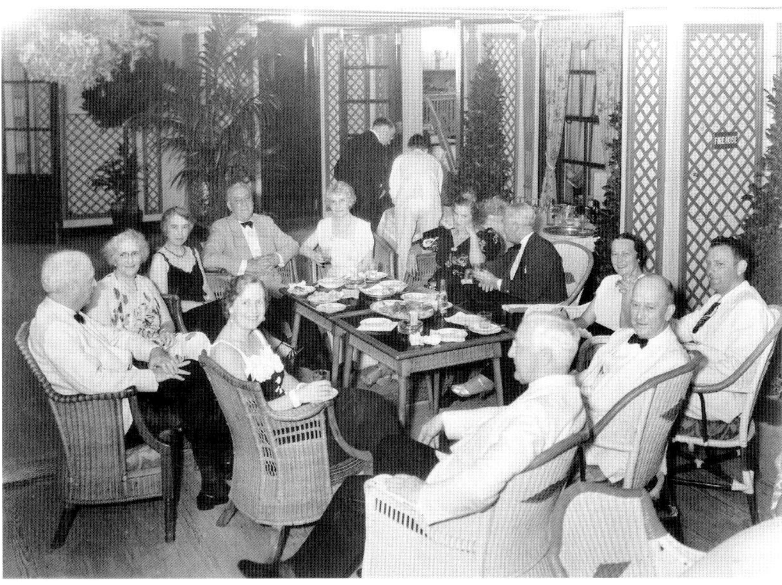

A group of passengers enjoys each others' company on one of the Verandah Cafes. Captain E.G. Diggle recalled of these matching rooms that during the evenings, 'multi-coloured fairy lamps shed their soft radiance on the forms of beautiful women gloriously gowned and handsome men in immaculate evening dress, who dance the hours away to the music provided by jolly orchestras'. (Ioannis Georgiou Collection)

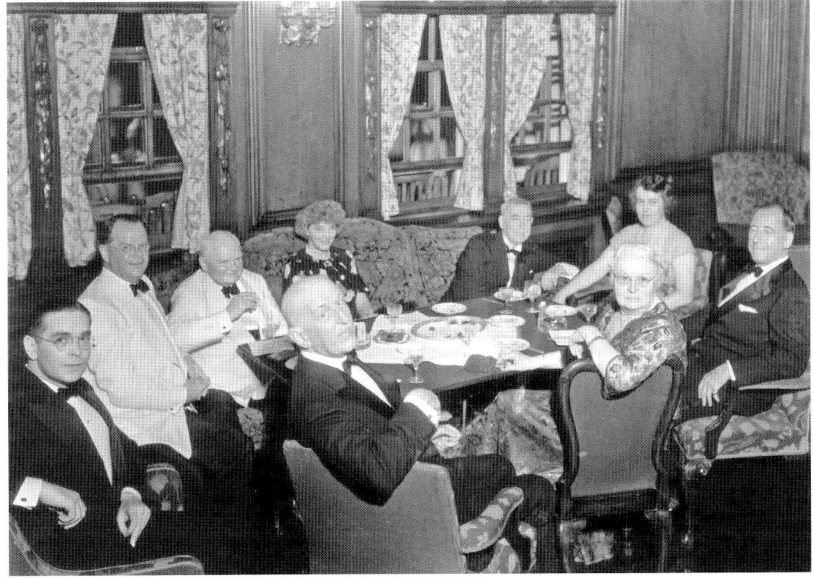

A group of people enjoy a meal just inside, with the windows behind them open to allow in the fresh sea air. (Ioannis Georgiou Collection)

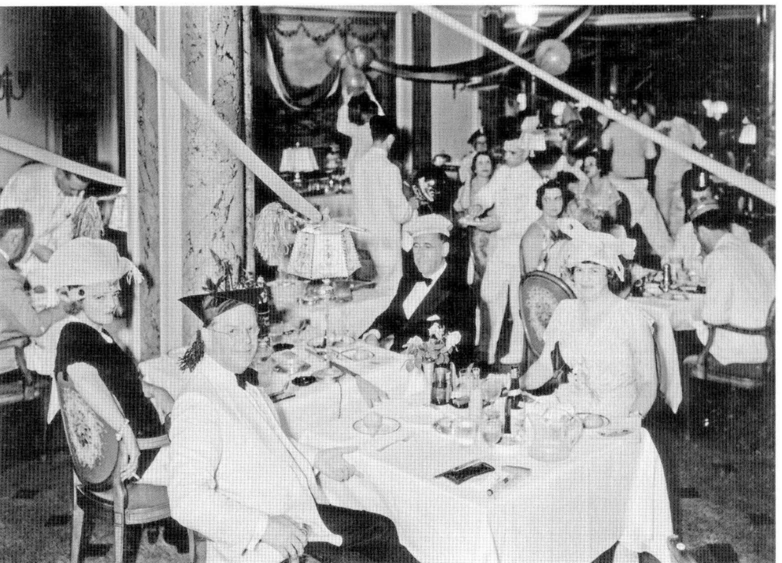

A silly hat party in the First-Class Restaurant. (Ioannis Georgiou Collection)

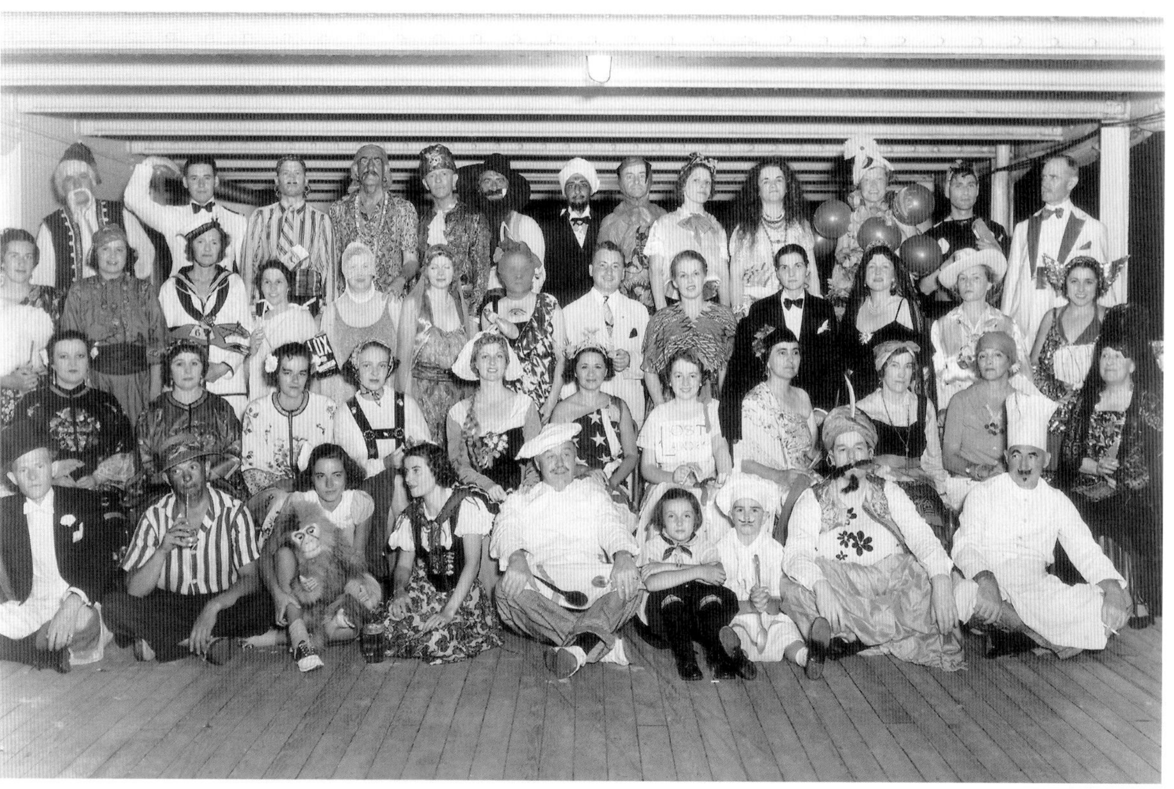

This is a fascinating group portrait, for everyone has dressed up in silly costumes. The degrees of success vary widely. Upon closer inspection, it also seems obvious that not everyone is the gender that you might at first conclude. There are chefs, bakers, pirates, prospectors, cowboys, sailors, a self-declared 'censor', Lady Liberty, a woman holding a shoe in a moderately threatening manner, another woman holding a sign for 'lost laundry' and a gentleman in a top hat with no trousers – among many others. (Ioannis Georgiou Collection)

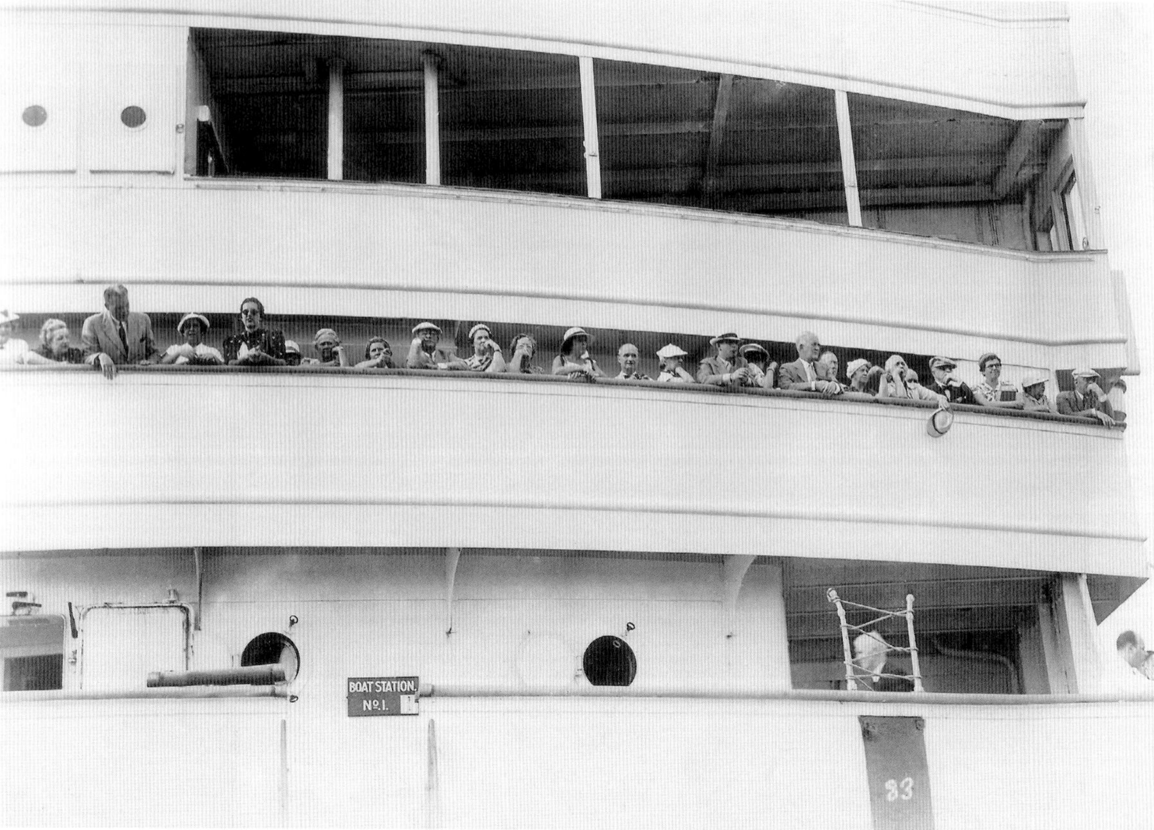

In this, the final Rio cruise photo, a line of passengers on A Deck, forward, is viewed from the forecastle. (Ioannis Georgiou Collection)

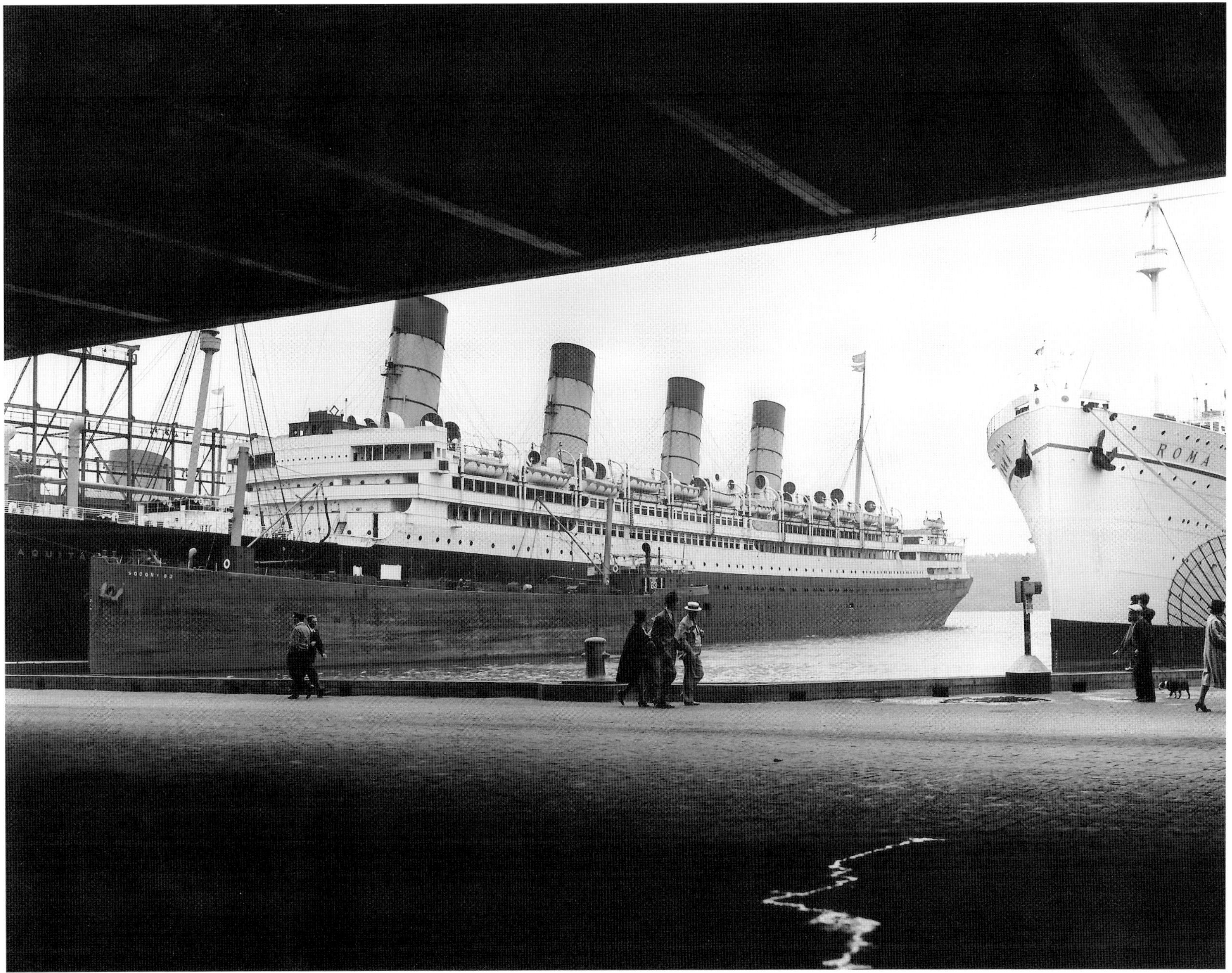

The *Aquitania* docked in New York on 29 August 1939. The scene is very quiet. Although a few pedestrians walk by on the shore, there is not a single person to be seen anywhere on the liner. The Cunard and White Star pennants fly from the mainmast. The *Roma* lies at the next pier north. This peaceful scene is particularly bizarre considering the fact that the world was on the precipice of the greatest war in history. (Clyde George Collection)

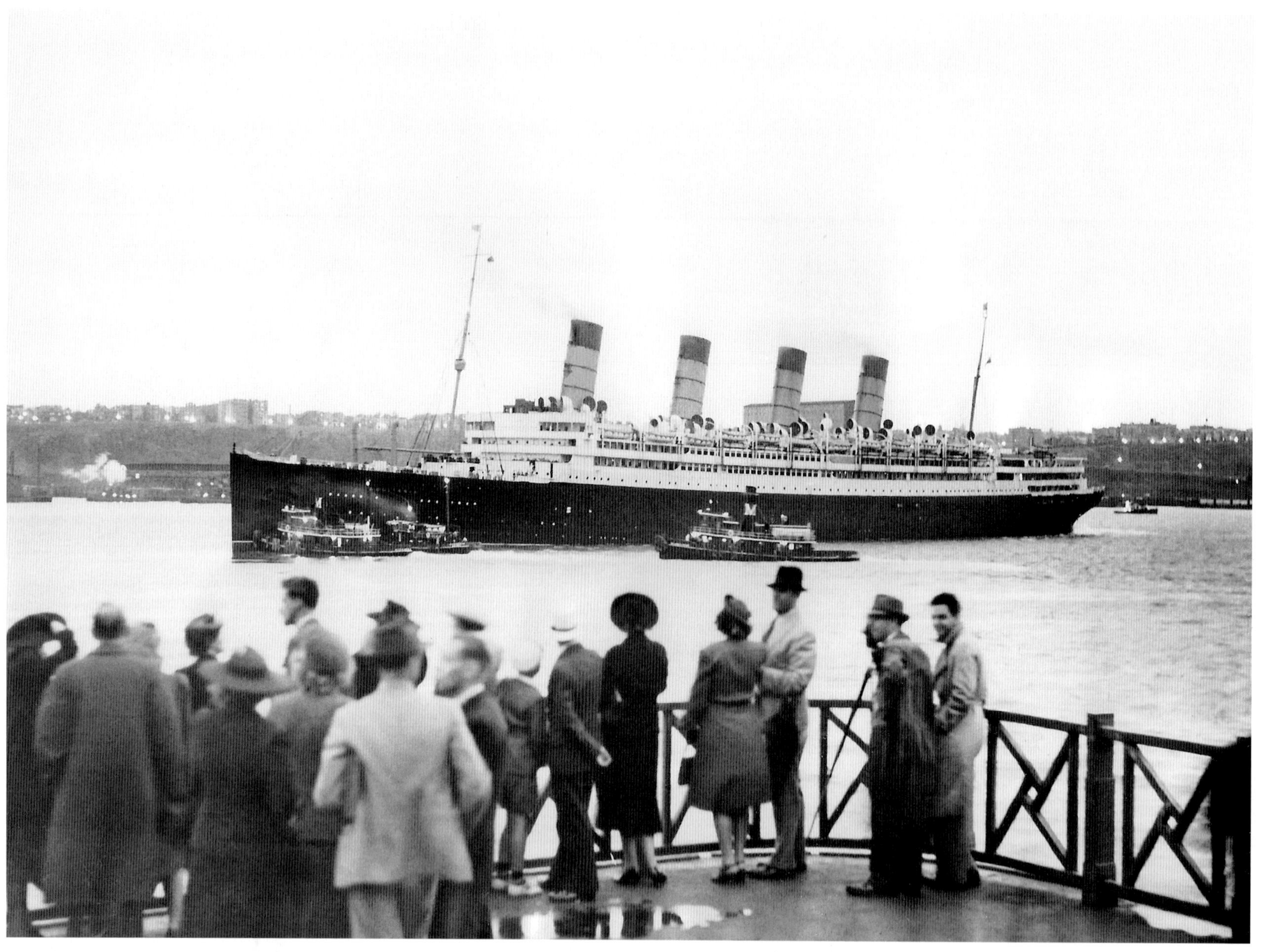

The *Aquitania* sails for England on 30 August 1939. She is seen off by a small knot of spectators, visible in the foreground. Europe would be engaged in a second world conflict before she had arrived there. (Clyde George Collection)

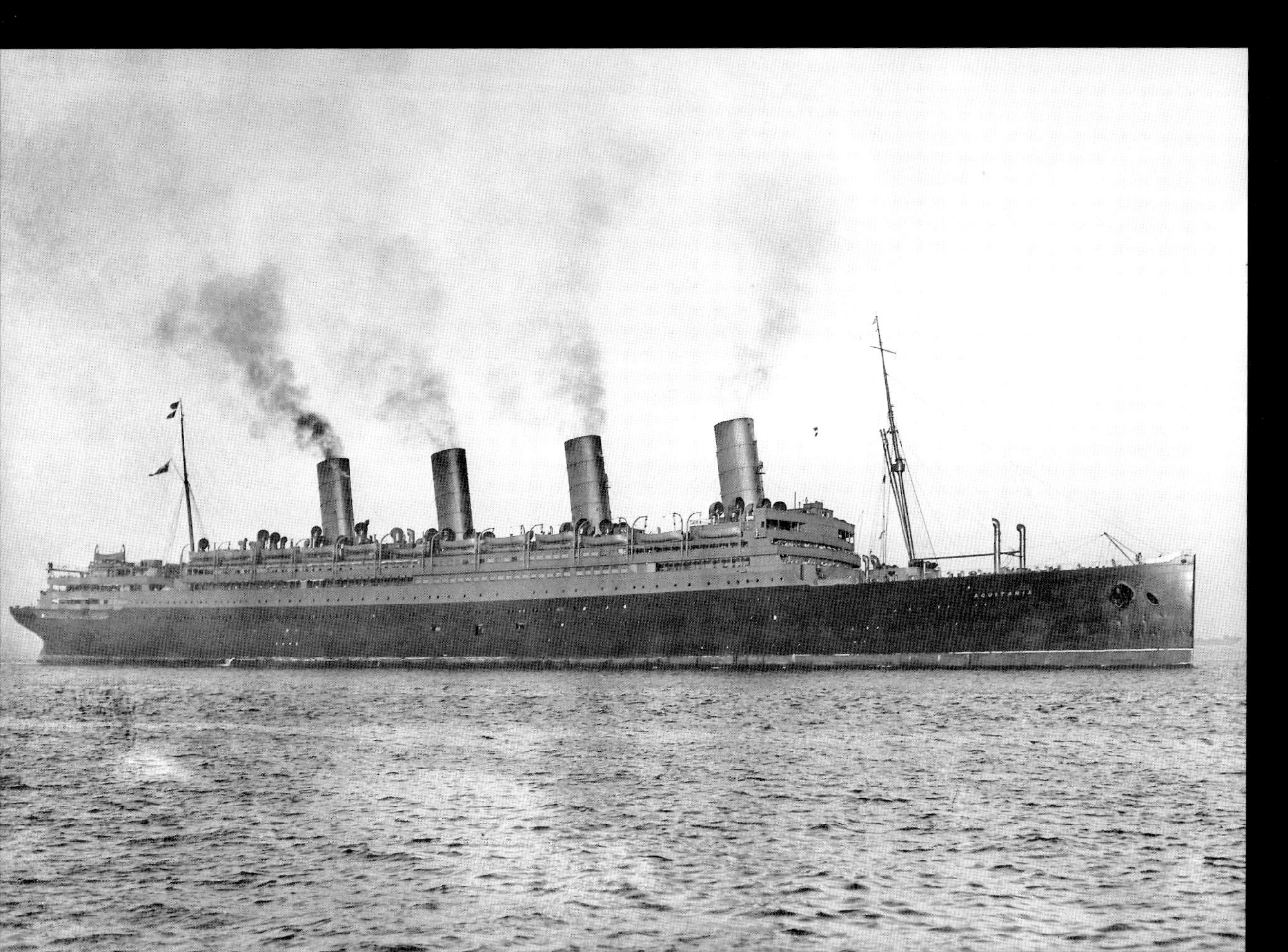

The *Aquitania* arrives in New York on 16 September 1939. It was said that she was surrounded by a 'special aura of mystery and adventure'. Her appearance was greatly altered: her white superstructure had been pained a deep grey, and all of her portholes had been blacked over. She looked 'dark and dirty' as she passed Staten Island. (Clyde George Collection)

Following the surrender of Germany in 1918, the Treaty of Versailles had imposed severe restrictions and economic burdens on the proud country, while seemingly blaming that nation alone for the war. Controversy immediately surrounded the terms of the treaty, which many felt sought punishment rather than reconciliation in Europe. Indeed, even France's Marshall Ferdinand Foch declared: 'This is not peace; it is an armistice for 20 years.' Similarly, the British economist John Maynard Keynes lambasted the economic terms of the treaty as 'outrageous and impossible', while predicting economic ruin for Germany and another war as a result.[148]

Unfortunately, these predictions proved prophetic. The treaty had the effect of creating extremely bleak financial conditions in Germany during the 1920s, a situation which was only exacerbated by the onset of the Great Depression. Unrest was rife, with the people desperately looking for a change. This created the conditions necessary for the rise to power of Adolf Hitler and the Nazi party. On 1 September 1939, Germany's invasion of Poland proved to be the opening move of the Second World War. By 3 September, Great Britain was at war with Germany once more. However, this time, the war would be much more difficult for the Allies, and there would be less surprise when civilians and civilian passenger liners became targets.

The outbreak of war would prove to be a new lease of life for the *Aquitania,* effectively ending any talk that she would be taken out of service or scrapped when the *Queen Elizabeth* entered service.

In late August 1939, with the build-up of German troops along the Polish border making war look inevitable, the *Aquitania*'s crew were ordered to extinguish her deck lights and to paint over or block her windows; her black hull was retained, but her upper works were painted in a very dark grey. All of this was to help prevent the vessel from being spotted by U-boats. Once war was declared, the ship, which was sailing east in the North Atlantic towards Southampton at the time, immediately increased speed, and began zig-zagging until it reached port.

As was the case at the onset of the First World War in 1914, many American citizens in Europe suddenly found themselves stranded in a war zone. A large number decided to return to the perceived safety of the neutral United States. The crews of American ships began flying prominent American flags and painting unmistakable large American flags on their vessels, in order to prevent them from being targeted by U-boats. These ships were quickly dispatched to bring fleeing American citizens home from Europe, but it would take time for them to arrive. Thus, when the Cunard Line announced that the *Aquitania* would depart on its next scheduled voyage despite the hazards, many American citizens seized the opportunity to book passage. The ship was armed with two 'defensive' 12-pound guns on her fantail.

Having learned the hard way with the loss of many civilian vessels during the First World War, including the *Lusitania*, the American ambassador to Great Britain, Joseph P. Kennedy, decided to take no chances. All of the American citizens boarding the vessel were herded into the Palladian Lounge. In this elegant and sophisticated setting, the realities of warfare were hammered home to the captive audience. The Consul General at the American Embassy in London, G.K. Donald, was sent by Ambassador Kennedy to deliver a blunt message: sailing on belligerent vessels, especially if they were convoyed, could lead to the opposing belligerent claiming the right 'to sink them without warning'. In other words: there were no guarantees of safety. Word had begun to spread about the torpedoing of the Cunard Line's *Athenia* on 3 September, only contributing to the misgivings and fears. Because of this, a number of American passengers – a Cunard official claimed it was only six people – lost their nerve and decided to disembark from the ship, opting instead to wait for the arrival of neutral American ships.

Setting sail on 9 September with 1,625 passengers, including 669 Americans, the *Aquitania* quickly slipped out to sea, omitting her customary stop at Cherbourg in France. This was done to avoid any U-boats which might have been lurking along the ship's expected route.[149] Despite the misgivings, *Aquitania* made it safely to New York on 16 September. The *Aquitania* spent part of September at Pier 90 in New York Harbor, awaiting a refit as a troop vessel. Fellow vessels lying in wait nearby included the *Queen Mary* and the interned French vessels *Île de France* and *Normandie. Aquitania* then returned to England.

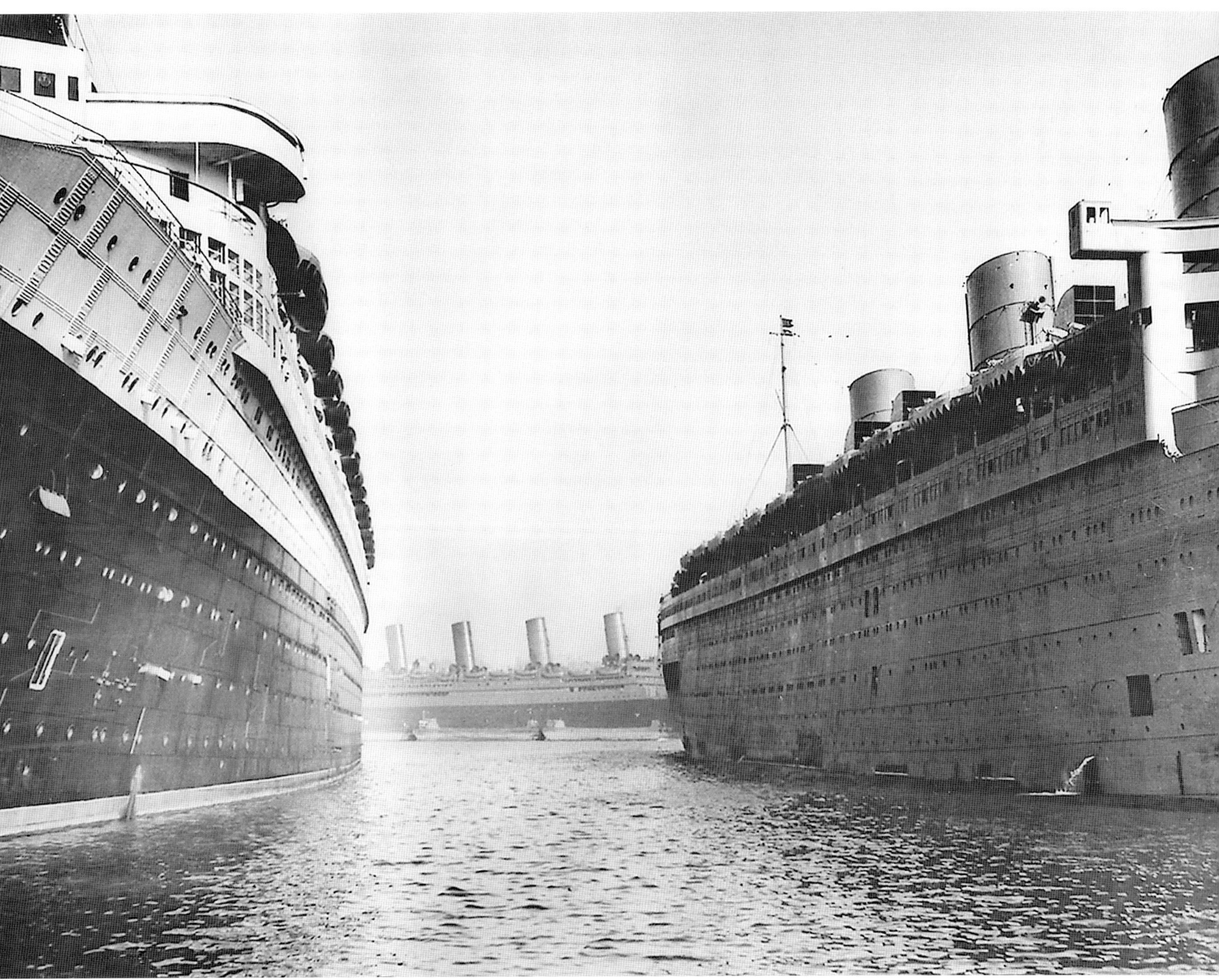

The *Queen Mary*, *Normandie* and a number of other great liners were tied up at Manhattan's Lower West Side Chelsea piers when the *Aquitania* arrived on 16 September. The *Aquitania* (centre) is framed by the SS *Normandie* (left) and RMS *Queen Mary* (right). The two British ships would both survive the war; the French liner had made her final crossing of the Atlantic. (Steven B. Anderson Collection)

Once again, *Aquitania* was tapped for government service. Already in possession of a distinguished war record, *Aquitania* was the sole remaining Atlantic liner that had served its country during the First World War. Despite being 25 years old, she was about to enter the most gruelling and distinguished phase of her career. On 18 November, *Aquitania* was officially requisitioned as a troop transport. She initially had berths for over 2,800 troops, although that capacity was increased later in the war, and she routinely carried over 3,000 per trip. These totals were a far cry from the over 6,000 troops she averaged on some voyages during the First World War. As part of the conversion, *Aquitania*'s hull was painted battleship grey. While much duller than the dazzle camouflage design that she sported during the latter days of the First World War, this new colour scheme did make the vessel less visible than the standard peacetime Cunard colours.

Due to her speed, the *Aquitania* was capable of 'usually proceeding without escort between North America and UK/Mediterranean'. She was also one of a select group of liners that were dubbed 'monsters' for their troop capacity. Some of the other vessels that earned this nickname included the *Queen Mary*, its new sister ship the *Queen Elizabeth, Mauretania* (1938), *Île de France* and *Nieuw Amsterdam*.[150]

Aquitania began transporting Canadian troops from Halifax to England in November 1939. The ship would be employed in this duty until being dry-docked late in February 1940. At that time, in addition to repair work, defensive weaponry, including two 6in deck guns and several smaller guns, were installed. *Aquitania* resumed service in March, at which point she was sent to the southern hemisphere to transport Australian and New Zealand troops. As she arrived, the Allies were forming several convoys that were designated by the moniker 'US'. *Aquitania* was assigned to the 'fast group' convoy, which was named US.3. Convoys could only travel as fast as their slowest ship, which is why the *Aquitania* was assigned to this elite group. Other ships

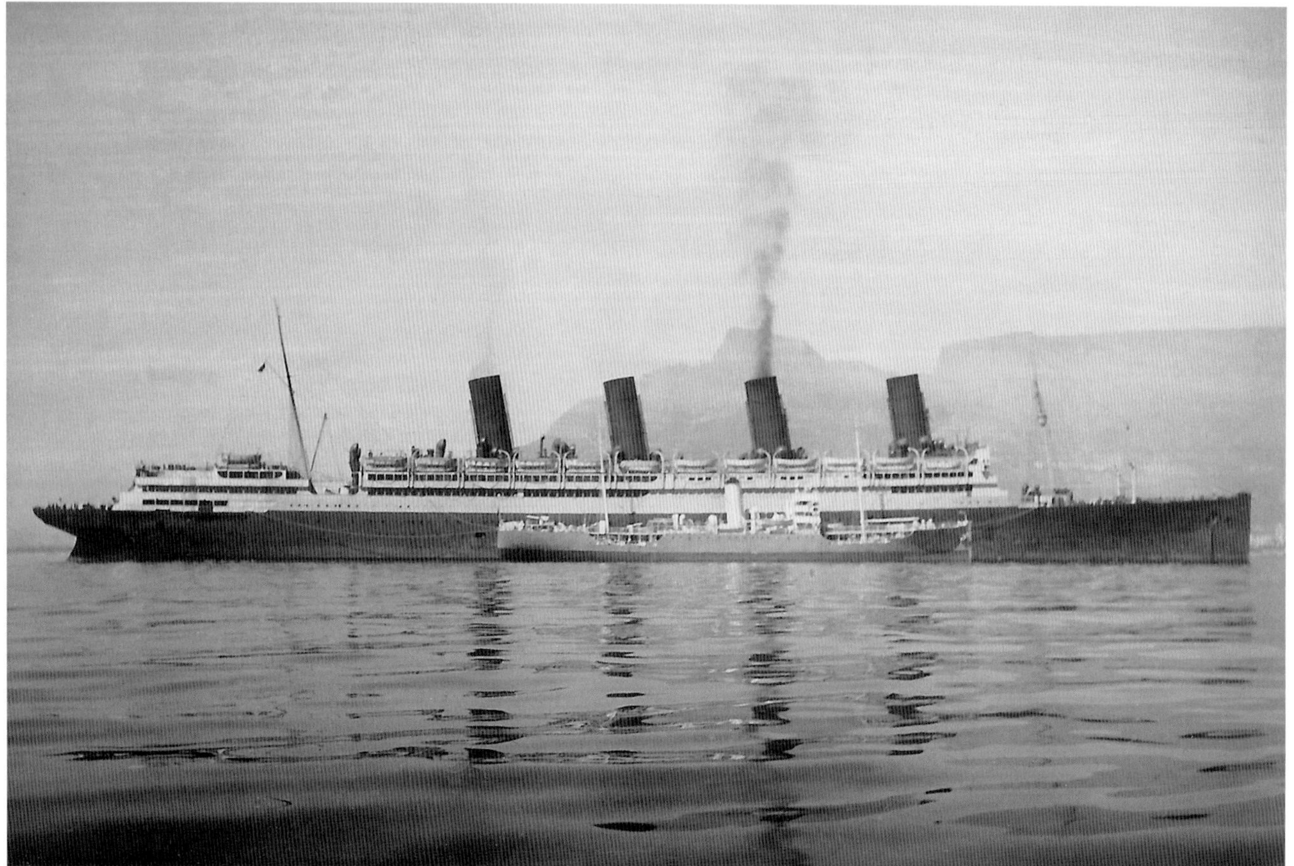

This photograph shows the *Aquitania* during her first visit to Capetown, South Africa. She was then sailing to New Zealand to bring troops back to the fronts. Unfortunately, she had run aground on 17 March as she was discharging the pilot. Unable to free herself, a tanker was brought alongside and some 1,800 tons of oil fuel were offloaded to the smaller vessel. She was refloated the following day and no serious damage was reported. (Clyde George Collection)

in US.3 included the *Queen Mary*, *Mauretania* (1938), *Empress of Britain*, *Empress of Canada*, *Empress of Japan* and the *Andes*.[151]

In May 1940, *Aquitania*, along with the *Empress of Britain* and *Empress of Japan*, sailed to Wellington Harbour to transport New Zealand troops.[152] As the war proceeded, *Aquitania* proved to be a workhorse once again. The ship was assigned to new routes and made numerous trooping voyages out of Sydney, including journeys to Madagascar, Bombay and Singapore. She also made two voyages between Pearl Harbor in Hawaii and San Francisco, California, before the year was out.

Aquitania turned out to be as popular with the troops she carried as she had been with civilians during peacetime. The vessel was older than many of the troops she carried. This led to her acquiring several nicknames that were clearly terms of endearment, such as 'The Grand Old Lady', 'Old Irrepressible' and 'Old Granny'. *Aquitania* often travelled alone, rather than as part of its designated convoy. Her top speed allowed her to travel solo without undue risk from Japanese submarines, rather than steaming at a lesser speed to allow slower vessels to keep pace, which would have placed her at greater risk.

The *Aquitania* was involved in a significant rescue late in 1941, and as a result would be embroiled in controversy for decades. Having departed Singapore on 19 November, the ship was steaming for Sydney, Australia. On 23 November, *Aquitania* was over 120 miles off the west coast of Australia when a crewmember spotted something strange in the water:

> Just before 0600 on Sunday 23 November, a cabin boy on the liner-transport *Aquitania* saw a low-lying raft bobbing on the pearly morning sea. The 26 men on the poorly equipped raft had seen her long ago, and were waiting anxiously for a sign that they had been noticed.

At first, the *Aquitania*'s crewmembers believed the men in the rafts were survivors of a German surface raider attack. It was also feared that the enemy could still be lurking in the area. Due to this, after quickly pulling the men aboard, *Aquitania* resumed its voyage to Sydney. *Aquitania*'s

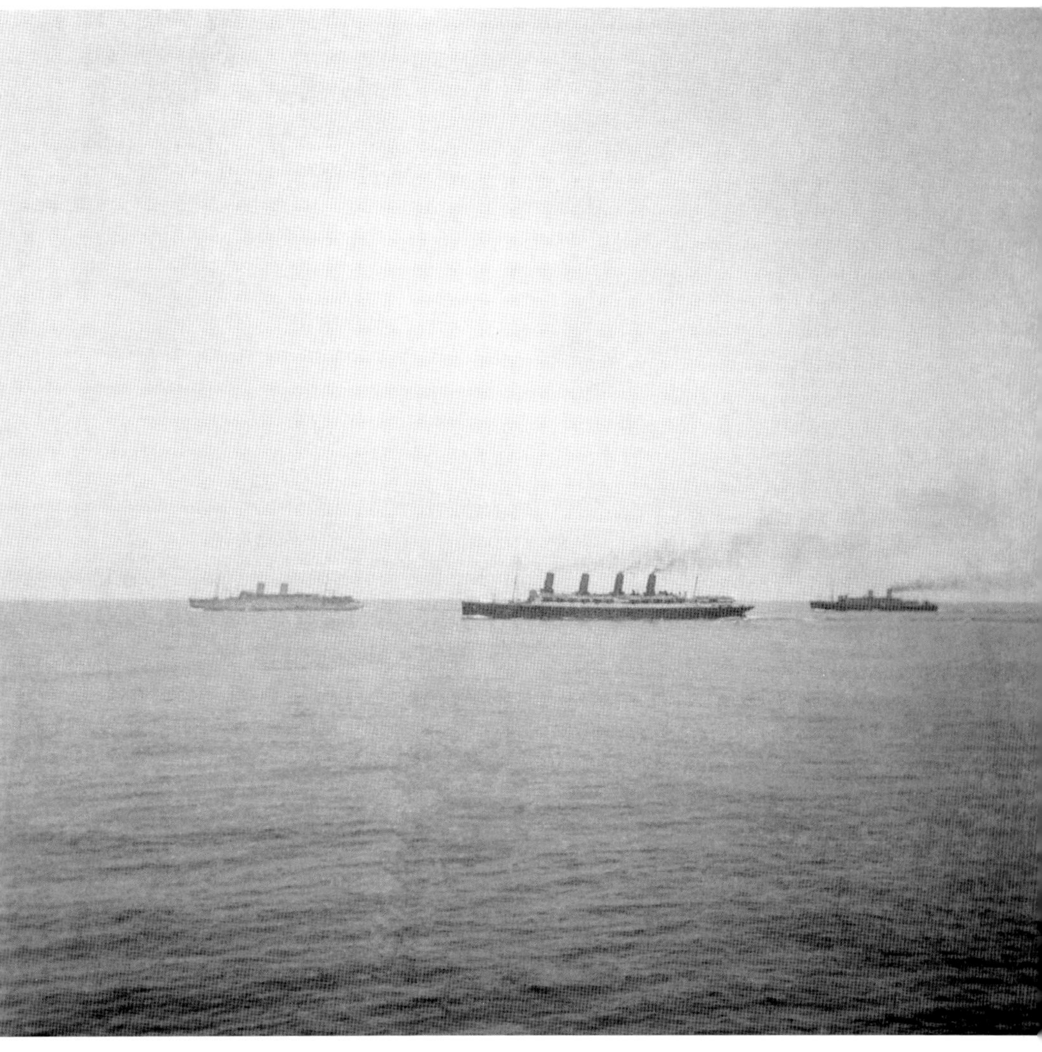

This image of the *Aquitania* (centre) was taken from the deck of the *Queen Mary* by a member of the Second Australian Imperial Force. The *Aquitania* was sailing as part of a convoy, which transported troops between Sydney, Australia and Africa. This voyage lasted from 26 December 1940 to around 10 January 1941. (Alexander Warton Collection)

commander was under strict orders to maintain radio silence to avoid being tracked by enemies. Thus, the crew's discovery was not radioed in until 27 November, when she arrived off Wilson's Promontory.

As it turned out, the survivors, rather than being the victims of a German raider, were actually the crewmembers of one. On 19 November, the German auxiliary cruiser *Kormoran* had encountered the Australian light cruiser HMAS *Sydney*. Catching the latter vessel by surprise, the *Kormoran* managed to sink the *Sydney* after a 30-minute surface battle, but was badly damaged in the exchange. With fires raging throughout the ship, the *Kormoran*'s crew was forced to scuttle their vessel before the flames could detonate the ship's magazine.

The *Aquitania* had stumbled across one of the lifeboats of survivors from *Kormoran*. Another lifeboat full of survivors was some distance away and actually witnessed the *Aquitania* picking up their comrades. However, the commander of the *Kormoran*, Theodor Detemers, was aboard this second lifeboat, and since the *Aquitania*'s crew hadn't sighted them, he had those aboard keep quiet and not draw attention to themselves, hoping to be rescued by a neutral vessel instead.

Out of 399 men aboard, eventually 319 were rescued. However, the exact opposite was true of those aboard the *Sydney*. None of its 645 crewmen were ever seen again. Following the rescue, some criticised *Aquitania*'s crew for having rescued German sailors, while none of *Sydney*'s crewmen survived.

Conspiracy theories flourished in the years following this incident, since many did not want to accept the German survivors' version of how the *Sydney* was lost, and because details of the sinking had been censored during wartime. Some critics of the official story began theorising that a Japanese submarine had actually sunk the *Sydney,* and that all of the survivors were killed in the water, or alternately, picked up by a Japanese vessel, and then later killed. They surmised that the true details of the incident were subject to cover-up.

Another, more extreme version of this conspiracy suggested that the Navy Office was aware of the *Sydney*'s loss earlier than they stated, and that they deliberately delayed the rescue. It was also alleged by several individuals that the *Aquitania* disobeyed orders and broke radio silence on 23 November, reporting to higher command their discovery of the *Kormoran* survivors and of the sinking of the *Sydney* that same day. However, the official record shows that the Navy Office was not even aware that the *Sydney* had been sunk until 24 November, when the British tanker MV *Trocas* picked up twenty-five additional survivors of the *Kormoran*. Immediately thereafter a search and rescue operation was launched to look for the *Sydney*.

Aquitania's crew would be subject to these conspiracy theories for years, despite adamant statements that they had maintained radio silence, and that they were not initially aware of the *Sydney*'s loss. The allegations have been investigated numerous times over the years, and

no convincing evidence to support their validity has ever surfaced. The theories, while probably being formulated to explain how the entire crew of the *Sydney* were lost, actually cast unfair suspicion and blame on the crew of the *Aquitania*; in reality, and in the finest traditions of the sea, they had risked their vessel's well-being to rescue the lifeboat's load of human cargo.[153]

The *Aquitania* had a close brush with disaster in June 1941, although nobody aboard the ship knew it at the time. On the night of 25/26 June, under the cover of darkness, the German minelayer *Adjutant* dropped ten magnetic mines, each containing 540kg of high explosives, at the entrance to Wellington Harbour. The *Aquitania* and other departing ships moored there were the intended targets.

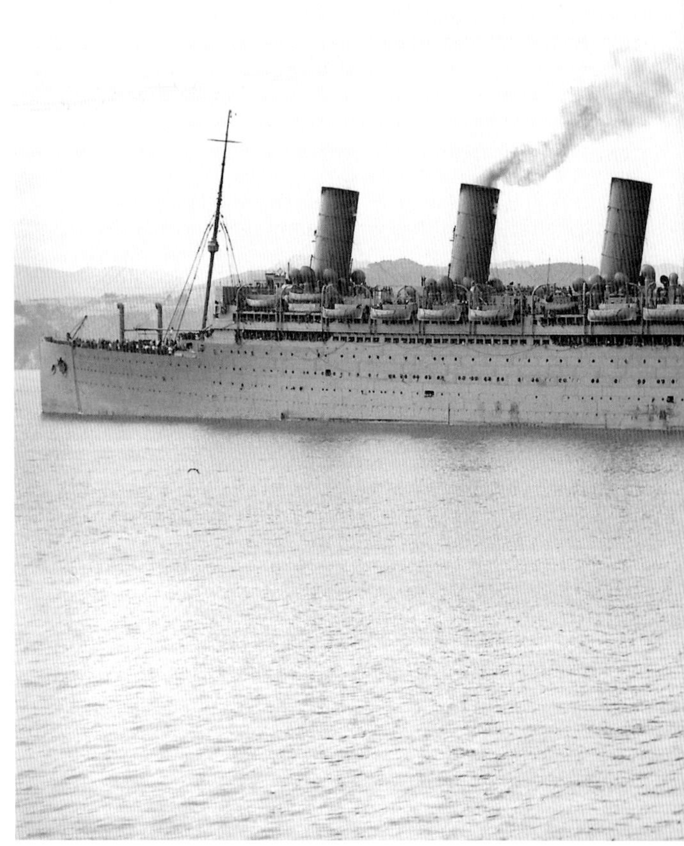

In September 1941, the *Aquitania*, painted battleship grey, departs Wellington Harbour, New Zealand, on a trooping voyage. (Steven B. Anderson Collection)

On the afternoon of 27 June, the *Aquitania*, with 4,000 New Zealand troops aboard, passed Pencarrow Head and headed out into Cook Strait. This course took the vessel directly over the location where the *Adjutant* had reportedly laid its mines. However, for reasons unknown, none of the mines detonated and the *Aquitania* escaped unscathed. None of the crew had any idea of the danger they faced that day, in what appeared to be a routine wartime voyage. In the decades since the war, naval searches have never located the mines on the seabed, and concerns have been raised about the potential threat to modern-day shipping that they could still pose.[154]

With the Japanese attack on Pearl Harbor on 7 December 1941 and the subsequent entrance of the United States

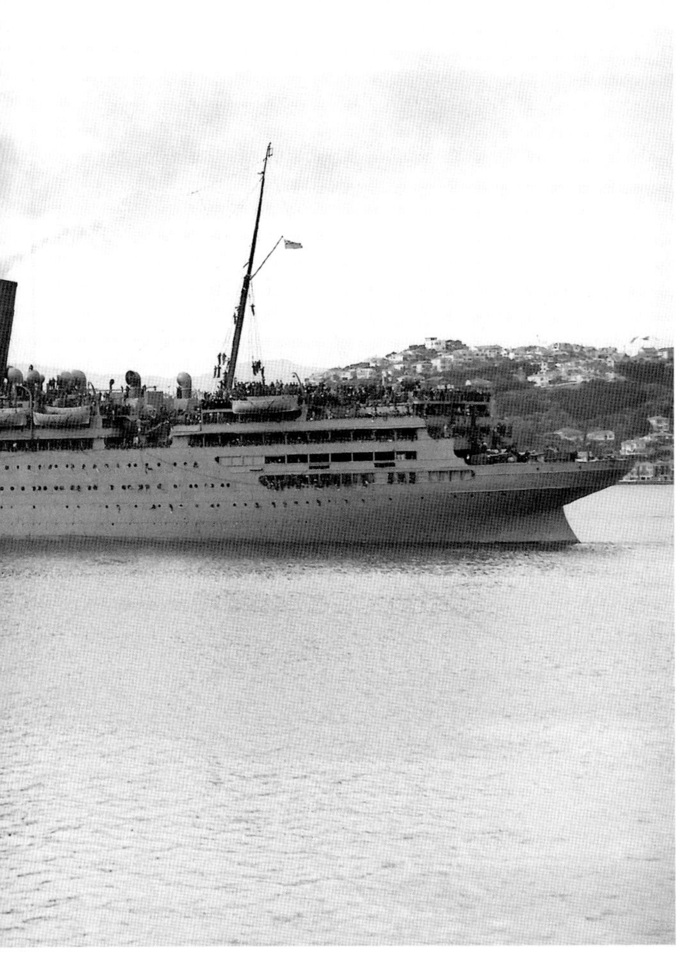

into the war, *Aquitania* was soon assigned to other duties. In January 1942, London had tentatively promised the *Aquitania* for February sailings from the west coast of the United States, since additional troops and supplies were badly needed in the Middle East. However, this plan went awry when *Aquitania* was dry-docked for repairs in Boston in February. After the repairs were completed, *Aquitania* was assigned to the Honolulu run. The Allies had determined that the ship's deep draft made its anchorage in Australian ports and intermediate ports hazardous.[155]

On 22 February, *Aquitania* departed Pearl Harbor with civilian evacuees, bound for San Francisco. On her return voyage to Hawaii, the ship transported American troops to the islands. Between March and April 1942, *Aquitania* transported 12,000 troops to Hawaii and many additional civilian evacuees to the United States.

On 21 September 1942, *Aquitania* departed New York, bound for Rio de Janeiro. The vessel was packed tight, with 7,000 troops aboard, along with necessary supplies and materials. Overloaded by about 3,500 tons and riding low in the water, strain was undoubtedly placed upon the ship's hull. Remarkably, despite the risks, *Aquitania* arrived safely.[156] The number of troops she carried is even more impressive when considering that her previous personal record, set during the First World War, was 6,900 on one crossing.

In January 1943, *Aquitania* was assigned to perhaps the most impressive convoy that she would ever be a part of. Sailing to Suez, the ship rendezvoused with the *Queen Mary*, *Île de France*, *Nieuw Amsterdam* and the *Queen of Bermuda*. More than 31,000 troops were berthed aboard the five ships. This impressive group set sail on 25 January, bound for Sydney. The slowest ship in the convoy was the *Queen of Bermuda*, with a top speed of 18 knots, which limited the pace at which they could sail. Because of this, the voyage would last a month. However, all of the vessels arrived safely in Australia on 27 February.

Aquitania departed for New York following this gruelling voyage, with a layover in Rio de Janeiro on the way. Following her arrival in New York on 4 May 1943, additional deck guns were installed, along with two-dozen smaller guns and four rocket launchers. The reason for this defensive upgrade was made clear when it was revealed that the *Aquitania* was being transferred to the North Atlantic route, to transport

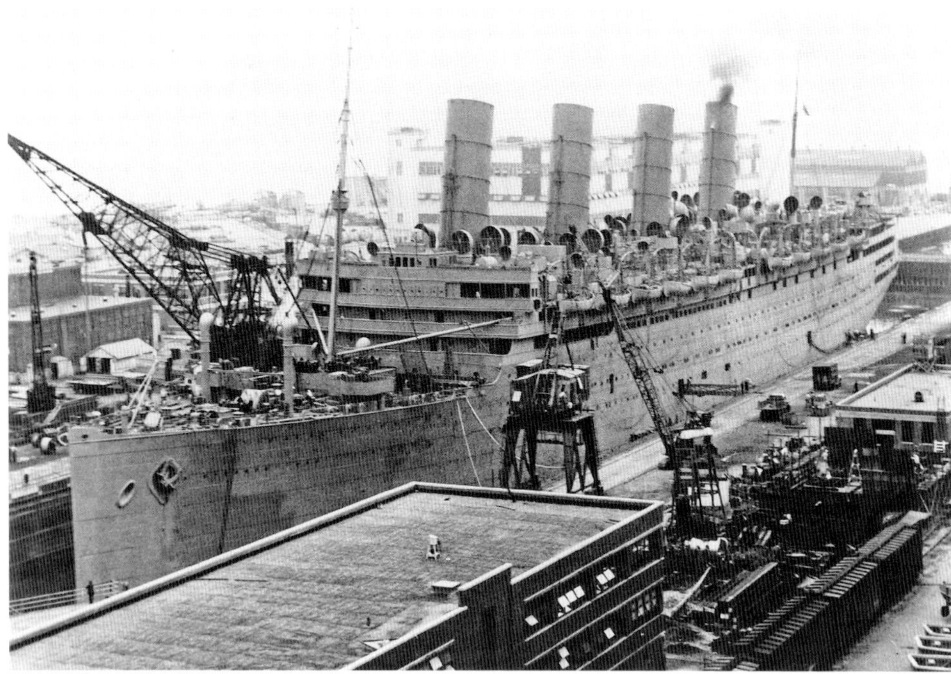

The *Aquitania* rests in the Boston Navy Yard dry dock in August or September of 1942. Here she received an overhaul, and an American gun crew was assigned to the liner, to work along with the British gun crew already aboard. (National Archives & Records Administration)

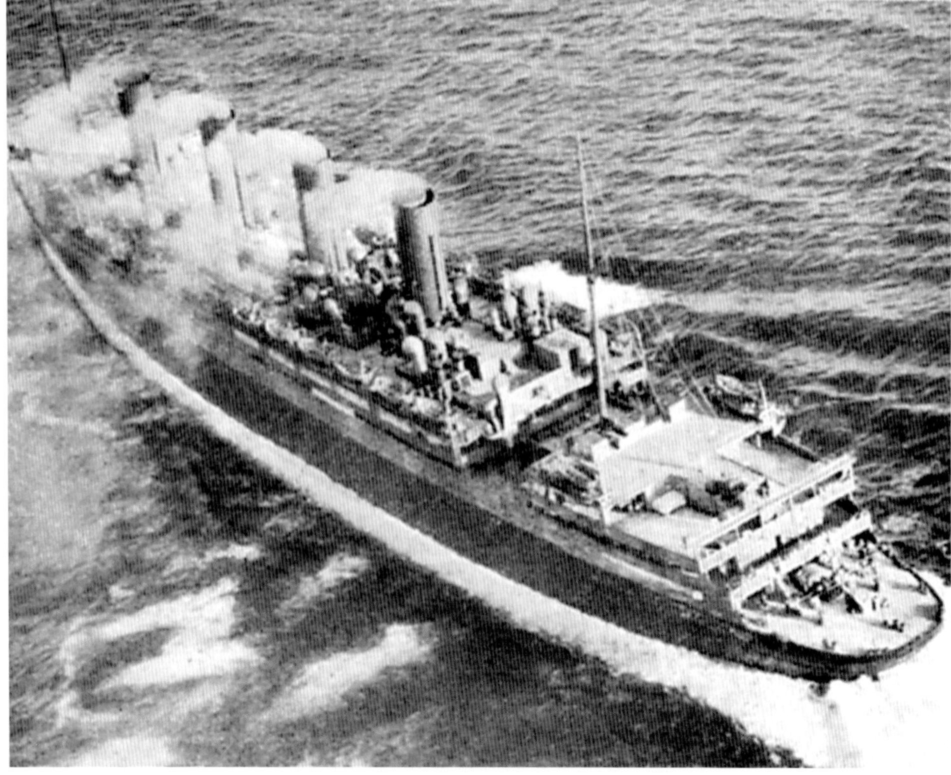

An aerial view of the *Aquitania* while she served as a troopship. (Ioannis Georgiou Collection)

troops between the United States and Europe. This new assignment would place the *Aquitania* squarely in harm's way from the U-boat wolf packs that stalked those waters. Her crew would need every advantage that they could get in order to survive this duty.

For the remainder of the war, *Aquitania* would remain on the transatlantic route, most frequently completing voyages between New York and Scotland. Ominously, Adolf Hitler had offered a bounty of a million dollars to any U-boat commander who could sink either the *Queen Mary* or *Queen Elizabeth*. The value of the monster transports was plain to see, even to the Nazi leadership. While the *Aquitania* did not have a price on her head, she was undoubtedly a high-priority target. Overall, *Aquitania* would make seventeen round-trip voyages in safety, a quite remarkable feat considering the horrific toll that U-boats inflicted upon Allied vessels during the war.

Henry W. Hudson, a sailor in the United States Naval Reserve, gave a detailed account of a wartime voyage aboard *Aquitania*. Serving in the Special Navy Advance Group 56, which was tasked with setting up a naval base hospital in England in advance of the Normandy invasion, Hudson's account begins with the departure from New York on 29 January 1944:

On the morning of the 19th sounds indicated to the seawise that we were getting underway and those who could find a place watched Manhattan recede from view. The P.A. announced that we were passing the [submarine] nets and that life preservers were to be worn at all times. No explanation was necessary.

Accommodations were not of luxury cruise character but ... were accepted with little griping. Another Navy outfit was aboard, officers and men who were to man LCGs [Landing Craft Guns] off the Normandy Coast. ... Enlisted personnel were berthed as low as 'I' deck and a continuous officer watch was required in each compartment. Clearance, between bunks, barely allowed a man to get in, much less turn or move about[,] and lighting was insufficient for reading. Officer staterooms housed two to thirteen in a room, dependent on size. In the two-officer rooms, one could stand or sit at a time, the other had to 'hit the sack', while those with larger numbers allowed more to stand but none to sit. Officers were berthed on 'B' deck, the nurses segregated (under Army guard, of course) on 'A' and they were crowded equally. Further, the nurses were allowed their quarters, the weather decks and Lounge and Library only. They could not go to other parts of the ship but could, and did, stand duty in the sick bay[,] as did most of the medical officers.

For the crew the covered decks, and for the officers the weather decks, were the only places for air and exercise. There was more room on the weather decks but they were also colder and on neither was any place to sit ... The number of officers, reduced by a third, would have crowded [the] Lounge and Library. As it was, a sardine would have had claustrophobia.

During daylight there was fresh air but no heat and the Atlantic is not balmy. After 'darken ship' there was no fresh air and such heat as was present came from animal sources. [W]ith the ship 'buttoned up' the aroma was reminiscent of a zoo.

For officers the two meals a day were fairly good, for the men less good. The two were adequate as part of breakfast could be hoarded and supplemented by canteen candy at noon. Then there were those who lost interest in eating, and a

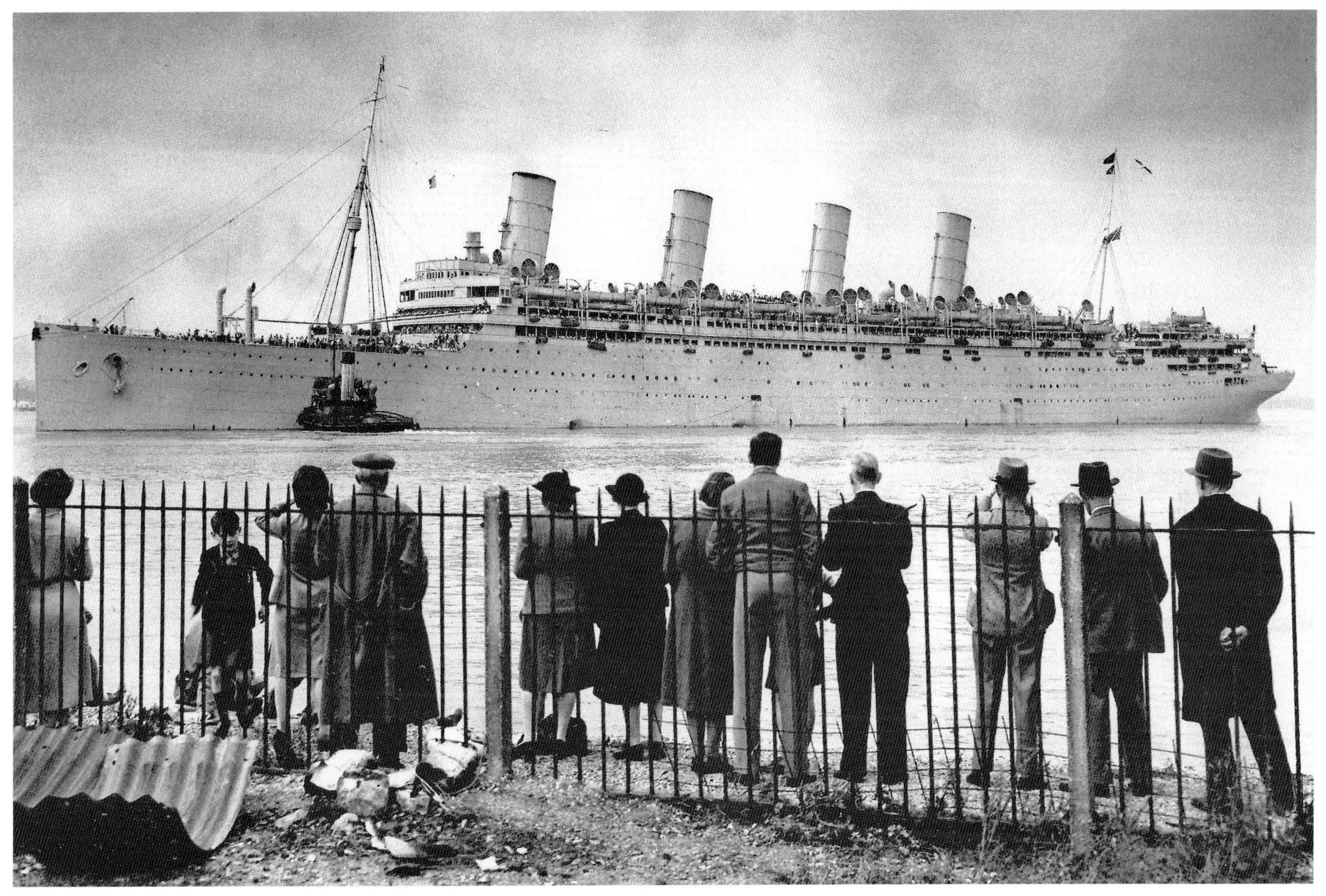

The troopship *Aquitania* impresses a small crowd of onlookers on shore, sometime in late 1944 or early 1945. The photograph shows the newly enlarged upper bridge, which was expanded around 1944. (Steven B. Anderson Collection)

few who lost interest in everything, as we struck a 48 hour stretch of heavy weather.

Once [*sic*: one] or more daily 'Action Stations' sounded and all hands assembled in passageways and by directions, which seemed to change each time, were moved ... to what was termed (facetiously it is presumed) boat stations. It was evidence that there were boats to accommodate only a fraction of those aboard ... These drills served useful purposes as fresh air and exercise for all hands and the consumption of time. Possibly lives might have been saved as a result had we encountered disaster.

Fortunately, we were not threatened until the day before making port. At about noon, just as drill was secured, 'Action Stations' sounded in earnest and was followed immediately by A.A. [anti-aircraft] fire. Three Jerry planes were overhead. One was believed to have been hit and disappeared. Evidently their mission was reconnaissance only. Personnel were held in berthing quarters for the next four hours but the anticipated attack did not follow. We would have been a rich haul for the Nazis.[157]

By the war's end in 1945, *Aquitania* had accumulated another spectacular war record. On 18 July – after the European conflict had ended, but before Japan had surrendered – Cunard's statements of accounts for the calendar year 1944 were released to the public. The company had made a profit for the year of £549,272 7*s* 10*d*, but during the war, they had lost the *Andania*, *Carinthia*, *Laconia*, *Lancastria* and *Laurentic*.

Meanwhile, the *Alaunia*, *Antonia*, *Aurania*, *Ausonia* and *Georgic* had all been acquired by the British Government. Cunard's chairman, Sir Percy Bates, in making his official statement on the year, took the opportunity to recount Cunard's contributions to the war:

I have no space for any long account of our proceedings during the war ... We have loaded 9,223,181 tons of cargo; collected £57,743,442 in freight; managed 30 Government ships, all without a commission of any kind for ourselves or a penny of expense beyond the audited disbursements. We found our pleasure and our duty to be the same. But we also have our pride. Up to May 31, 1945, the Company's own fleet had carried on all routes and in all directions a grand total of 2,473,040 troops, of which 1,243,538 were carried in the *Queen Mary* and *Queen Elizabeth* ... I like to believe that these two ships shortened the war in Europe by a whole year.[158]

Even compared to the larger, faster *Queens*, *Aquitania*'s record stood as quite an accomplishment. Not only was she the only passenger liner to have served in both world wars, but she had sailed 526,264 miles during the Second World War, carrying some 334,586 troops during that time[159] – or, by Sir Percy's method of calculation, shortening the war by three months through her efforts alone. As a basis of comparison, she had carried 'only' 120,000 troops during the entirety of her First World War service. This was a remarkable achievement for any passenger liner, much less one that was over three decades old by the end of the war, and which had seen much rough duty and hard use in those years.

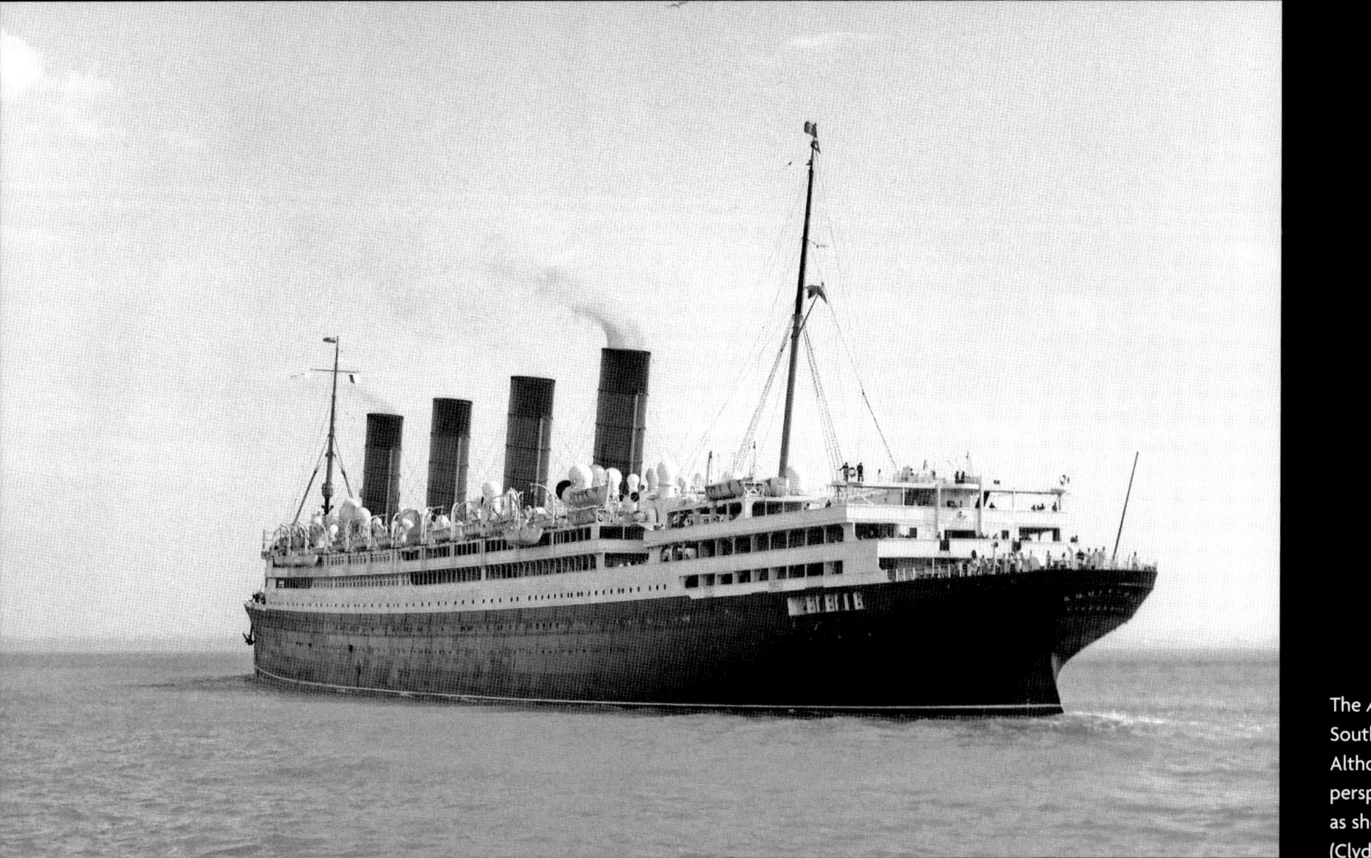

The *Aquitania* departs
Southampton on 30 May 1948.
Although 34 years old, from this
perspective she looks as splendid
as she did on her maiden voyage.
(Clyde George Collection)

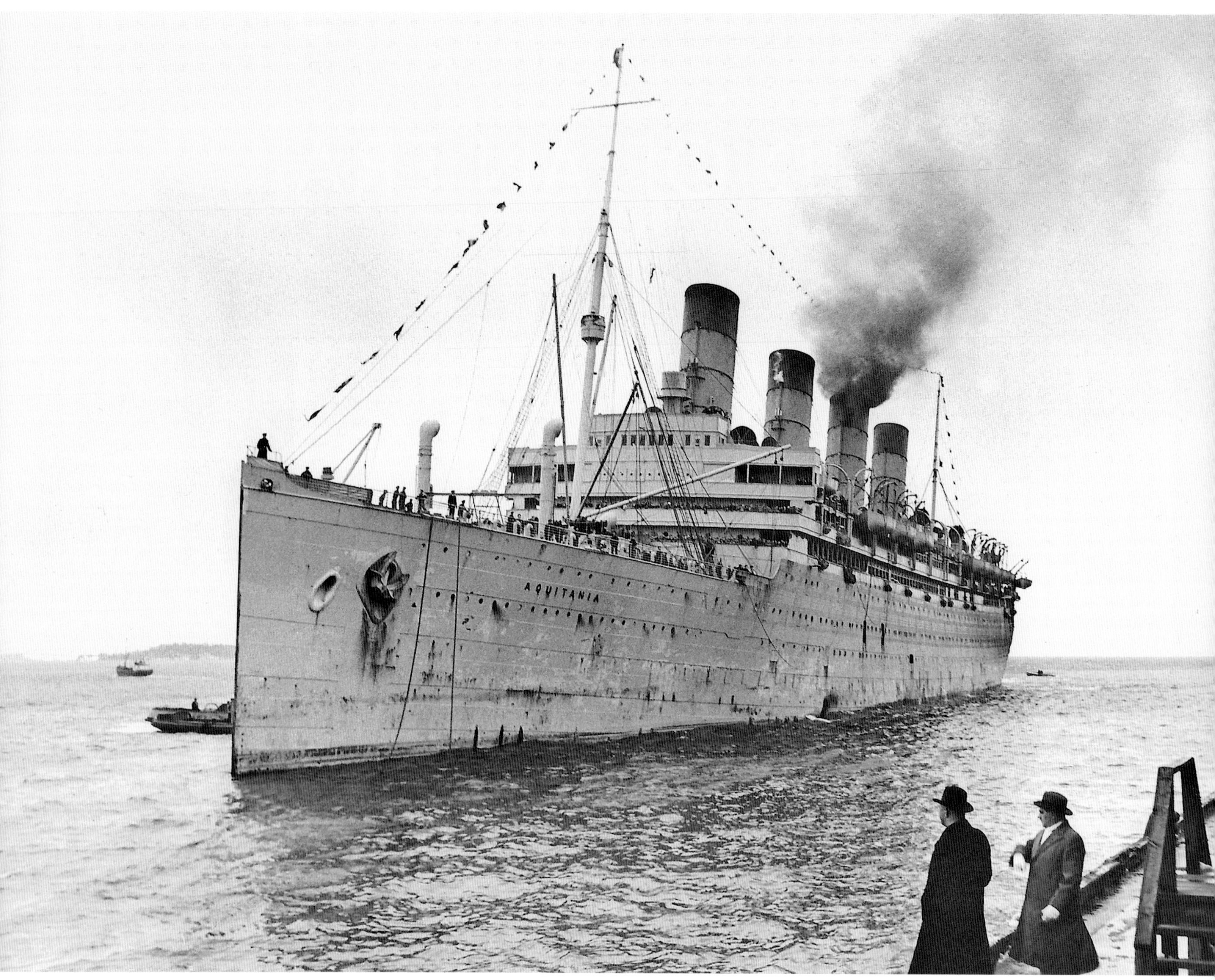

Aquitania arrives in Halifax on 4 February 1946. Her funnels have been returned to their civilian livery, although her hull retains its grey wartime scheme and shows symptoms of age and wear. This was her first trip carrying English war brides to their Canadian husbands. (Clyde George Collection)

Once the conflict in Europe ended on 8 May 1945, great numbers of American troops had to be repatriated to their home countries, or redeployed to the Pacific Theatre, which raged on until mid-August. For this effort, the *Aquitania* and *Queen Elizabeth* had been loaned to the United States. By October, *Aquitania* was returned to the British, and thereafter began a gruelling voyage to Sydney, repatriating Commonwealth troops back to Australia and New Zealand.

In November, while making a refuelling and revictualling stop at Capetown, South Africa, 3,000 troops departed, swarmed over what tugs and launches were alongside the liner and went ashore, despite the fact that they had not been given shore leave. In the end, the liner's skipper was forced to broadcast orders by radio, repeated throughout the city by loudspeaker, for the men to return to the ship. The affair caused no small amount of political backlash, with everyone denying knowledge of why shore leave had not been granted and why the ship had not been allowed to dock rather than be serviced at anchor in the roadstead.[160]

In January 1946, *Aquitania* returned to the Atlantic, repatriating Canadian troops along with their 'war brides' and children, via Halifax, Nova Scotia. When she arrived in that port on 10 April carrying Viscount Alexander, the new governor-general of Canada, it was in the middle of a blizzard. Cheering crowds came out to greet him, but the driving snow and high winds had wreaked havoc with the festivities.[161] After that trip, she underwent a refit where some of her accommodations were restored to peacetime configuration. First Class could now accommodate about 400, Tourist Class was partially restored and Third Class spaces retained their dormitory-style bunks.

Shortly after the war, Queen Elizabeth herself toured the *Queen Elizabeth* at Southampton, just before that liner made its first commercial crossing in October 1946. The *Aquitania* was nearby, and the Queen had lunch aboard the older vessel. Although rationing was still in effect on many 'luxury' foods, including white bread – which would not be available in England again until 1953 – the luncheon on the *Aquitania* was typical deluxe fare,

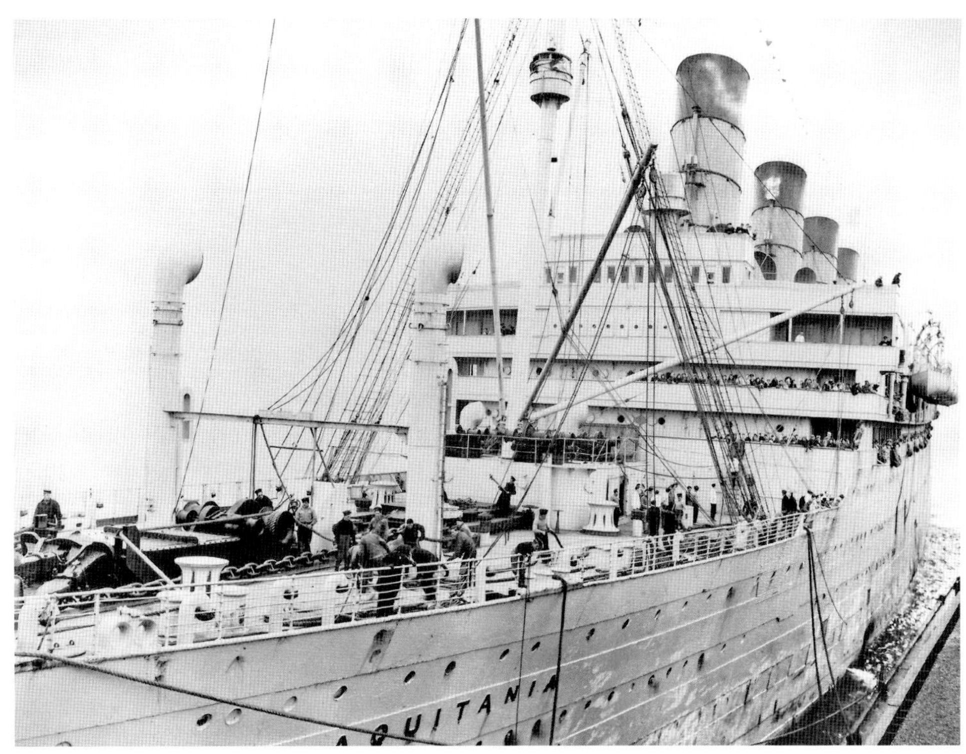

The first of three photos showing the *Aquitania* arriving in Halifax on 2 March 1946, carrying another load of war brides. Here the liner approaches the dock and crewmen work to feed hawsers down to the dock to secure her. One hawser is already fast. (Clyde George Collection)

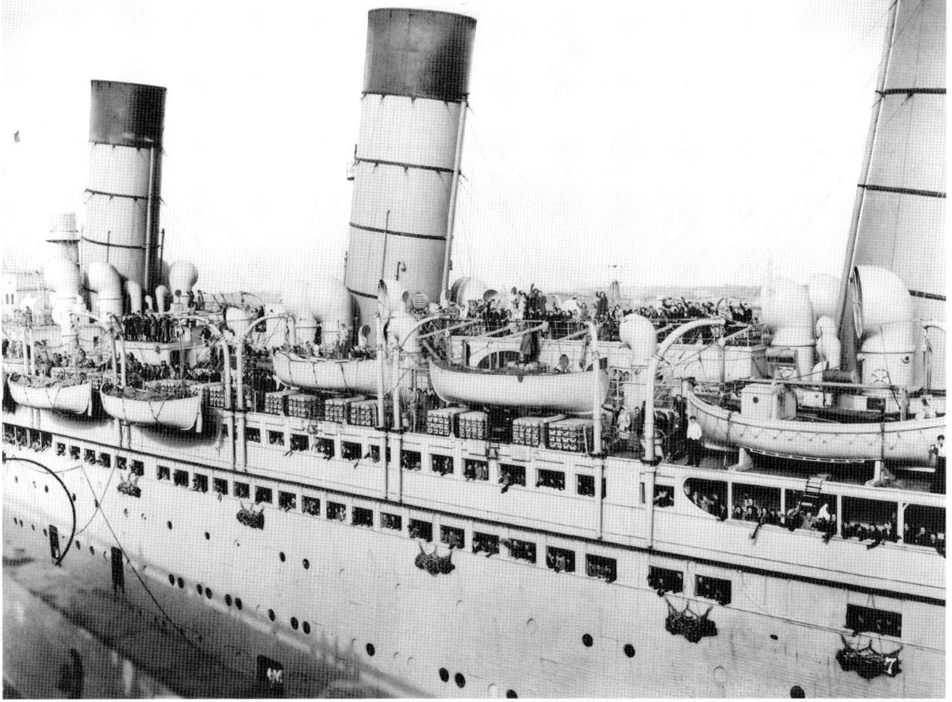

The upper decks of the *Aquitania* are packed with war brides. (Clyde George Collection)

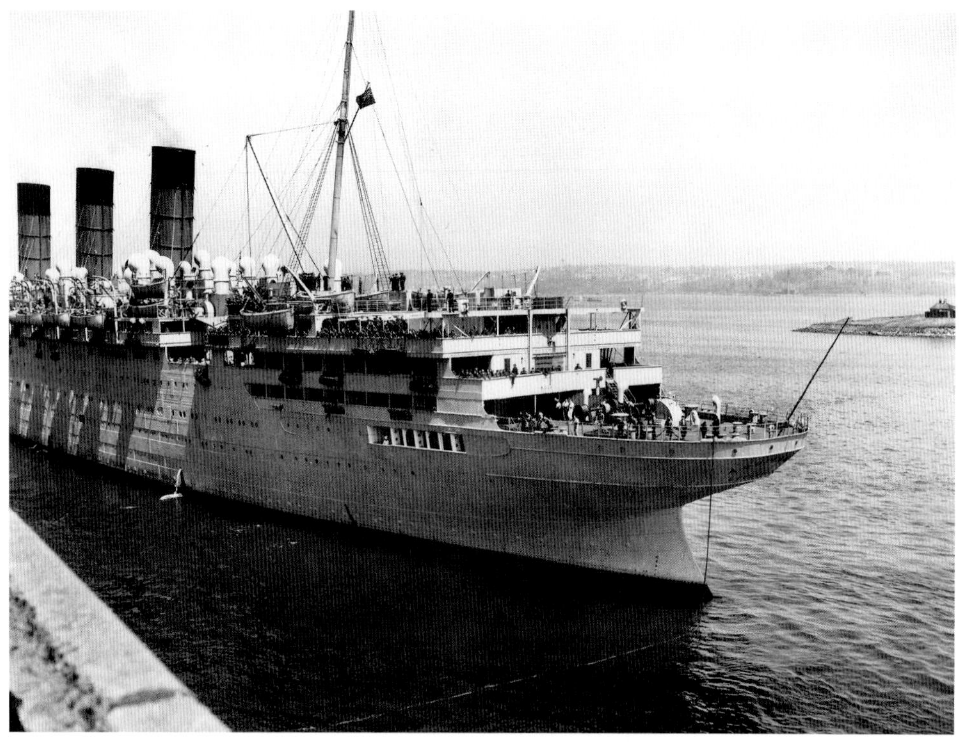

The stern of the *Aquitania* alongside the dock in Halifax. (Clyde George Collection)

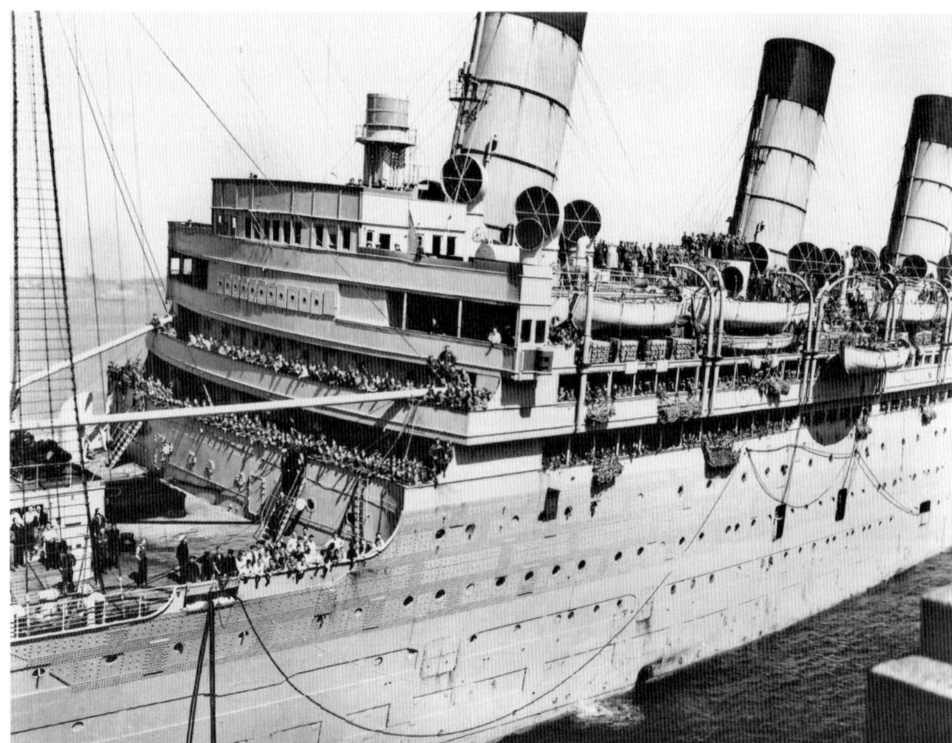

A batch of troops returns home to Halifax in the summer of 1946. (Clyde George Collection)

free of all rationing restrictions. When later asked what impressed her most about that lunch, the Queen replied: 'The white rolls.'[162]

While the ship was clearly showing her age, her condition was not as bad as is commonly believed. As 1946 drew to a close, Cunard were trying to piece together a peacetime transatlantic service. Some officials felt that the ship – both in engines and in hull – was still sound enough to allow her to be restored for civilian service for a few more years.[163]

Today, one oft-told tale endures: many books claim that, because of the badly deteriorated condition of her decks, a piano had either come crashing through a deck, or nearly had – and the details vary with each retelling. Where this had happened, and when, is also open to interpretation: some choose to add the detail that the piano had threatened to fall 'through the roof of one of the dining rooms' from the deck above, while 'a corporate luncheon'[164] was being held on the ship.

The origins of this tale – in any way, shape or form – are uncertain; Captain Harry Grattidge, true to his 1914 promise, had finally been appointed as the *Aquitania*'s master in early 1948; he published his extensive memoirs in 1956 under the title *Captain of the Queens*. While allegedly the original source, his text tells a different story:

She *was* a little old-fashioned, bless her, and by this time, sailor-men from New York to Southampton were calling her 'The Grand Old Lady,' for you could scarcely have called her 'The Ship Beautiful.' She was now the last four-funneled ship afloat, rust and paint flaked from her hull and the day had long gone when passengers clamored to travel in her because they believed the more funnels the greater the speed. But it was my sneaking conviction that her Carolean smoke room, oak-paneled like a London club, with its gorgeous painted ceiling, was still the finest afloat.

... [My] time with the *Aquitania* could not have lasted. Her hull was still firm and sound, but her decks had seen many better days. One rainy afternoon, not long after I had left her, one of these split. The resulting deluge of water almost washed

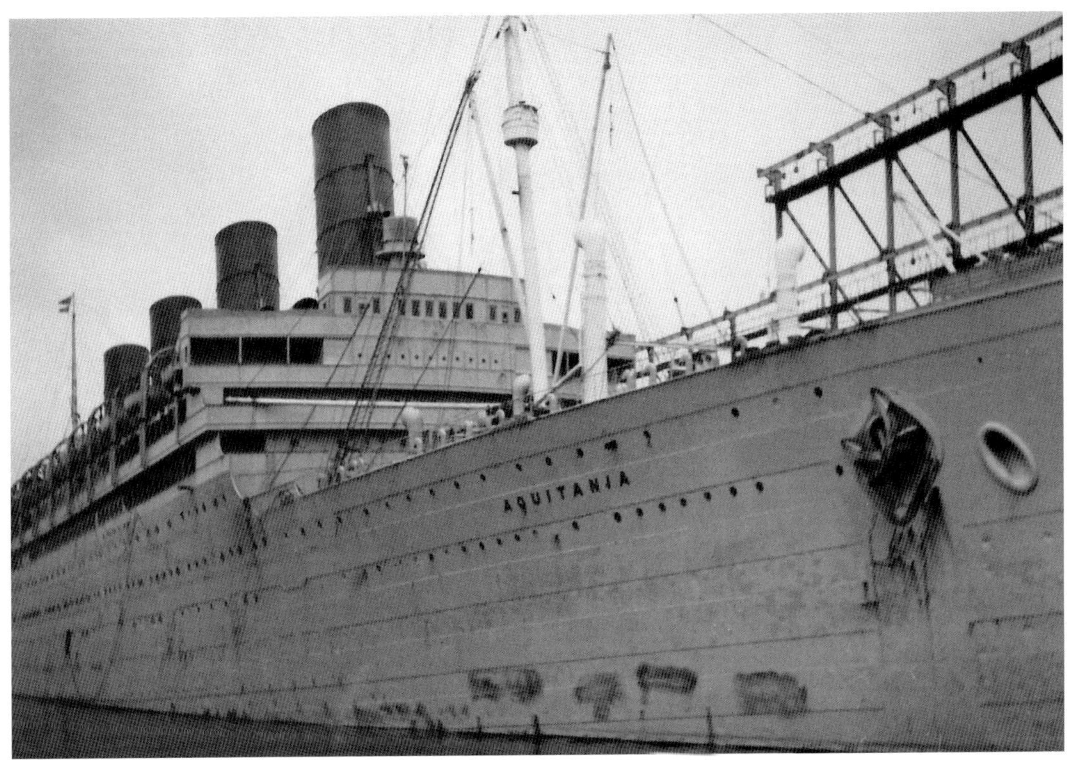

This photograph was purportedly taken during the *Aquitania*'s last stay in New York. (Clyde George Collection)

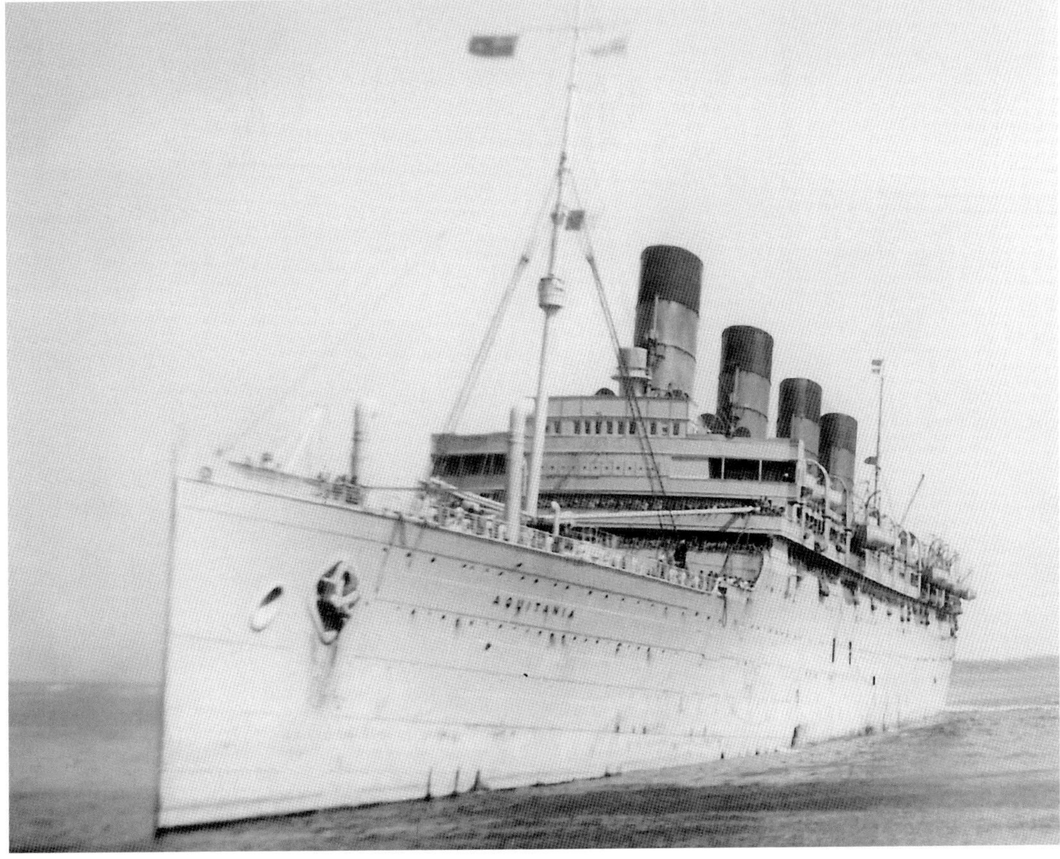

A splendid 1947 view of the liner arriving in Halifax. (Clyde George Collection)

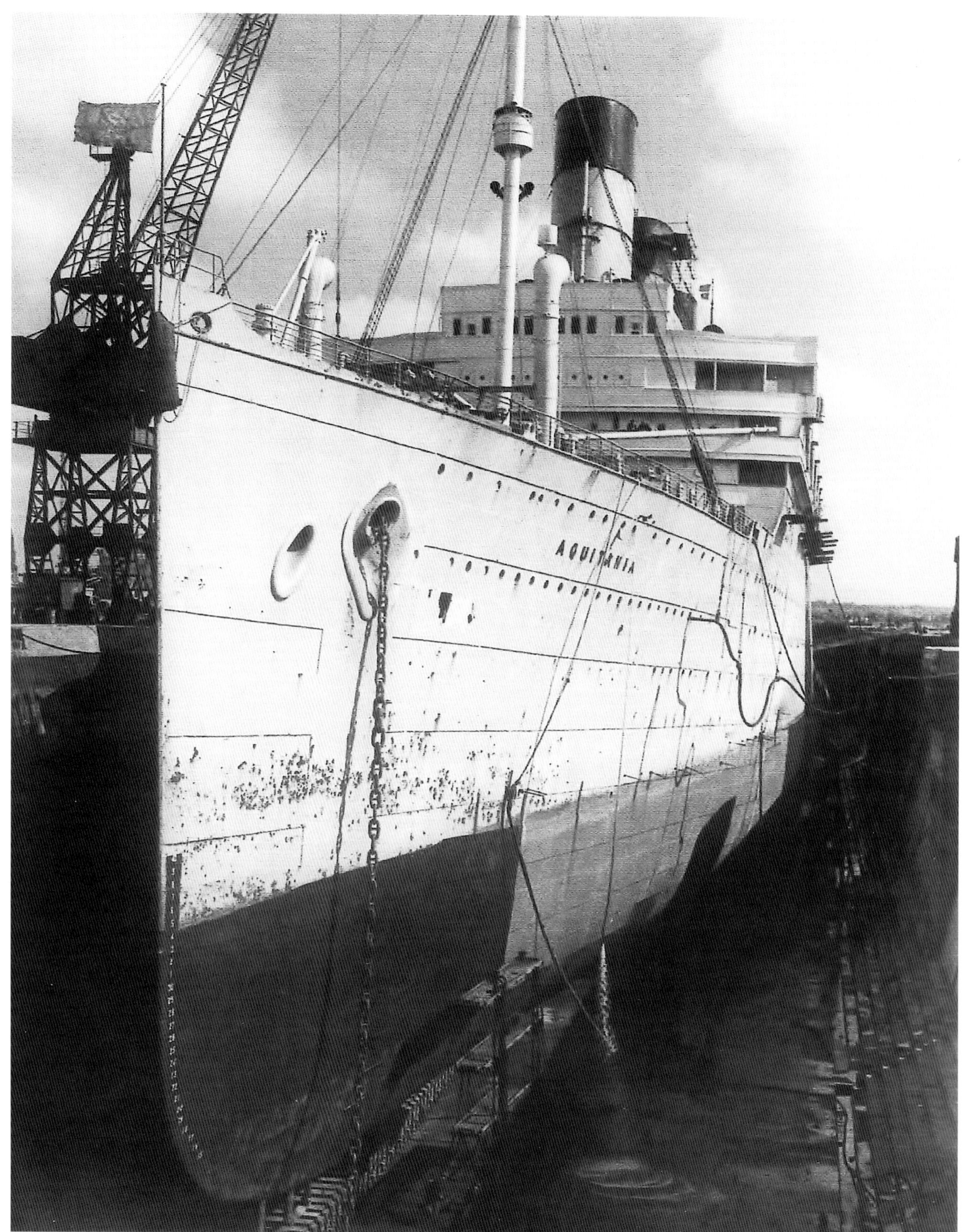

In March 1948, the *Aquitania* was undergoing a refit in the King George V Graving Dock in Southampton. This bow view shows her underwater surfaces being painted in civilian colours once again. (Clyde George Collection)

out the Board of Trade surveyors, the men who give the certificate of seaworthiness, who were lunching below. After that there were no more certificates of seaworthiness.[165]

While not the story of a wayward piano, there were similarities: a weakened deck that split, interrupting a luncheon of henceforth soggy Board of Trade surveyors. Yet even Grattidge's telling was second-hand, as it occurred after he had left the ship. In lieu of first-hand accounts proving the tale, the piano story would appear to be just another oft-recycled myth.

What is certain, however, was that the ship's future was not going to be a long one. *Aquitania*'s contract with the Canadian Government would expire on 16 March 1948, at which time the liner was to be released from her duties with the Ministry of Transport. A month before that deadline, it was noted: 'Her testimonial will read: "Conduct excellent, hull and engines in first-class condition. Age 34." ... But *Aquitania*'s 23 knots cannot match the "*Queen*" ships, and she is not worth restoring to Atlantic luxury standards. ... [And] nobody seems to want her except the scrap-hungry steel barons.'[166] However, on 19 February, it was announced the Canadian Government had signed an agreement with Cunard to use the ship for the emigrant trade to Canada for a year. By 25 May, the ship had been restored to her civilian colours for the first time since August 1939.[167]

Externally, the ship sparkled following the March 1948 refit. While her passenger spaces were not fully reconditioned, things were improved from the previous refit, and she could now carry some 650 in First Class. In her engineering spaces, a great deal of money was spent to keep her equipment functioning; however, her No. 2. Boiler Room was shut down to conserve costs, thereby reducing her speed to 18.5 knots.[168]

The Canadian charter was quite successful, and was renewed for 1949. Her passenger lists in 1948 and 1949 were longer than they were during the 1930s; when the two *Queens*, then back in civilian service on the New York route, were held up by a dockers' strike, it was said that the *Aquitania* could not take any of their passengers, since she was already fully booked.[169]

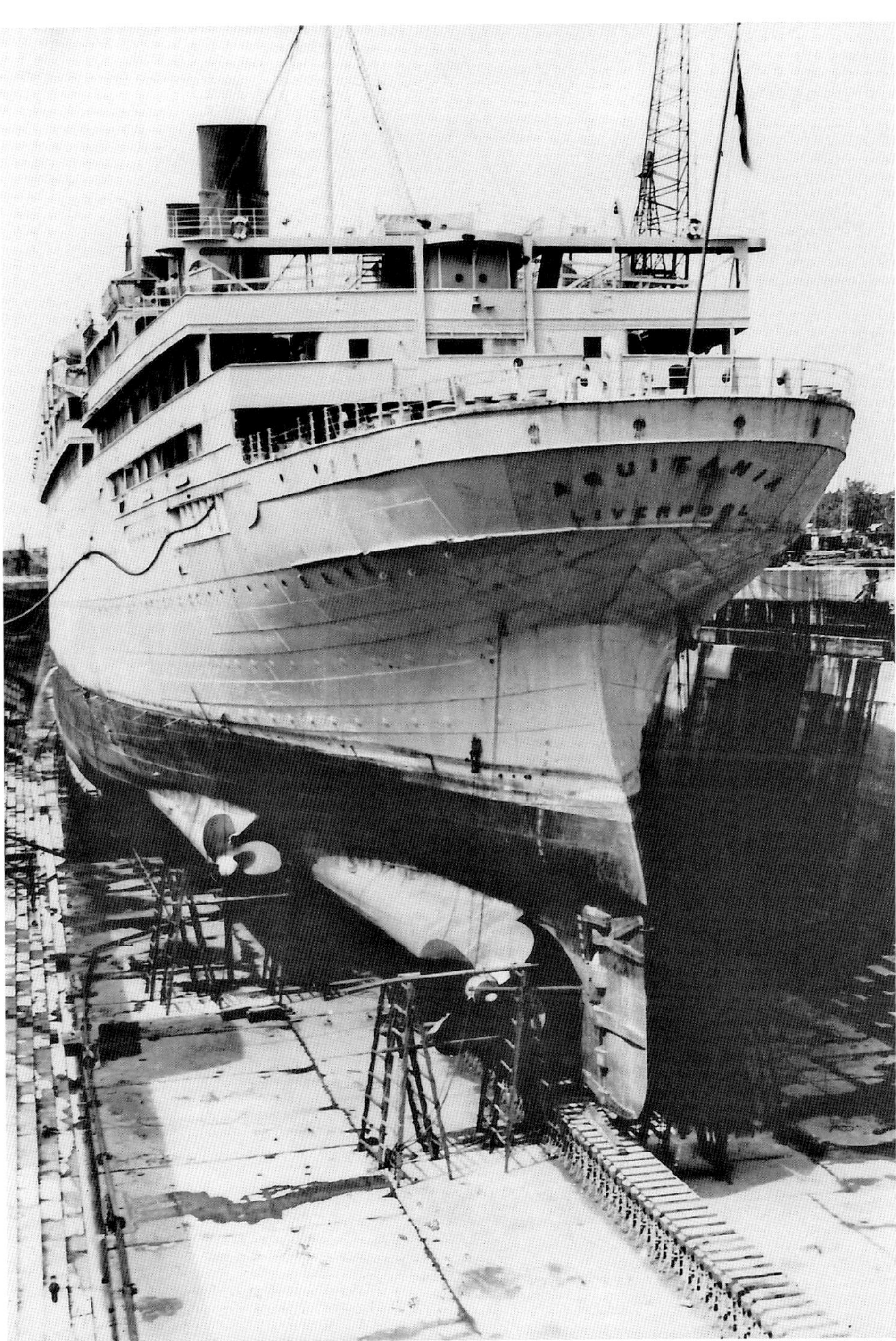

A stern view of the *Aquitania* in dry dock in early 1948. (Clyde George Collection)

On 25 October 1949, the *Aquitania* sailed into a gale off the English coast. The coaster *Yewbank*, off Cornwall, radioed an SOS. Several vessels, including the Cunarder, made for the crippled ship. Despite her 35-year-old hull, *Aquitania* made it safely to port. However, the end was

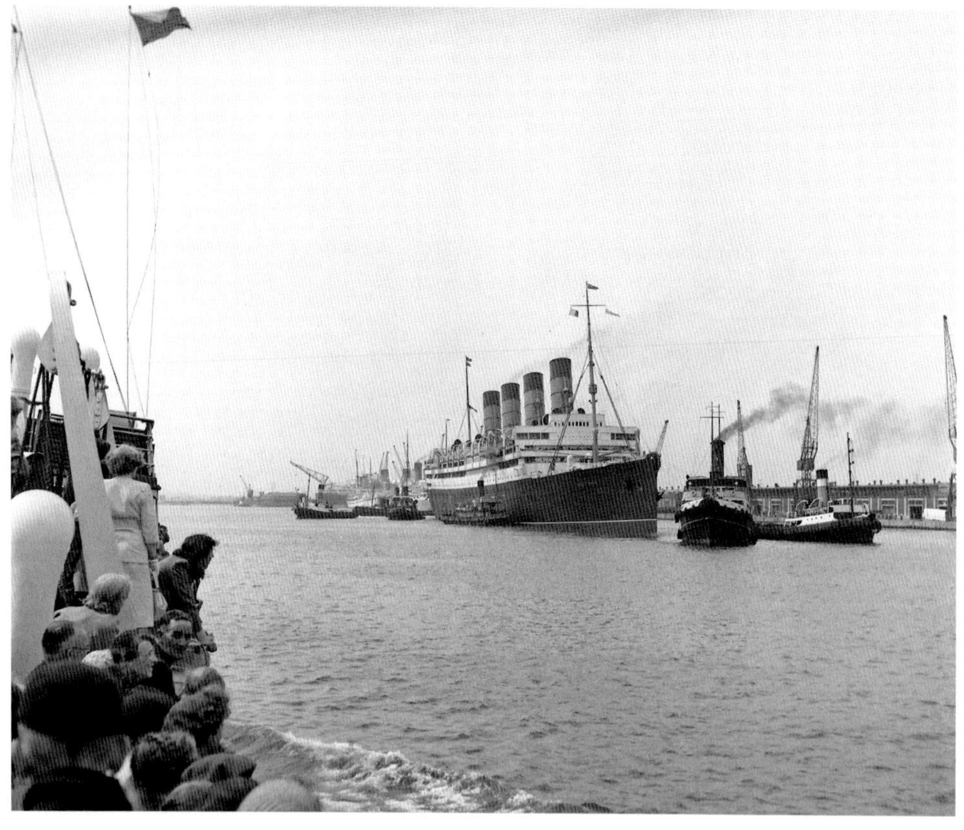

The *Aquitania* steams out of Southampton's Western Docks on 30 May 1948. She is shepherded by at least five tugs. This view was taken from a smaller vessel, and the liner's fame is evidently a big enough incentive to draw many over to the vessel's side to watch her slip by. (Clyde George Collection)

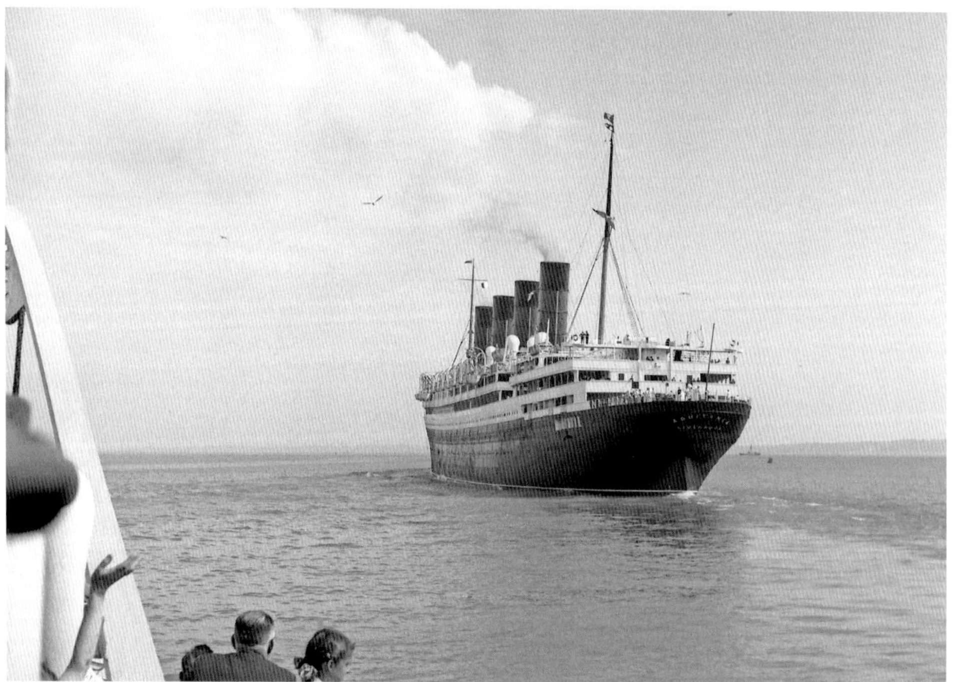

A fine stern view of the liner departing Southampton on 30 May 1948. (Clyde George Collection)

near. On 9 November, under the headline, 'The fourth funnel may vanish', it was reported:

> The last of the four-funnel liners, the 44,786-ton *Aquitania*, may go to the ship-breakers soon. After an Atlantic voyage, which starts from Southampton on Tuesday [15 November], she will be surveyed by engineers who will report on her future.[170]

It turned out to be her last round-trip transatlantic voyage. The results of the survey showed that keeping the ship in service any longer would require greater expense then anyone was willing to pay. Thus, on the evening of 14 December 1949, Cunard announced that the *Aquitania* would be withdrawn from service. The *New York Times*, rather uncharacteristically, waxed nostalgic:

> It must have been with a deep nostalgic sigh that many persons here read of the retirement from active service of the Cunard White Star liner *Aquitania*. No ship in modern maritime history has had a more honorable and distinguished record. She was one of the queens of the transatlantic fleet.
>
> But to those in New York she was something else. She was a recurrent adornment to the waterfront skyline. Her famous four stacks were there after her sister ships disappeared. She was unmistakable; dignified, proud, often sedate, always the great lady. Those who traveled on her know of her comfort; those who watched her know of her beauty. She was a great and a proud ship, and it is sad to think that we may not again see her coming up the bay.[171]

The ship's furnishings were removed to Shed 108 during January, preparatory to being auctioned off along with her fittings in some 2,000 lots. The event, which lasted a week, was held beginning on 13 February. Bargain-hunters were largely disappointed, for the bidding was brisk: some £6,000 was realised on the first day alone. The liner herself was sold to the British Iron & Steel Corporation Ltd for £125,000 to be broken up at Faslane, Scotland.

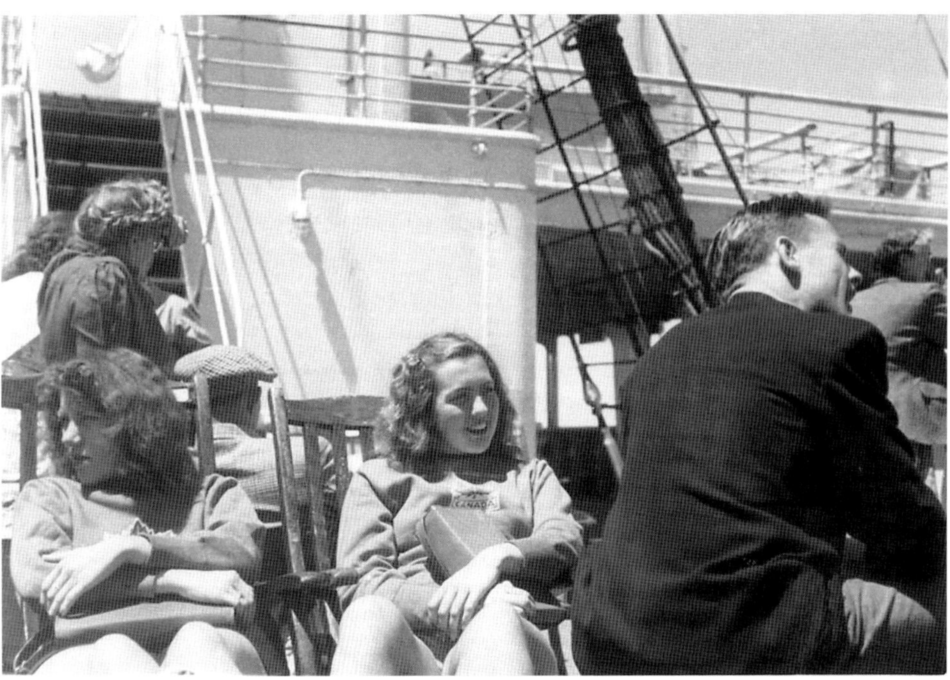

Two girls from the Canadian Olympic team enjoy time on deck during their 1948 crossing on the *Aquitania*. (Clyde George Collection)

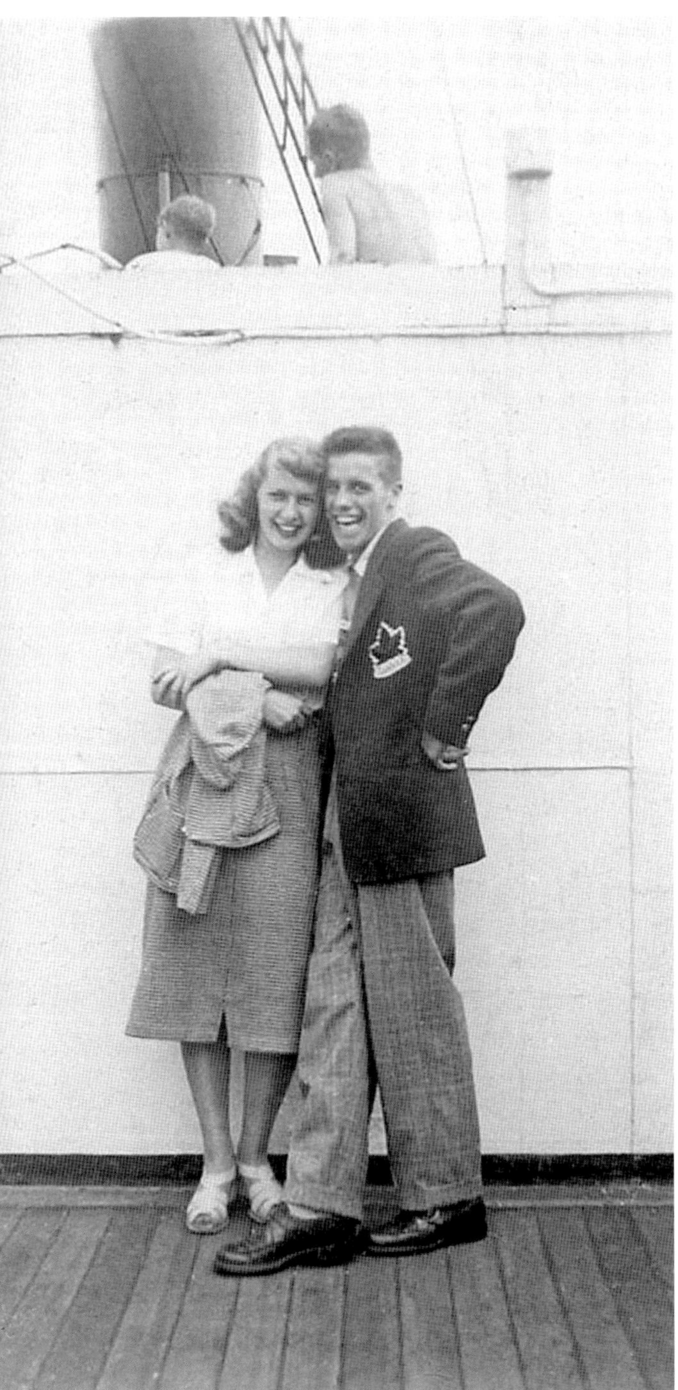

In 1948, the Canadian Olympic Team sailed on the *Aquitania*. They departed Halifax bound for the games on 14 July, and returned there on 25 August after the London games had ended. Here a pair pose on deck. (Clyde George Collection)

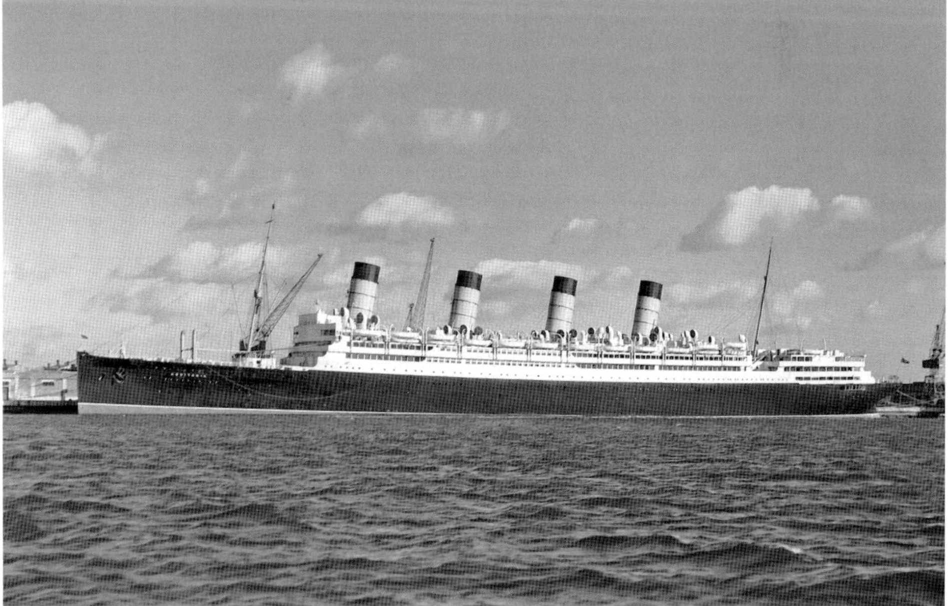

The *Aquitania* in Southampton during August 1948. (Clyde George Collection)

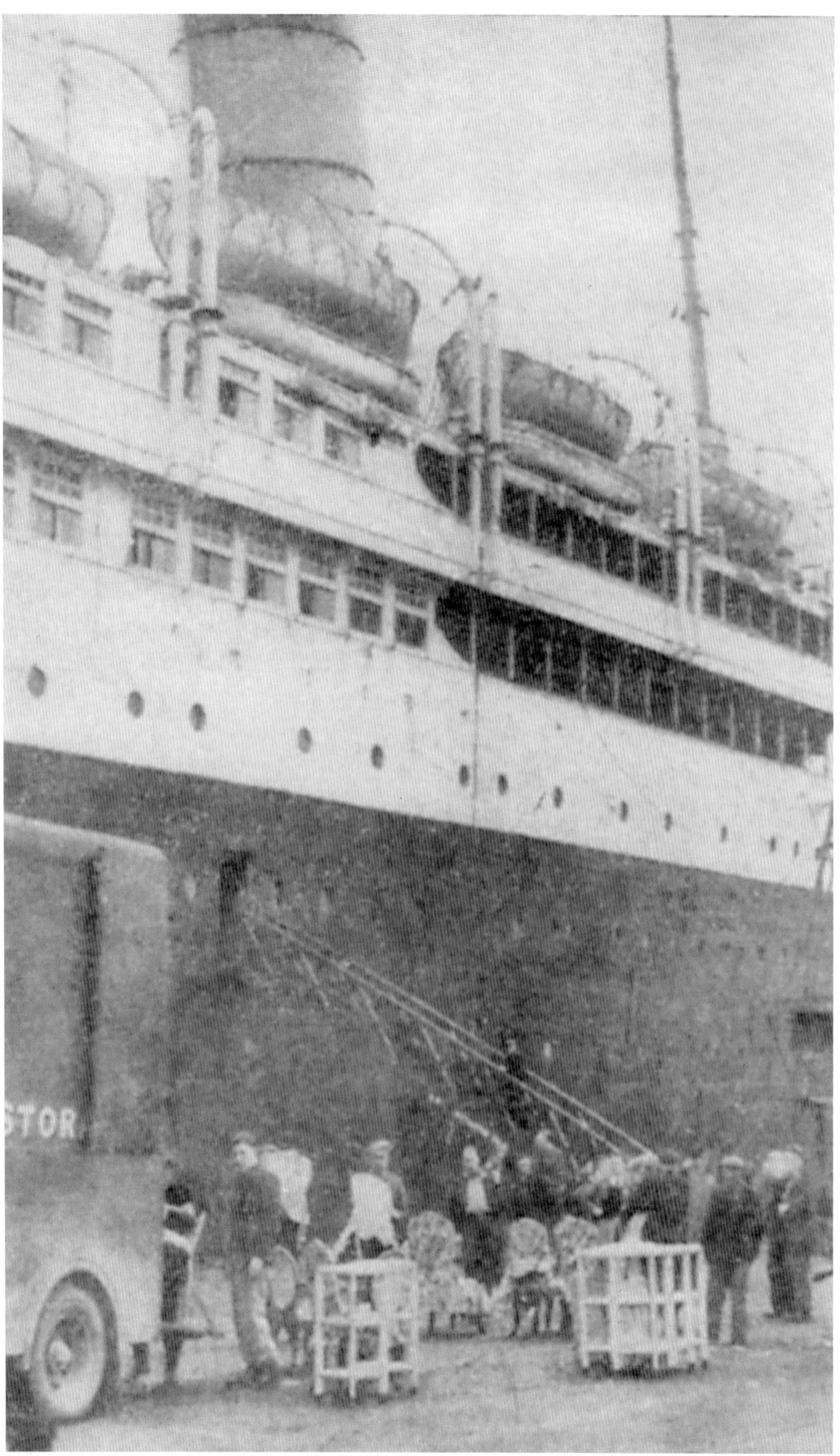

The ship's colours were struck for the last time on 18 February, and standing on deck for the ceremony, Captain Woollatt bade goodbye to many members of her crew, some of whom had been with the ship for as long as thirty years. On Sunday, 19 February, the time arrived for the great old *Aquitania* to make her last departure from Southampton. It was said:

Right to the last Southampton was reluctant to see her go. Fog swept up the Solent during the night, and the last journey, due to begin at 9 a.m., was delayed by four hours.[172]

Another witness felt that the old ship was reluctant to leave the port, rather than the other way around. Yet once the weather had cleared, the ship cast off and moved toward open water for the last time. People turned out in droves to witness the end of an era. Every ship she passed gave her a noisy send-off, their sirens and flags signalling a fond farewell. For the trip, the old liner suffered the indignity of being categorised as a cargo ship; the twenty-five journalists who would accompany her on her 443rd and final voyage were not considered passengers; they were signed on as 'shilling-a-month supernumerary seamen'.[173]

The ship's interior spaces, once warm, cosy refuges from the chill North Atlantic weather, had been stripped of their luxurious fittings, their personalities. They were cold, dark and presented a gloomy atmosphere as sombre as a grand but now empty house. Her great suites were vacant, littered with trash from the workmen who had removed the fittings. Her chief engineer, John Moffatt, was still aboard. At age 60, he had been a strapping young man when he had sailed on her maiden voyage in 1914 as a junior sixth engineer. He had accumulated over twenty years with the ship, and now had the sad task of bringing her in for the last time.

That evening, off Land's End, as she ran north to the Bristol Channel, the *Aquitania* encountered a moderate swell.

Furniture and fittings were offloaded from the *Aquitania* in December 1949 and January 1950. The items were stored in a shed on the dockside, where the final, depressing auction was eventually held. (Ioannis Georgiou Collection)

It was said that 'the old liner creaked and groaned as she took the Atlantic rollers on her port side. But the last 10 to 15 hours were glorious from a weather point of view', which was followed by a 'wonderful display of northern lights, the skies being tinged with silver and reddish gold'. As the ship arrived at the Clydeside, the 'sun shone brightly and the sparkling waters of the Firth were flanked by green and snow-capped hills. Many who were making their first visit to the Clyde sang its praises, and even the hardened native had to admit that the estuary was being seen in its pleasantest mood.' The old ship had averaged an unhurried pace of 14.14 knots over the 580-mile journey.

As the *Aquitania* moved upstream, several passing vessels sent messages of farewell, ranging from a 'somewhat terse "R.I.P."' to a more convivial 'We are very proud to have met you.'[174] The Clyde pilot, who had come aboard at Southampton to make the trip, guided the ship up the Clyde with the assistance of several tugs. She inched forward cautiously, the largest ship ever to enter Gareloch. By 3.35 p.m., the *Aquitania* let go her anchors, finally pulling into the quay. Then, at 4.10 p.m., on Tuesday 21 February 1950, the telegraph handles on the bridge were swung to the 'Finished With Engines' position for the last time. Her prestigious career – encompassing over 3 million miles and having carried over a million passengers over portions of five different decades – had finally concluded, less than 20 miles from where she had been built.

It truly was the end of an era. The world had changed completely from the days of late 1909, when Cunard's Board of Directors had first conceived of building her. Just three round-trip voyages into her career in the summer of 1914, things had begun to unravel in earnest on the world scene. The *Aquitania*'s story was that of a ship that had been designed for one world, but which had spent all but a few days of her life in a completely different one. Yet, through two global conflicts and some of the most changing decades the world had ever seen, the *Aquitania* adapted well to transient circumstances, and hers was, ultimately, one of the greatest success stories ever seen on the North Atlantic.

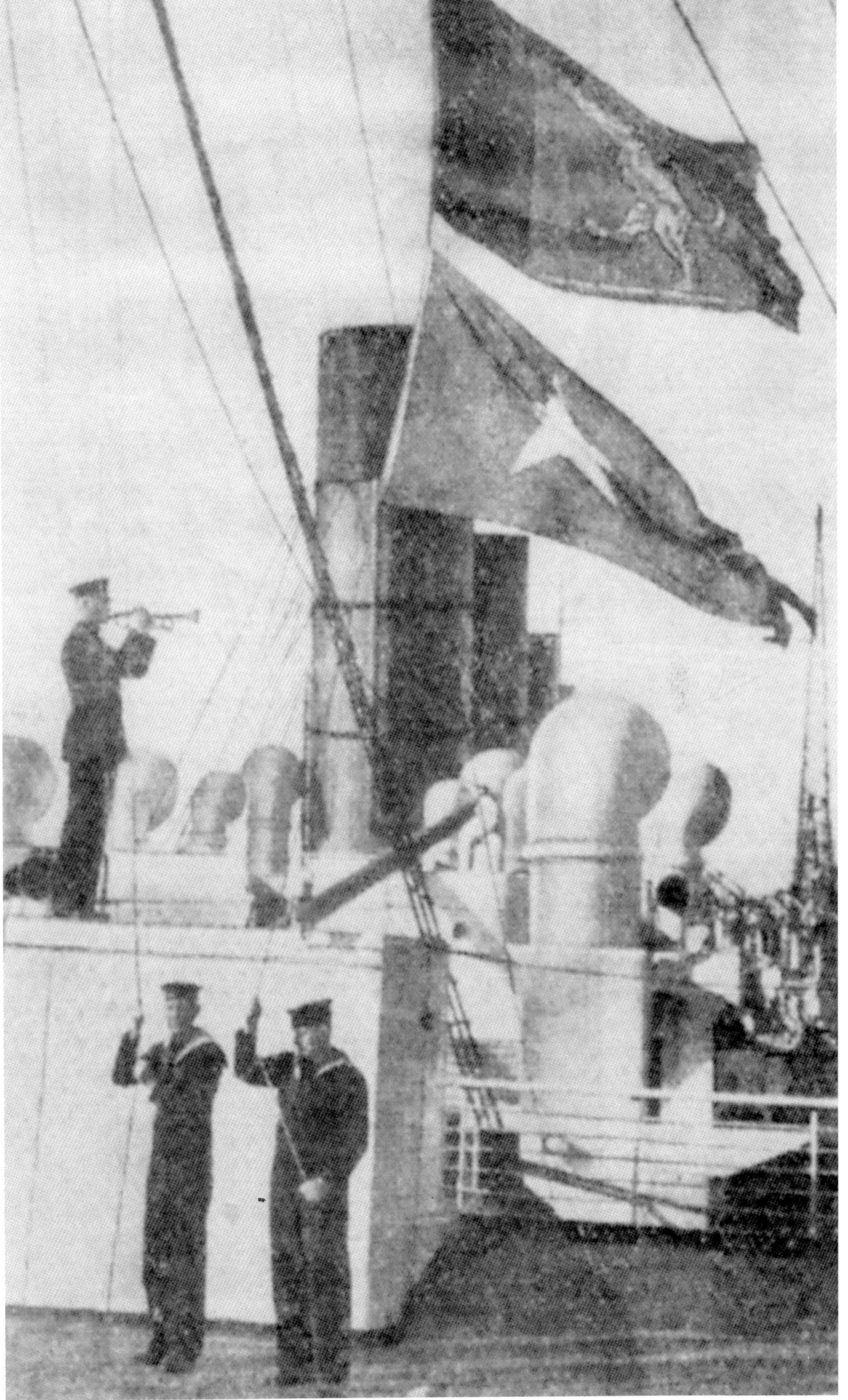

Aquitania's company flags are struck for the last time on 18 February 1950. (Ioannis Georgiou Collection)

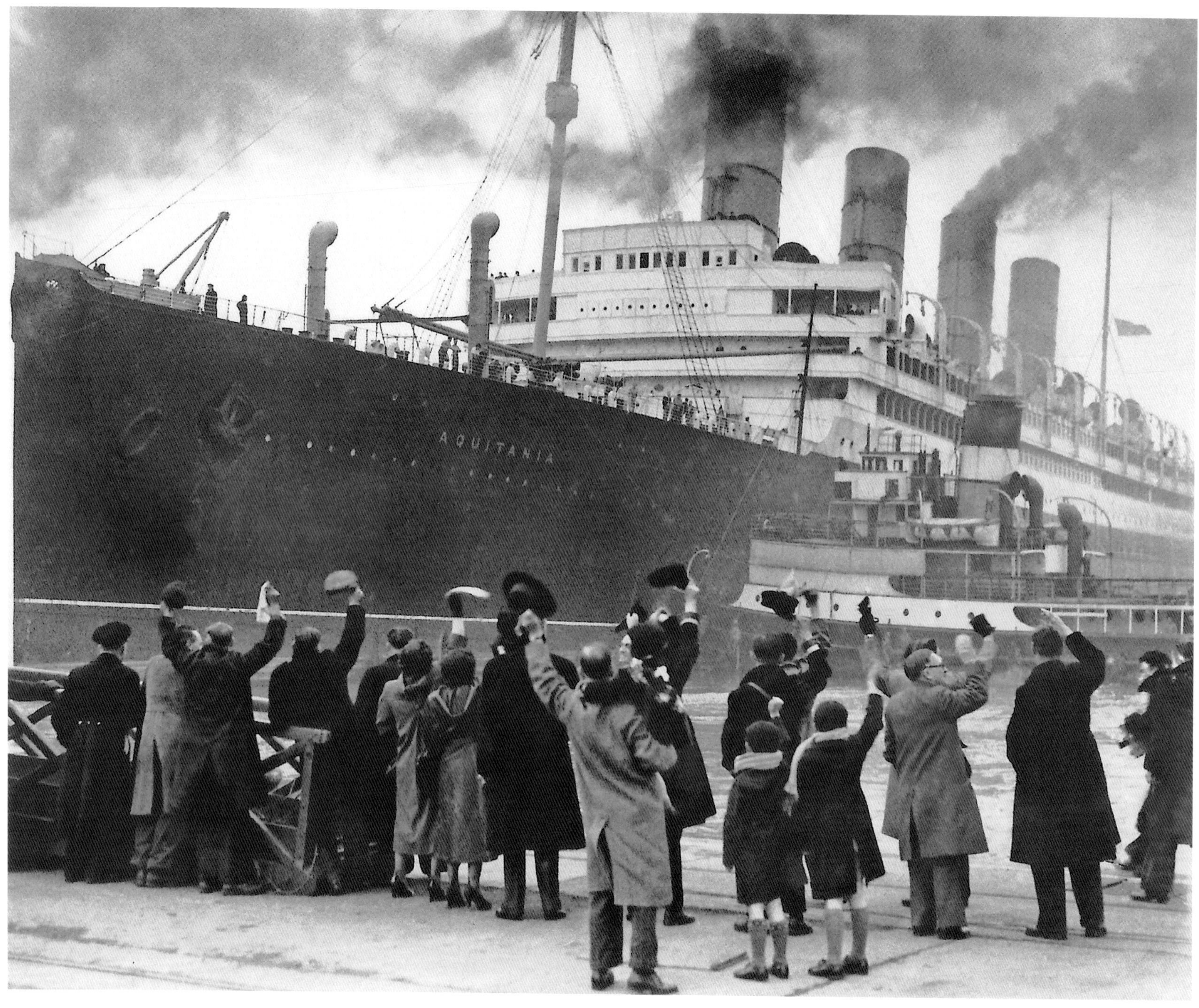

On 19 February 1950, one of the greatest Atlantic liners ever to sail the seas leaves her home port of Southampton for the last time. She is seen off by a crowd of well-wishers. It was the end of an era. (Clyde George Collection)

Aquitania had aged gracefully and, in some respects, even improved with the passage of years. She had overcome great dangers from the elements and from enemies. She sailed through waters that she had never been intended to traverse, in both peace and conflict. She had carried passengers looking for a way to relax through tropical waters on grand, extended cruises to exotic destinations, and had carried people from all walks of life: immigrants, teachers, magnates, debutantes, politicians and ambassadors, famous sportsmen and sportswomen, stars of the stage and theatre, and even thousands of troops heading to war. Romances had blossomed aboard her, stars had been discovered, reporters had swarmed over her looking for a great quote or photograph from some famous person or other.

Her most famous captain, Sir James Charles, had put on legendary feasts in her Louis XVI Restaurant for his distinguished guests, demanding that these select passengers don only the finest of clothes when he spent time with them. His great ship had become so beloved to him that the strain of leaving her seems to have quite literally finished him, yet he was not the only one who had grown attached to it. From the crewmen who had stayed with her for five, ten, twenty years and more, to the passengers who chose to take passage on her time and again because they connected with her on a profound and deep level, she was a liner that evoked deep attachment, sentiment and loyalty. Even after her four-funnel profile had become passé and been replaced by more modern-looking, and by some definitions infinitely less regal, two-and three-funnelled liners, those who caught sight of her reliably steaming through Southampton Water, into or out of Cherbourg Harbour, or passing by the Statue of Liberty in New York, couldn't help but feel a flash of warm recognition – the kind one feels at catching sight of an old and much-loved friend.

Most importantly, during her breathtaking thirty-six years of existence, she had always delivered her passengers and crew safely through a storm-tossed sea to the other side of the Atlantic. She had never held the title of 'world's largest', 'world's longest' or 'world's fastest' ship. However, by the time her career had come to its conclusion, the *Aquitania* could, arguably, have laid claim to the title of 'world's greatest ship'.

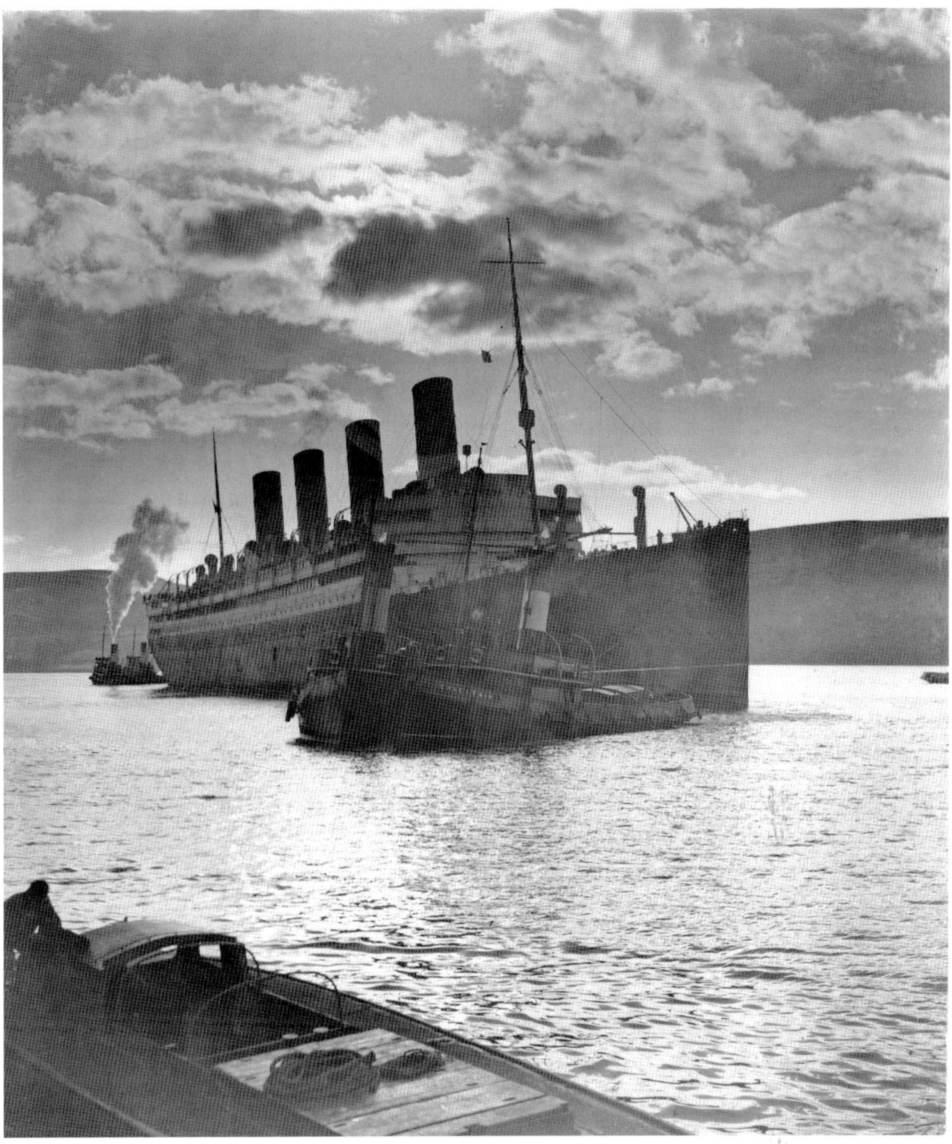

Aquitania arrives in Faslane, Scotland, and approaches the berth where she will be dismantled on 21 February 1950. (Clyde George Collection)

ENDNOTES

1   *The Unseen Lusitania*, pg. 10.

2   'The Dinner at Lord Pirrie's in Summer 1907: Just a Legend?', Gunter Babler (*Titanic Post*, Swiss Titanic Society, 2000.); *On A Sea of Glass*, pp. 17–8, 380.

3   Contrary to popular misconception, 'gross tonnage' is a term used to measure enclosed volume, not weight. A ship's weight is referred to as 'displacement' because it is measured by the weight of the water their hulls displace. This figure varies during a ship's voyage, as fuel and stores are consumed.

4   *TDE*, 24/03/1911, 4.

5   *The Engineer*, 29/05/1914, pg. 587.

6   *Ships For A Nation*, pg. 131; UCS 1/74/6.

7   *The Unseen Mauretania (1907)*, pg. 27, 55; *Lusitania: An Illustrated Biography*, 2 vols (The History Press, 2024 & 2025).

8   *Ships For A Nation*, pg. 132.

9   Chirnside, pg. 8.

10   *The Shipbuilder*: Aquitania, pg. 33.

11   *Daily Consular and Trade Reports*, pg. 233.

12   Diggle, pp. 63–4.

13   *The Engineer*, 31/01/13, pp. 127–8.

14   *NYTimes*, 20/04/13.

15   *Ships for A Nation*, pg. 132.

16   *TDE*, 15/04/13.

17   *IME*, 7.

18   *NYTimes*, 20/04/13.

19   *The Engineer*, 03/05/12, pg. 468.

20   The figure in the text, 1,496, represents the best-documented study of the ship's passenger and crew list ever completed. There had been 2,208 people aboard when the ship departed her final port of call; 1,496 perished in the sinking, and 712 survived. For further details on this, please see *Titanic: A Centennial Reappraisal*, Chapter 4 and Appendices A–F.

21   *The Ocala Evening Star*, 29/05/14, 1.

22   For a detailed study of Cunard's reports on the *Olympic*, please see *RMS Olympic*, Appendix III.

23   *Romance*, pp. 19–21.

24   *The Engineer*, 03/05/12, pg. 468.

25   *The Engineer*, 31/01/13, pg. 127, 128.

26   *TDE*, 22/04/13, pg. 4.

27   *TDE*, 22/04/13, pg. 4.

28   *The Engineer*, 25/04/13, pg. 447.

29   *NYTrib*, 20/08/13, 7.

30   *TDE*, 11/11/13, 5.

31 *TDE*, 10/01/14, 5.

32 *Ships for A Nation*, pg. 134; SCA Report for 31/03/14, unlisted documents.

33 *IME*, Aug 1914, (vol. 19) 363.

34 *NYTimes*, 07/06/14.

35 *NYTrib*, 07/06/14, pg. 11.

36 *TDE*, 11/05/14, pg. 1.

37 *TDE*, 11/05/14, pg. 1.

38 *Romance*, pg. 40

39 *Ships for A Nation*, pg. 135; ULA, AC 14/46, 606. Of this figure, £1,413,152 was paid to John Brown (some £1,070 less than the contract price), with the remaining £271,531 being paid for public rooms, architects' fees, engine work, final items such as linen, crockery and cutlery, and interest at 4.5 per cent.

40 *NYTrib*, 31/05/14.

41 Press accounts printed immediately after the liner arrived in New York cited 1,019 passengers on board: 334 First Class, 213 Second and 472 Third.

42 *Evening Star*, 31/05/14, 2.

43 Some newspaper accounts provide an average speed of 23.17 knots, while others cited 23.10 for the crossing. However, the official abstract of log for the voyage cited 23.17. The figures provided here are from the official log abstract. Oddly, the average daily speeds provided by the log, if added up and then averaged out, amount to 23.13 knots.

44 *TDE*, 06/06/14, 1, 5.

45 *TDE*, 06/06/14, 1, 5.

46 *TEW*, 05/06/14, 1; *The Sun*, 06/06/14, 3.

47 *The Sun*, 06/06/14, 3.

48 *TEW*, 05/06/14, 1.

49 *NYTrib*, 09/06/14, 3.

50 Grattidge, pg. 67.

51 *TWH*, 21/06/14, 20.

52 Further details on the *Aquitania*'s early technical problems can be found in *Changes*.

53 House of Commons Reports.

54 *A Merchant Fleet At War*.

55 *The Unseen Mauretania*, pg. 93.

56 *The Parliamentary Debates* [Official Report], pp. 1557–60.

57 *Ambulance Company 113, 29th Division*.

58 *King's Complete History of the World, Visualizing the Great Conflict in All Theaters of Action, 1914–1918*.

59 *The World War and Historic Deeds of Valor, Vol. 6*. The entire *Shaw* incident was censored from the public initially. However, when the war ended, one crewmember from the *Shaw* claimed that a submarine periscope had been spotted on the *Aquitania*'s port side, and that the commander had tried to rush toward it across the *Aquitania*'s path. However, Commander Glasford's account of the incident does not support this, instead stating that the jammed helm was the cause of the collision.

60 *TDE*, 24/02/19, 4.

61 Chirnside, pg. 35; *Atlantic Liners of the Cunard Line*, pg. 103.

62 *Pacific Marine Review*, June 1922, pg. 337.

63 *NYTimes* and *NYTrib*, 01/03/19, 1. The owners of the *Lord Dufferin*, Gaston, Williams & Wigmore Ltd, brought suit against the Cunard Line for the loss. The partially submerged wreck of the freighter was eventually salvaged.

64 *NYTrib*, 21/07/19, 7; Chirnside, pg. 35.

65 *TDE*, 08/09/19, 5; *NYTrib*, 13/09/19, 18.

66 *NYTrib*, 19/09/19, 12.

67 Spedding, pp. 165–6.

68 *Times*, 29/06/20; *TST*, 06/08/20.

69 *IME*, October 1920, 818.

70 *Scientific American*, 14/08/20.

71 *IME*, August, 1920, 635.

72 *IME*, October, 1920, 821.

73 *TNG*, 31/07/20, 137.

74 *Times*, 29/06/20; *TST*, 06/08/20.

75 Spedding, pg. 168.

76 Spedding, pp. 168–9.

77 *Times*, 29/06/20; *TST*, 06/08/20.

78 *TNG*, Vol. 99, 31/07/20, 137.

79 *TDM*, 19/07/20, 4; *NYTrib*, 20/07/20, 2.

80 *NYTrib*, 25/07/20, 14.

81 Some period sources refer to this engineer as Seymour Barkway.

82 *TNG*, 31/07/20, pg. 137.

83 Chirnside, pg. 35, endnote 15. Exactly when the bilge keels were added is unclear: if the Frahm's tanks were turned into oil tanks during this stay in New York, then it is possible that the bilge keels were installed during the ship's dry-docking in Liverpool before the crossing. It is also possible that they were installed at a later date. Up to now, we have been unable to ascertain the precise timing.

84 *TNG*, 07/08/20.

85 The *Sunday Express*, 17/07/21, 7.

86 Chirnside, Appendix Five.

87 Chirnside, pg. 39. It was Mr. Chirnside's research that unearthed the total number of passengers carried in each class, as well as the total revenues generated by each class. A similar, if rather rough, study of the *Olympic* and *Titanic*'s 1911–2 potential capacity for generating revenue done in 2002 would indicate First Class could contribute between 58 per cent and 68 per cent of the total revenue, while Third Class would contribute only 9–21 per cent of that total. See *Titanic Passenger Ticket Revenues*, by Captain Charles B. Weeks.

88 *Daily News*, 31/03/31; *TDM*, 20/04/31.

89 For a thorough comparison of many of the *Aquitania*'s changed interior spaces, please see *Aquitania 'Dossier'*, by Mark Chirnside.

90 Valentino, pp. 90–4.

91 strictly-vintage-hollywood.blogspot.com/2010/05/valentinos-letters.html.   The letter went up for auction in 2010: www.liveauctioneers.com/item/7456691_rudolph-valentino-autograph-letter-signed.

92 Valentino, pp. 90–4.

93 Spedding, pg. 251.

94 Spedding, pp. 196–7.

95 Spedding referred to this ship's original name as *Fokker*, which was mistaken. The *Foca I* had been built in Norway in 1917, and was 111ft in length.

96 See also *Shackleton: An Irishman in Antarctica*, pg. 179; *Shackleton: By Endurance We Conquer*, Chapter 40; and *Acts of Occupation: Canada and Arctic Sovereignty, 1918–25*, pg. 104, 116, 125.

97 *NYTrib*, 31/01/21, 20.

98 Spedding, pg. 197.

99 *NYTrib*, 27/02/21.

100 *Musical America*, 19/03/21, 26.

101 Letter sold at auction: store.paulfrasercollectibles.com/Ernest-Shackleton-Handwritten-Personal-Letter-p/pt206.htm.

102 Spedding, pg. 199.

103 *Eyewitness Accounts: Shackleton's Last Voyage*.

104 Spedding, pp. 201–2.

105 *NYTimes*, 17/03/22; *NYTrib*, 18/03/22.

106 Spedding, pg. 243, 244.

107 *NYTimes*, 30/12/77; *NYTrib*, 12/04/22.

108 Spedding, pp. 182–3.

109 Spedding, pg. 243, 244.

110 Spedding, pg. 252.

111 *NYTrib*, 26/02/22, 11.

112 *TWH, The Washington Times, NYTrib*, 01/10/22.

113 *TDE*, 25/01/24, 1.

114 *TDM*, 29/01/25, 3; *TDE*, 18/02/26, 1.

115 *NYTrib*, 29/04/22, 7.

116 *NYTrib*, 06/06/22; *NYTimes*, 29/06/22; *Pittsburgh Post-Gazette*, 11/10/35; *Reading Eagle*, 02/10/38.

117 *TDE*, 29/12/23, 7.

118 Spedding, pg. 269; *NYTimes*, 24/05/22.

119 *NYTimes*, 10/11/23, 26/11/23; *TDE*, 28/11/31.

120 *NYTimes*, 21/08/26

121 *TDE*, 25/05/25.

122 *TDE*, 06/01/38, 6.

123 *The Sun, The Globe* 21/08/23, 07/09/23; *NYTimes*, 22/08/23, 08/09/23.

124 *TDM*, 03/01/33.

125 *TDM*, 05/06/24.

126 *TDM*, 09/08/27, 10; *TDE*, 08/08/27, 9.

127 Although many accounts say that Captain Charles retired to his cabin, some period, and very detailed, press accounts specifically state that he did not go to his cabin, but instead went to the Chart Room and laid down on a couch there.

128 *NYTimes* 05/01/28, 07/01/28, 09/07/28; *The Binghamton Press*, *The Brooklyn Daily Eagle*, *The Auburn Citizen*, *TDE*, *NYTimes* 16/07/28; *The Singapore Free Press*, 17/07/28. Diggle, pg. 76– 77.

129 Private correspondence provided by Mike Poirier.

130 *NYTimes*, 08, 12, 25, 30/04/31; 02, 06/05/31.

131 *NYTimes*, 04/07/31, 22/12/31.

132 *TDM*, 04/01/32, 21.

133 *TDE*, 01/04/82, 25.

134 Cunard advertising, as seen in *The Milwaukee Sentinel*, 11/06/29.

135 *NYTimes*, 10/06/36; *TYP*, 11/06/36.

136 *TYP*, 04/08/37.

137 *TDE*, 05/05/37.

138 *TDM*, 22/04/39.

139 Chirnside, pg. 71.

140 *TYP*, 12/04/35, 12.

141 Period newsreel narration.

142 *TDE*, 12/04/35, 13.

143 Chirnside, pg. 63; 'Aquitania: "The Grand Old Lady" Dossier', www.markchirnside. co.uk/Aquitania_OldLady_Dossier.htm.

144 *NYTimes*, 17/02/37, 30/03/37.

145 *NYTimes*, 29/03/37.

146 *NYTimes*, 31/03/37.

147 *Philadelphia Inquirer*, 05/12/37.

148 *Versailles and After, 1919–1933*.

149 *NYTimes*, 17/11/39.

150 *United States Naval Administration in World War II, History of Convoy and Routing*.

151 *Royal Australian Navy, 1939–1942*.

152 Ibid.

153 *Report on the Loss of HMAS* Sydney.

154 *The Evening Post*, 03/03/2001. Theories about why the mines did not detonate include the mines being duds, or laid further offshore than reported in the *Adjutant*'s logs.

155 *Global Logistics and Strategy, 1940–1943*.

156 Chirnside.

157 *The Story of SNAG 56*.

158 *TYP*, 18/07/45, 5.

159 *TDE*, 18/02/48, 2.

160 *TDE*, 13/11/45, 4.

161 *TDE*, 11/04/46, 3.

162 *The Chicago Tribune*, 01/09/53.

163 Chirnside, pg. 84.

164 The text, never directly quoted but instead paraphrased, varies slightly with each retelling. This particular phrasing comes from the ship's Wikipedia page, but is repeated elsewhere in print and online.

165 Grattidge, pp. 203–04.

166 *TDE*, 18/02/48, 2.

167 Chirnside, pg. 85.

168 Chirnside, pg. 86.

169 *TDE*, 24/11/48, 1.

170 *TDE*, 09/11/49, 1.

171 *NYTimes*, 17/12/49.

172 *TDE*, 20/02/50, 5.

173 *TDE*, 20/02/50, 5.

174 *The Glasgow Herald*, 22/02/50.

BIBLIOGRAPHY

Books

Arnold, Alan, *Valentino* (Library Publishers, 1954) (cited as *Valentino*).

Cavell, Janice and Noakes, Jeff, *Acts of Occupation: Canada and Arctic Sovereignty, 1918–25* (UBC Press, 2011).

Chirnside, Mark, *Aquitania: The Ship Beautiful* (The History Press, 2008, First Edition) (cited as Chirnside).

Chirnside, Mark, *RMS Olympic: Titanic's Sister* (The History Press, 2015, Second Edition) (cited as *RMS Olympic*).

Coakley, Robert W. and Leighton, Richard M., *Global Logistics and Strategy, 1940–43* (Office of the Chief of Military History, 1955).

Diggle, Captain E.G., *The Romance of the Modern Liner* (Sampson Low, Marston & Co., 1930, reprinted by Patrick Stephens Ltd, 1989) (cited as *Romance*).

DuPuy, William Atherton, *The World War and Historic Deeds of Valor,* Vol. 6 (National Historic Publishing Association, 1919).

Fitch, Tad, Layton, J. Kent and Wormstedt, Bill, *On A Sea of Glass: The Life & Loss of the RMS Titanic* (Amberley Books, 2015, Third Edition).

Fitch, Tad G. and Poirier, Mike, Into *the Danger Zone: The Lusitania, First Battle of the Atlantic and Liners During the Great War* (Blurb Books, 2023).

Gill, G. Hermon, *Royal Australian Navy, 1939–1942* (The Griffin Press, 1957).

Grattidge, Harry, *Captain of the Queens* (E.P. Dutton & Co., Inc., 1956) (cited as Grattidge).

Halpern, Sam et al., *Titanic: A Centennial Reappraisal* (The History Press, 2012, First Edition).

Hansard, *House of Commons Reports,* vol. 139, cc673–4 (Queen's Printer, 1904).

Henig, Ruth, *Versailles and After, 1919–1933* (Methuen & Company, Ltd, 1984).

House of Commons, *The Parliamentary Debates [Official Report]* (H.M. Stationary Office, 1916).

Hudson, Henry W., *The Story of SNAG 56* (Harvard University Printing Office, 1946).

Hurd, Archibald, *A Merchant Fleet At War* (Cassell and Company, Ltd, 1920).

Johnston, Ian, *Ships For A Nation: John Brown & Company, Clydebank* (Argyll Publishing, 2008).

King, William C., *King's Complete History of the World, Visualizing the Great Conflict in All Theaters of Action, 1914–1918* (History Associates, 1922).

Layton, J. Kent, *Lusitania: An Illustrated Biography* (Amberley Books, 2015, Second Edition).

Layton, J. Kent, *The Unseen Mauretania (1907): The Ship in Rare Illustrations* (The History Press, 2015).

Layton, J. Kent, Fitch, Tad, Poirier, Michael, Lynskey, Tom and Rourke, Levi, *Lusitania: An Illustrated Biography*, 2 vols (The History Press, 2024 & 2025)

McCart, Neil, *Atlantic Liners of the Cunard Line* (Patrick Stephens, Ltd, 1994).

Ocean Liners of the Past: The Cunard Quadruple-Screw Atlantic Liner Aquitania, No. 3 in a series of reprints from *The Shipbuilder* (Patrick Stephens, Ltd, 1971) (cited as *Shipbuilder*).

Robinson, Ralph J., *Ambulance Company 113, 29th Division* (The Lord Baltimore Press, 1919).

Sauder, Eric, *The Unseen Lusitania: The Ship in Rare Illustrations* (The History Press, 2015).

Shackleton, Jonathan et al., *Shackleton: An Irishman in Antarctica* (University of Wisconsin Press, 2002).

Smith, Michael, *Shackleton: By Endurance We Conquer* (Oneworld Publications, 2014).

Spedding, Charles T., *Reminiscences of Transatlantic Travellers* (T. Fisher Unwin, 1926).

Wild, Frank, *Eyewitness Accounts: Shackleton's Last Voyage* (Amberley Publishing, 2015).

Primary Source Documents

Navy Department, *United States Naval Administration in World War II, History of Convoy and Routing*, Washington, D.C., 1939–1945.

Records of the Upper Clyde Shipbuilders (cited as UCS).

Sheffield City Archives (cited as SCA).

University of Liverpool Archives (cited as ULA).

Periodicals & Newspapers

Periodical dates are listed as day/month/year.

The Auburn Citizen

The Binghamton Press

The Brooklyn Daily Eagle

The Chicago Tribune

Daily Consular and Trade Reports, Vol. 4, Issues 231–307, Department of Commerce and Labor, Bureau of Foreign Domestic Commerce, October–December 1912.

The Daily Express (cited as *TDE*)

The Daily Mirror (cited as *TDM*)

Daily News

The Engineer

The Evening Post

The Evening Star

The Evening World (cited as *TEW*)

The Glasgow Herald

The Globe

International Marine Engineering (cited as *IME*)

The *Milwaukee Sentinel*

Musical America

The Nautical Gazette (cited as *TNG*)

The New York Times (cited as *NYTimes*)

The New York Tribune (cited as *NYTrib*)
The Ocala Evening Star
Pacific Marine Review
Philadelphia Inquirer
Pittsburgh Post-Gazette
Reading Eagle
Scientific American
The Singapore Free Press
The Straits Times (cited as *TST*)
The Sun
The Sunday Express
The Times (cited as *Times*)
The Washington Herald (cited as *TWH*)
The Washington Times
The Yorkshire Post (cited as *TYP*)

Websites & Online Research Papers

Australian Joint Standing Committee on Foreign Affairs, Defence and Trade. *Report on the Loss of HMAS* Sydney, 1999, museum.wa.gov.au/explore/sydney

Mark Chirnside's Reception Room, www.markchirnside.co.uk

Chirnside, Mark, *Aquitania: Changes to a Design* (cited as *Changes*), www.titanic-model.com/articles/aquitania_maiden_voyage/aquitania_maiden_voyage.shtml

Chirnside, Mark, *Aquitania 'Dossier'*, www.markchirnside.co.uk/rms_aquitania_interior.html

The *Titanic* Research & Modeling Associaton, www.titanic-model.com

Weeks, Captain Charles B., *Titanic Passenger Ticket Revenues*, www.encyclopedia-titanica.org/titanic-passenger-revenue.html

ALSO AVAILABLE

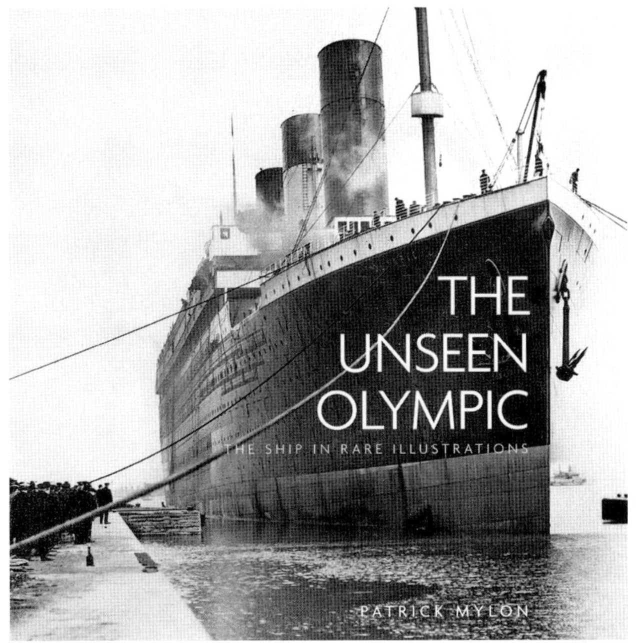

9780750982672

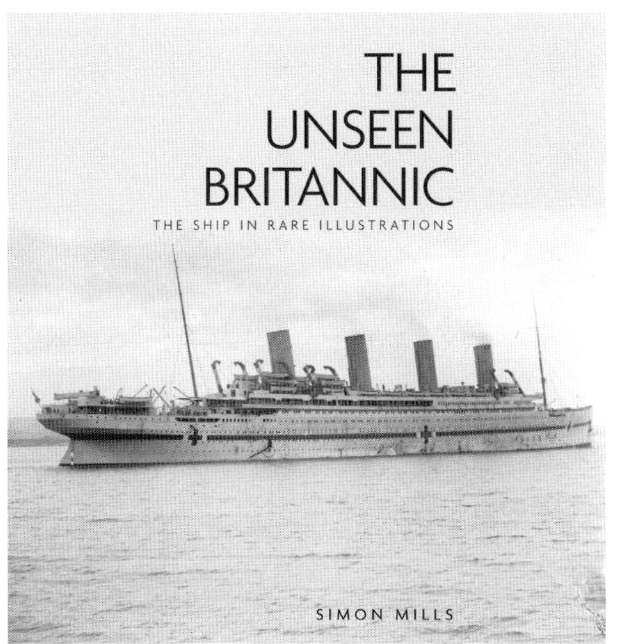

9780750993777

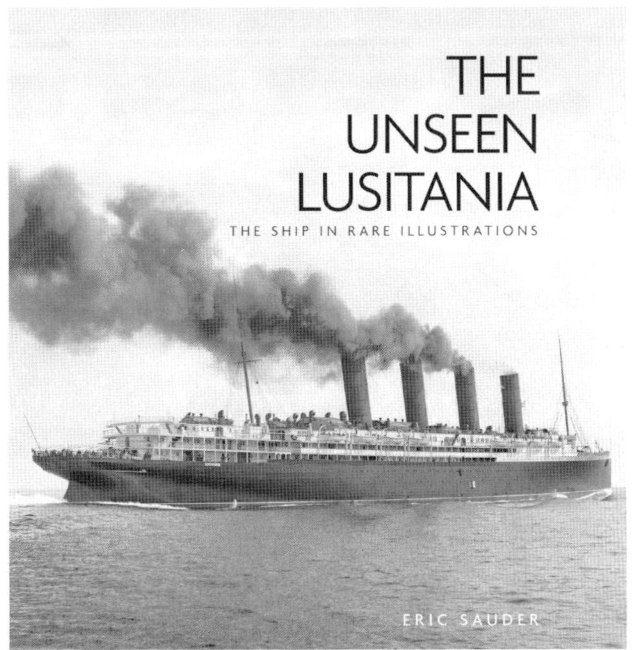

9780750998871

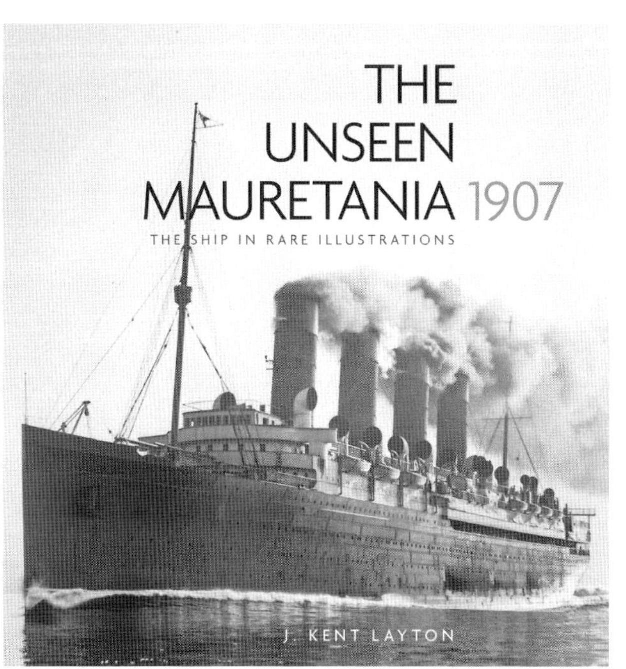

9780750996549

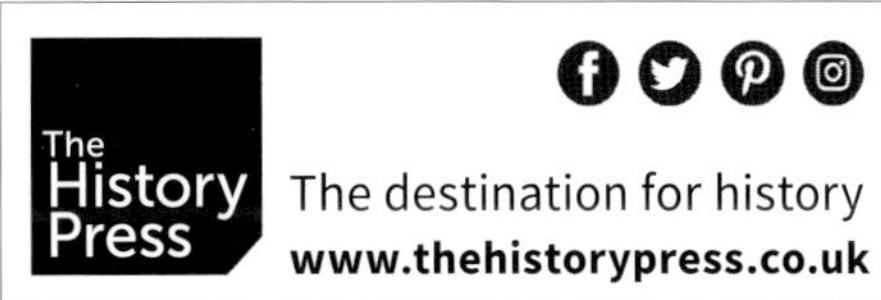